Green Energy and Technology



For further volumes:
http://www.springer.com/series/8059

Fabio Orecchini · Vincenzo Naso

# Energy Systems in the Era of Energy Vectors

## A Key to Define, Analyze and Design Energy Systems Beyond Fossil Fuels

Springer

Prof. Fabio Orecchini
Centro Interuniversitario di Ricerca
Per Sviluppo Sostenibile (CIRPS)
Sapienza Università di Roma
Piazza San Pietro in Vincoli 10
00184 Rome
Italy
e-mail: fabio.orecchini@uniroma1.it

Prof. Vincenzo Naso
Centro Interuniversitario di Ricerca
Per Sviluppo Sostenibile (CIRPS)
Sapienza Università di Roma
Piazza San Pietro in Vincoli 10
00184 Rome
Italy
e-mail: vincenzo.naso@uniroma1.it

ISSN 1865-3529
e-ISSN 1865-3537
ISBN 978-1-4471-2713-0
ISBN 978-0-85729-244-5 (eBook)
DOI 10.1007/978-0-85729-244-5
Springer London Dordrecht Heidelberg New York

British Library Cataloguing in Publication Data
A catalogue record for this book is available from the British Library



*Cover design:* eStudio Calamar, Berlin/Figueres

Printed on acid-free paper

Springer is part of Springer Science+Business Media (www.springer.com)

# Contents

# Chapter 1
# Energy Systems

## 1.1 The Energy Cycle

### *1.1.1 Introduction*

The *energy cycle* concerns energy evolution in its passage from natural availability to the most suitable form for end use, meaning the form that allows the desired useful effect and entails the disposal of its unexploited portion.

*Energetics* is the branch of learning that studies the forms and quantities through which nature provides energy to human beings, and the sources (called *primary sources*) through which this availability is carried out. *Energy systems* are those devices in which, through **transformation and conversion processes and flows**, energy is processed until reaching **end uses** and thence the **useful effect**. Obviously, as shown in Fig. 1.1, the energy cycle—in its form commonly used nowadays—entails an interaction with the environment as well as the production of waste and pollutants.

This cycle is shown in detail in Fig. 1.1, with the *general reference scheme*. It shows the so-called **primary** (or **natural**) forms of energy, the **energy flows**, which include the **transformation from primary energy into secondary forms of energy**, through their processing and physical or chemical transformation, that is to say the transformation into **electric energy** and the energy of **end use**, until achieving the **useful effect**.

By following the indications of the arrows shown in the figure, the possible pathways of the energy cycle can be identified. It is therefore shown how some sources of primary energy are *directly employable for end use* (line (a); for instance: the natural gas or coal used for domestic heating, the wind energy or hydropower directly used in mills since ancient times, or the wind energy used for sailing purposes).

More often, the forms of energy available in nature must be adapted to the demand coming from end users that is, to say transformed and/or converted. In this

F. Orecchini and V. Naso, *Energy Systems in the Era of Energy Vectors*,
Green Energy and Technology, DOI: 10.1007/978-0-85729-244-5_1,

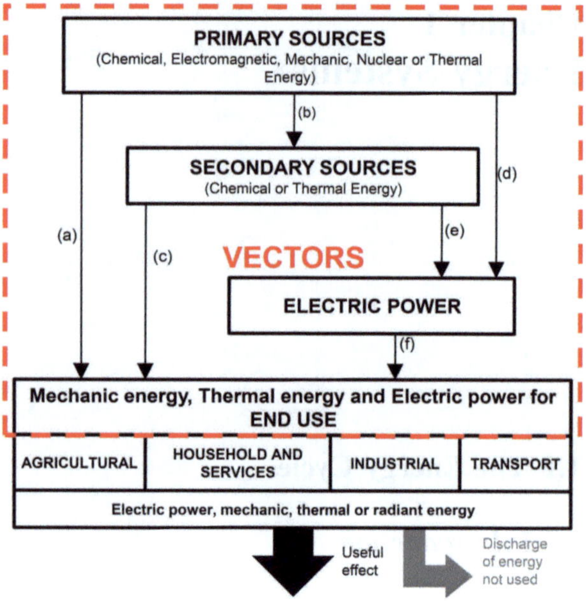


**Fig. 1.1** General reference diagram of the energy cycle

way, a passage occurs from *primary forms* (i.e., the potential chemical energy contained in liquid hydrocarbons) to *secondary forms* (line (b); i.e.: oil refining products) and they are directly employed for end use (line (c)).

In other cases, some primary sources (line (d)), such as natural gas and coal, or secondary sources, can supply the thermoelectric plants that generate electric power (line (e)). In turn, the latter is employed by end users (line (f)); in this final process of the cycle, energy allows the achievement of the desired *useful effect*, whereas the portion not used is generally released into the environment.

In this chart, particular attention is paid to *electric power*, due to its peculiarity and the priority it is gaining in the entire energy framework, compared to the other forms of energy used.

It is appropriate to underline that the definitions herein introduced differ from those that several authors adopted on the basis of other nomenclatures. For instance, some prefer to define "secondary energies" those types of energy that allow to transfer energy in space or in time, with a view to make it available wherever and whenever needed. In this framework, the name *energy vector* is ascribed to this type of function.

For instance, when hydrogen production is finally obtained—having considered that it is a very good fuel not directly available for use in nature—by using the waste heat of thermoelectric power plants, a sort of energy recovery through heat "accumulation" in hydrogen is carried out. In fact, the *hydrogen energy vector* will collect the waste thermal energy that would otherwise be lost. The availability of hydrogen so produced, in fact, will make energy available wherever and whenever needed. As a matter of fact, through this process, the waste thermal energy will be transferred over time by means of the hydrogen energy vector.

In the next few paragraphs, an analysis is made of the main usable natural forms of energy. Subsequently, the units of measurement used in the energy sector will be dealt with, since some of them—although commonly used among operators—are not the ones adopted by the International System. As a matter of fact, as we will see more in detail afterwards, these particular units were selected for their characteristics and size, which make their use handy in the analysis of energy flows.

Finally, an analysis of natural energy sources will be made, preceded by general considerations on global energy flows on the Earth, seen as an energy system interacting with the surrounding space.

### *1.1.2 Energy and Power: Natural Forms of Energy*

*Energy*, according to the most widespread definition, is the *capacity of a system to do work*. As regards the unit of time, *power* is similarly defined. The "energy problem" consists in making energy available at the place, time and conditions requested by users; and this for the entire time needed. It is therefore even more difficult to find a solution to the "power problem", having considered that the quantity of energy demanded may vary minute after minute. Solving the energy problem from the quantity view point and in a given period of time may not be sufficient.[1]

The energy available in nature—and which human beings avail themselves of—can be distinguished into different heterogeneous forms. The forms of energy presently used are the following:

- Chemical
- Electrical
- Electromagnetic
- Mechanical
  - Kinetic
  - Potential and pressure
- Nuclear
- Thermal

This classification is important since it provides the choice of the form of energy to be employed, according to its intended use.

[1] This concept can be more easily understood by means of an example. Let's consider the case of electric utilities: since it is known that electric power is stored with difficulty and expensively, it is not easy for Utilities to avail themselves of all the necessary energy. Supplying the electric power demanded by users will entail the ability of providing, moment by moment, the exact quantity of *energy requested* that is, to say supplying the *power demanded second after second*.

In fact, according to the characteristics and the attitude of each form of energy that is converted into useful work to be destined to a particular end use, it is possible to adopt each time the most convenient energy source, that is to say the conversion process that is most suitable to pursue the objectives set.

For instance, in all the applications in environments with no free oxygen—as space and submarine applications—it is little convenient, if not impossible, to use the chemical energy developed from conventional combustion reactions, namely those that develop heat through exothermic oxidation reactions, entailing the availability of oxygen. In these cases, other conversion processes are privileged or, alternatively, other forms of energy: a technically interesting solution may be represented by nuclear power, supported by closed-cycle thermodynamic plants, or by the resort to special systems for energy conversion, such as some so-called *direct conversion* devices that will be described below.

Here it follows a brief description of the characteristics and properties of the different forms of energy available in nature.

#### 1.1.2.1 Chemical Energy

It is inborn in the bond strengths existing at a molecular and atomic level in some particular substances. It can be defined either as *potential*, underlining the intrinsic character that can be expressed through suitable reactions in the form of thermal energy, or *electric*. In the former case, *exothermic reactions* (in general, oxidation) occur, and the quantity of heat provided, called *reaction energy* or *reaction heat*, corresponds, in an absolute terms, to the enthalpy variation pertaining to each reaction. Conversely, the generation of electric power is due to the variation of free energy relating to oxidation–reduction reactions or to the concentration gradient of different substances in given solutions (in this case, reference is made to *free energy of mixing*). The chemical energy that is mainly used is the one contained in fossil fuels.

#### 1.1.2.2 Electric Power

This form of energy, utterly linked to the development of the human society, is produced by the movement of free electrons in conductors. Although it is incorrect to refer to "electric power"—since, in addition to the electric field, a magnetic field is always involved and therefore the definition of "electromagnetic energy" should apply—this distinction is however suitable to define the type of energy used for its electric features only. The electric power available in nature (for example through the phenomenon of thunderbolts) is not directly exploitable at the present state of technology; *it is therefore necessary to produce electricity artificially, converting other forms of energy available in nature into electric power.*

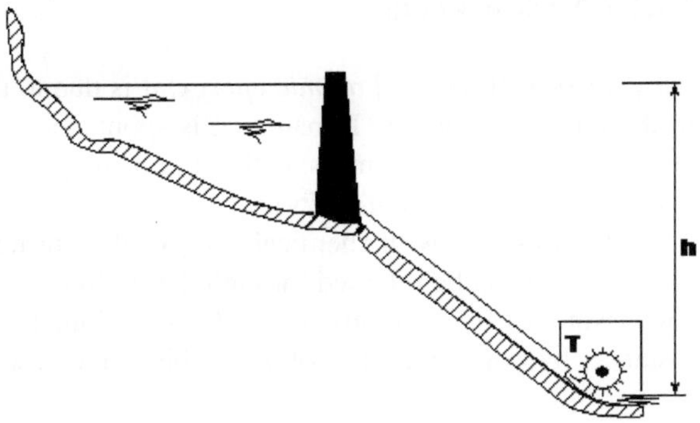


**Fig. 1.2** Water stored in a mountain basin with potential energy proportional to the level *h* of difference in height of turbine *T*

#### 1.1.2.3 Electromagnetic Energy

Linked to the interaction between electric and magnetic fields, it is a kind of energy conveyed without the support of any physical means, and therefore also in vacuo, where it is practically free from dissipation phenomena. In nature it is mainly present in solar radiation, which predominantly provides for the energy supply of our planet.

#### 1.1.2.4 Mechanical Energy

It is available in nature in two different forms: potential and pressure, and kinetic.

Potential and Pressure Mechanical Energy

It derives from the gravitational pull acting on the Earth; it is owned by bodies at rest and depends on the position ("height") that they have in that pull, in relation to a conventional position which is ascribed the reference potential energy content. The water stored in a basin, for instance, has a potential energy proportional to hits height, *h*, compared to a reference point (i.e., the discharge of a turbine downstream) (Fig. 1.2).

A potential energy is also pressure energy, strictly linked to the homologous status variable. Also in this case, the reference to a hydraulic example can be useful, in the sense that there is always a component of this form of energy in a moving fluid.

Kinetic Mechanic Energy

It is linked to the movement of bodies (solid, liquid, and gaseous bodies); it is owned, for instance, by the wind (in this case, wind energy), by the water of rivers (hydropower) and by the sea (energy of wave motion, tides), etc.

#### 1.1.2.5 Nuclear Energy

Also (incorrectly) called *atomic energy*, it is due to the cohesive strengths present in the nucleus of atoms. In nature it is spontaneously released by elements, generally of a large atomic mass, called *radioisotopes,*[2] which emit radiations and/or particles (***natural radioactivity***).

Furthermore, as with chemical energy, also for nuclear power a *potential* form exists, which can be released through the fission of nuclei of heavy atoms (such as some isotopes of uranium) or the fusion of nuclei of light atoms (such as, for instance, hydrogen and its isotopes); this is the case of ***artificial radioactivity***.

#### 1.1.2.6 Thermal Energy

It is the energy contained in any body depending on the fact that their temperature is higher than absolute zero and is due, from the microscopic viewpoint, to the energy (kinetic and potential) linked to the state of motion present in the matter at a molecular, atomic or subatomic level.

In the thermodynamic approach, thermal energy is considered as the less "noble" form of energy, since all the other forms tend to almost entirely transform themselves into it, whereas the opposite phenomenon is not so spontaneous. As stated in the Second Law of Thermodynamics, in fact, *a free thermodynamic transformation whose only result is to transform heat into another form of energy is not possible*.

An example of thermal energy available in nature is ***geothermal energy***.

### *1.1.3 Units of Measurement in Energy Systems*

On the basis of the nature itself of energy and the use that human beings make of it, energetics and any of its related subjects range among fields that are even too far from the merely engineering framework, adopting each time economic, political and social values.

The principles of the *International System (SI) of units of measurement* in the energy sector are shown in

Table 1.1, Whereas

Table 1.2 sums up the main units *derived* from it.

[2] Radioisotopes are the radioactive isotopes of some chemical elements. The word "isotopes" indicates the atoms of chemical species that have the same atomic number, namely that own the same number of protons; for this reason—and this is the etymological explanation—they are positioned at the same level in the scale of elements. However, they have a different number of neutrons in the nucleus, therefore a different atomic weight. Most of elements are available in nature in the form of mixtures of isotopes with constant composition.

**Table 1.1** Main units of measurement referred to energetics in the international system (S.I.)

| Magnitude | Unit of measurement | |
|---|---|---|
| | Name | Symbol |
| Length | Metre | m |
| Mass | Kilogram | kg |
| Time | Second | s |
| Intensity of electric current | Ampere | A |
| Thermodynamic temperature | Kelvin | K |
| Quantity of mass | Mole | Mol |
| Light intensity | Candela | Cd |

Owing to its marked interdisciplinary character, as well as to the huge range of values encountered among the energy values involved, the general procedure in the world of trade and information is, however, to contravene the S.I. principles by using units of measurement originally adopted in other sectors (economic, commercial, etc.), or in any case not formally accepted in official standardised systems.

The most important unit in the energy sector, besides the units of general use (J, cal, erg), is the *ton of oil equivalent* (TOE), used to quantify energy consumption at a microscopic level, in industrial, transportation, civil and residential use. It is based on the characteristic of petroleum of having a nearly constant calorific value[3] and is conventionally adopted as amounting to 10,000 kcal/kg. The energy content ascribed to the TOE is then $10^7$ kcal.

Another unit of measurement sometimes used is the *ton of coal equivalent* (*TCE*). After its spreading in the 70s', however, the TOE was preferred over it due to the higher commercial spreading of oil in the present energy market. A practical limitation to the use of TCE derives from the fact that, differently from petroleum, pit coal has a variable calorific value, as well as its carbon content varies according to the place of origin and to the composition of the fossil fuel.

Besides these units of measurement, the *oil barrel*[4] is often used to develop energy balances based on which the price of crude oil is indicated, as well as the *barrel of oil equivalent* (BOE), a unit of measurement similar to those described above.

Among the most widely used energy units there is also the kWh (kilowatt-hour), a tipical electric unit of measurement (corresponding to the energy developed by a machine providing 1 kW of power per hour) and the BTU (British Thermal Unit),[5] mainly used in the field of thermotechnics.

---

[3] This is the *lower heating value*: it is obtained by deducting from the *higher heating value* the heat deriving from water condensation that is formed during combustion, as well as the latent heat of the humidity contained in the fuel. The *higher heating value* is the quantity of heat, expressed in kilocalories, generated by 1 kg of fuel by burning entirely, at a constant pressure, when the products of combustion are brought to 0°C and 760 mmHg. Frequently, in case of gaseous fuels, instead of referring to kg, reference is made to the normal cubic meter ($Nm^3$).

[4] A barrel of oil corresponds to 158.98 l.

[5] A BTU corresponds to the quantity of heat necessary to increase the temperature of a pound of water (1 lb ≅ 4.44 kg) by 1 degree Fahrenheit (1° F ≅ 0.55°C).

**Table 1.2** Main energy-related units of measurement "derived" from the S.I.

| Magnitude | Unit of measurement | |
|---|---|---|
| | Name | Symbol |
| Frequency | Hertz | Hz |
| Force | Newton | N |
| Pressure | Pascal | Pa |
| Energy, work, quantity of heat | Joule | J |
| Power, energy flow | Watt | W |
| Quantity of electricity, electric charge | Coulomb | C |
| Electric voltage, electric potential, electromotive force | Volt | V |
| Electric resistance | Ohm | Ω |
| Electric conductance | Siemens | S |
| Electric capacitance | Farad | F |
| Magnetic induction flux | Weber | Wb |
| Magnetic induction | Tesla | T |
| Electric inductance | Henry | H |
| Flux of light | Lumen | lm |
| Illumination | Lux | lx |
| Activity | Becquerel | Bq |
| Absorbed dose/Radiation dose | Gray | Gy |
| Equivalent dose | Sievert | Sv |

The use of so many units of measurement is due to the variety of energy applications characterised by very different orders of magnitude Table 1.3 shows that, in order to represent on the scale of Joule the entity of all energy phenomena concerning the "Earth system", over 50 orders of magnitude are needed [1].

Table 1.4 shows the conversion factors relating to the units introduced. Finally, Table 1.5 shows the conversion factors of the main units of measurement used for the measurement of calorific values

## 1.1.4 Energy Availability in Nature

The analysis of energy availability in nature starts from the global energy cycle of the system in which human beings operate the *Earth system*.

**Table 1.3** Comparative scales of some energy-related measurements (from [1])

| Energy phenomenon | Energy (J) |
|---|---|
| Fission energy of a U-235 atom | $1.6 \times 10^{-12}$ |
| Energy of one ton of coal | $2.9 \times 10^{10}$ |
| Energy of one ton of oil | $4.5 \times 10^{9}$ |
| Energy equivalent of one gram of matter | $9 \times 10^{13}$ |
| Energy consumption all over the world (2004) | $5.04 \times 10^{20}$ |
| Daily solar energy received by the Earth | $1.49 \times 10^{22}$ |
| Daily emission of solar energy | $3 \times 10^{32}$ |

**Table 1.4** Conversion factors of the main energy units of measurement

| | J | Cal | toe | tce | bop | Wh | BTU | Erg | eV |
|---|---|---|---|---|---|---|---|---|---|
| J | 1 | $2.39 \cdot 10^{-4}$ | $0.023 \cdot 10^{-9}$ | $0.034 \cdot 10^{-9}$ | $0.16 \cdot 10^{-9}$ | $2.78 \cdot 10^{-4}$ | $0.94 \cdot 10^{-3}$ | $10^{7}$ | $6.24 \cdot 10^{18}$ |
| Cal | $4.186 \cdot 10^{10}$ | 1 | $0.092 \cdot 10^{-6}$ | $0.142 \cdot 10^{-6}$ | $0.68 \cdot 10^{-6}$ | 1.16 | 3.95 | $4.186 \cdot 10^{10}$ | $2.61 \cdot 10^{22}$ |
| toe | $4.537 \cdot 10^{10}$ | $10^{7}$ | 1 | 1.55 | 7.37 | $1.26 \cdot 10^{7}$ | $4.28 \cdot 10^{7}$ | $4.537 \cdot 10^{16}$ | $2.83 \cdot 10^{29}$ |
| tce | $2.93 \cdot 10^{10}$ | $7 \cdot 10^{6}$ | 0.645 | 1 | 4.74 | $8.14 \cdot 10^{6}$ | $2.76 \cdot 10^{7}$ | $2.93 \cdot 10^{17}$ | $1.83 \cdot 10^{29}$ |
| bop | $6.17 \cdot 10^{9}$ | $1.47 \cdot 10^{6}$ | 0.135 | 0.21 | 1 | $1.71 \cdot 10^{6}$ | $5.8 \cdot 10^{6}$ | $6.17 \cdot 10^{16}$ | $3.85 \cdot 10^{28}$ |
| Wh | $3.6 \cdot 10^{3}$ | 0.86 | $0.079 \cdot 10^{-6}$ | $0.12 \cdot 10^{-6}$ | $0.58 \cdot 10^{-6}$ | 1 | 3.4 | $3.6 \cdot 10^{10}$ | $2.24 \cdot 10^{22}$ |
| BTU | $1.06 \cdot 10^{3}$ | 0.25 | $0.023 \cdot 10^{-6}$ | $0.036 \cdot 10^{-6}$ | $0.171 \cdot 10^{6}$ | 0.29 | 1 | $1.06 \cdot 10^{10}$ | $6.61 \cdot 10^{21}$ |
| erg | $10^{-7}$ | $2.38 \cdot 10^{-11}$ | $0.022 \cdot 10^{-16}$ | $0.034 \cdot 10^{-16}$ | $0.16 \cdot 10^{-16}$ | $2.77 \cdot 10^{-1}$ | $0.94 \cdot 10^{-10}$ | 1 | $6.24 \cdot 10^{11}$ |
| eV | $1.6 \cdot 10^{-19}$ | $3.83 \cdot 10^{-23}$ | $3.68 \cdot 10^{-30}$ | $5.44 \cdot 10^{-30}$ | $2.56 \cdot 10^{-29}$ | $4.45 \cdot 10^{-23}$ | $1.5 \cdot 10^{-22}$ | $1.6 \cdot 10^{-12}$ | 1 |

**Table 1.5** Conversion of the units of measurement relating to calorific values

| | J/kg | kcal/kg | BTU/lb |
|---|---|---|---|
| J/kg | 1 | $2.39 \times 10^{-4}$ | $4.29 \times 10^{-4}$ |
| kcal/kg | $4.19 \times 10^{3}$ | 1 | 1.82 |
| BTU/lb | $2.33 \times 10^{3}$ | $5.5 \times 10^{-1}$ | 1 |

**Table 1.6** Classification of primary energy sources according to the criterion of renewability

| Energy flows of the earth system | Non-renewable energy sources | Renewable energy sources |
|---|---|---|
| Solar radiation | Petroleum | Human and animal energy |
| | Coal | Wood/biomass |
| | Natural gas | Energy crops |
| | Bituminous schist | Solar energy |
| | Oil sands | Wind energy |
| | | Sea waves |
| | | OTEC |
| | | Saline gradient of the sea |
| | | Hydropower |
| Gravitational energy | | Tides |
| Internal energy of the earth | Natural radioisotopes | Geothermal energy |
| | uranium, thorium, lithium | |
| | hydrogen isotopes | |

The figures and energy flows making up the Earth's energy balance are shown in Sankey diagram (Fig. 1.3), in which these flows are represented through arrows whose width indicates the total contribution of each single item [1].

It is therefore evident that the energy balance of our planet is mainly based on the *energy flow coming from the Sun*, accounting for 99.98% of the total.

The mean energy flow coming from the Sun is called *solar constant* and corresponds to the energy that, outside of the atmosphere (namely without atmospheric absorption) and at the average distance of the Earth from the Sun, passes through a surface of 1 $m^2$, placed perpendicularly to the direction of the flow, in the interval of 1 s. Measurements made through artificial satellites have indicated for the solar constant a value of 1.353 $kW/m^2$, with a $\pm$ 3% variability due to the

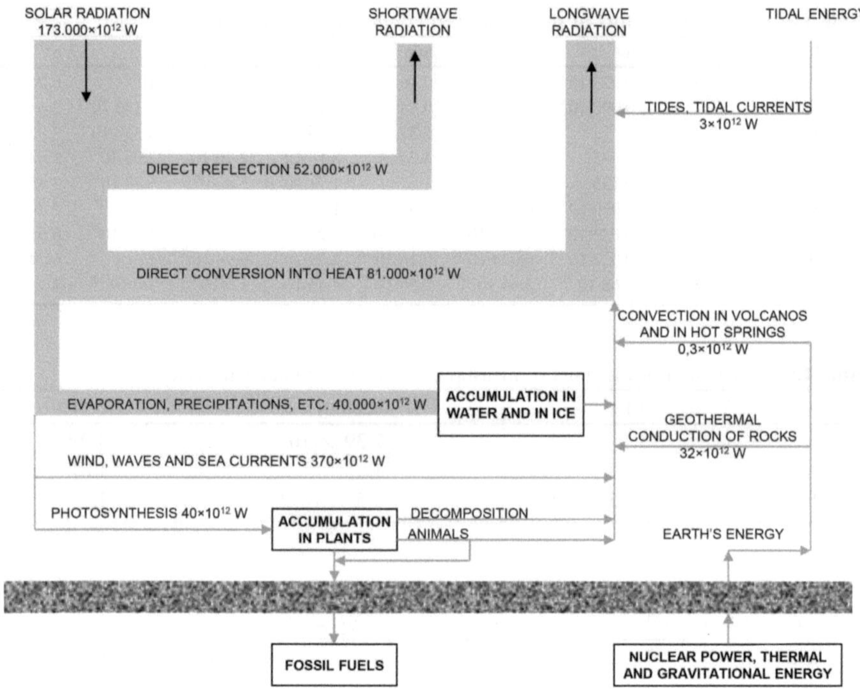


**Fig. 1.3** Global energy balance of the earth [1]

ellipticity of the Earth orbit.[6] Conversely, on the average, the value of incident radiation on the terrestrial surface, considering the effect of absorption of the atmosphere, amounts to nearly 1 kW/m$^2$. In any case, the amount of total mean power intercepted by the Earth is estimated at nearly $173{,}000 \cdot 10^{12}$ W.

Besides the direct solar energy, the other sources contributing to the Earth's energy balance include the energy coming from the *gravitational interaction* (involving the triad Earth, Sun and Moon, with a prevalence of the Earth-Moon interaction), as well as the *thermal energy present inside of the Earth*, conveyed to the surface through conduction and convection. A further contribution finally comes from *natural radioisotopes* (radioactive elements), mainly responsible for the phenomenon of *geothermics*.

The Sankey Diagram in Fig. 1.3 shows how a remarkable portion of the incident solar energy ($52{,}000 \cdot 10^{12}$ W) is directly returned to the external space, whereas the remaining portion proceeds towards the Earth's surface, passing through the atmosphere.

[6] As a matter of fact, the causes of the variability of solar flow are manifold: the irregularity of the solar surface, the different emissions by the star, the rotation of the Sun and of the Earth around their axis, the variability of the Sun-Earth distance, etc.

Please note that the energy flows referred to are intended in a time unit, therefore "power" will more precisely be referred to (See Sect. 1.1.2). In this connection, it is suitable to underline the critical aspect of the energy issue: ***although, at any moment, the power available in nature is,*** as described more in detail afterwards, ***at least four orders of magnitude higher than the present needs of the entire world population, it is technically and/or economically impossible to meet energy demands, at any moment and for each user, with the only power that is directly and instantaneously available in nature from the Sun or from the other renewable sources mentioned above.***

***Meeting energy needs is, therefore, a problem relating to the availability in space and time rather than being a problem of quantity. The issue to be solved is that of making available to individual users the suitable quantity of energy (and in the suitable form) they instantly need.***

The Earth's energy balance will be analysed in detail in the next chapters.

## *1.1.5 Classification of Energy Sources*

After considering in which *forms* energy is available in nature and what are the *sources* that ensure the total energy supply of the Earth system, an "operational" classification of the energy utilisation cycle is needed, indicating which *energy sources allow to meet the energy needs of the human society, and which is the contribution that each of them can offer*. To do so, it is necessary to refer back to the diagram of Fig. 1.1, which will represent the starting point for the more detailed classifications hereunder.

In fact, if from a global viewpoint the Sun represents the most substantial energy source for the Earth system, different ways are in place for its processing, storage and availability throughout our planet. This storage allows the availability of new forms of energy that, together with those already seen in the global balance, are defined as ***primary energy sources***. They originate, either directly or indirectly, the energy continuously provided to users for the so-called ***end use***.

However, in order to contribute to end use, primary energy sources cannot be directly used in the forms and quantities available in nature, but rather they *have to be modified and adjusted to the demand*. In addition, the *problems of transportation and cost-effectiveness of exploitation* have to be solved, as well as meeting *technological and environmental needs*, supplying the type, the nature and the quantity of instant energy demanded by users.

### 1.1.5.1 Primary Energy Sources

In the field of energetics, different subdivisions of primary sources have been proposed, which have had several evolutions over time. In order to highlight the results achieved by the energy market through technical and commercial

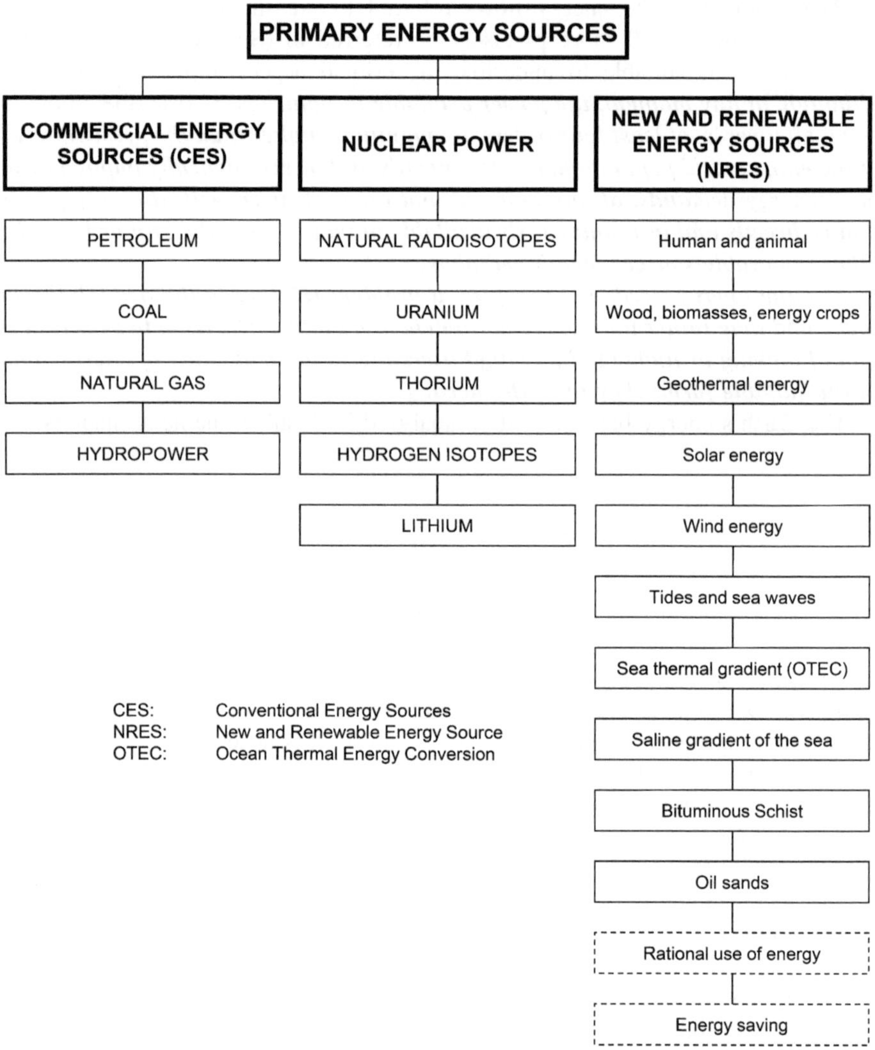


**Fig. 1.4** Classification of primary energy sources

evolutions, and to underline the incidence of use of the different forms of energy, it is suitable to know the usual subdivision of the various primary energy sources into ***conventional, nuclear, new and renewable energy sources*** (Fig. 1.4).

Conventional Energy Sources **(CES)** are characterised by the most extensive use and by a well-established structure in the energy market. Therefore, these sources are commonly defined as ***commercial*** sources.

***Nuclear energy*** is considered separately since, although its role in the energy market is largely well-established and has major possibilities of use, it shows problems that need to be analysed separately. Furthermore, its use as primary source is practically destined to the production of electric power alone (electro-nuclear plants) and propulsion—naval in particular, both surface and submarine—or, eventually destined to space use.

Finally, the New and Renewable Energy Sources ***NRES*** have to be considered: *new*, due to their new (or renewed) introduction in the technical practice; *renewable*[7] since they are characterised by relatively short regeneration time (Table 1.6).

NRES are available in nature with a higher variety of forms compared to commercial sources.

A direct availability of *thermal energy* comes from the sun, from the Earth's endogenous fluid, and from the marine thermal gradient. Wood, biomass, energy crops, the saline gradient of the sea, bituminous schists and oil sands provide potential *chemical* energy. The remaining sources mainly concern *mechanic energy*.

Particular attention must be paid to the *rational use of energy* and *energy saving,* which in modern terminology fall within the most correct definition of ***energy efficiency***. Obviously, these are not actual forms of energy, but their adoption is seen as a virtual energy source, since it allows to reduce the consumption of other forms of energy that have to be employed to achieve the useful effect. This concept is underlined by "measuring" the energy saved – given the same result obtained – in "**negawatt**" (energy that *was not* consumed).

A further classification of some primary energy sources that in a sense represents a variant of the previous one, is shown in Fig. 1.4, which highlights the renewable character of these sources and their position in the energy balance of the Earth System (Table 1.6).

## 1.2 Closed Cycles of Energy Resources

The reference scheme for reaching the sustainability of Energy Systems is referred to as "Closed Cycles of resources". What yesterday seemed impossible, today is the clear and declared objective: to realise energy systems that do not consume resources and do not produce waste. Energy systems based on what we call "Closed cycles of energy resources" can be set-up and run by human technical skills in the twenty first century.

Zero-consumption, zero-waste: this is the objective, together with high efficiency conversion technologies and flexibility of energy end uses.

---

[7] Renewable sources are those sources with a relatively short recovery period, and in any case comparable to the duration of the utilisation cycle (which guarantees their practically unlimited availability over time). Non-renewable sources, on the contrary, are those sources in which the recovery cycle has a duration largely higher than the utilisation cycle. Usually, non-renewable sources include fossil fuels, since the fossilisation cycle necessary for the formation of fossil fuels has a duration amounting to millions of years (which makes these primary energy sources exhaustible or non-renewable from a practical viewpoint). Even nuclear fuels, deriving from natural raw materials which cannot be regenerated with the present technological processes, have to be considered as non-renewable sources.

## *1.2.1 The "Measurable" Definition of Sustainability*

For ages the human society has been consuming resources that it is not able to reproduce. Its activity can be considered as still based on "open cycles", starting from a condition of natural environmental balance that brings to environmental imbalance. The general challenging purpose of scientific and technological research towards sustainability appears extremely clear if based on this assumption: to identify and implement new social and economic systems based on "closed cycles" of resources.

The challenge of closed cycles allows to propose a new definition of sustainability, able to indicate the way leading to—as well as stating the general principles of—sustainable development.

In the energy sector, the implementation of the sustainability scheme of closed cycles shows its real feasibility, together with its theoretical inspiring role. Energy vectors—some of which are well-known and established like heat-transfer fluids, electricity, or others that are not yet as used and widespread as hydrogen—allow to close the energy resource cycle in this field by effectively meeting the zero-consumption and zero-waste objectives through viable technological solutions.

### 1.2.1.1 Open Cycles, Closed Cycles and a New Definition of Sustainable Development

Any technological cycle that brings either a product or a service to our everyday life presently consists of a sequence of activities. From a wider perspective, these activities include:

- extraction of resources
- storage
- transportation
- processing/transformation
- production
- storage and distribution
- use
- waste formation
- partial recycling of materials and product reuse
- waste disposal into water, air and ground.

This cycle, commonly used and apparently normal, implies all the limits that the sustainability concept has brought to evidence in our current development model. It might represent, in a close future, the main element leading to the end of our civilisation.

By only considering the first and the last rings of the chain, it is immediately evident how human beings take any matter—which is in a condition of natural equilibrium with the environment—and transform it into useless waste—which unfortunately, in most of cases, is even dangerous for the mankind, the wildlife, and the environment as a whole [2, 3]. Human activities consume resources that they are not able to reproduce. This is an *open cycle*: a cycle that starts from a status of environmental balance and ends with a condition of environmental imbalance. This cycle "consumes" resources, and "produces" waste.

The human society as seemed to accept this flow as inevitable, mainly because till now human economic development was almost entirely based on such an apparently simple and convenient solution. During the last two centuries, humankind has built its industrial evolution on the exploitation of raw materials which, through production processes, end products and waste disposal, imposed open cycles on the environment. It is usually known where these cycles begin (usually in a mine shaft, or in a well), but most of times it is not exactly known where they end – water, air, ground (in a specific natural reservoir, such as a landfill, or randomly all over the planet, as uncontrolled waste).

The era of open cycles must come to an end. It has been an era of rapid development, but only for a portion of the human population. Consequently, the possibility of finding resources to be consumed and land destined to waste disposal was huge if compared to the percentage of world population that was undergoing development. In the last decades this system evolved towards more and more complex cycles, and now includes recycling and reuse practices. In this way, the main cycle has been enlarged by adding secondary cycles, able to extend the useful life of materials and products. Such extension in useful life is obtained by spending energy and by progressively deteriorating the quality of original materials and products. This can be considered as an excellent practice, but not a final solution: it still brings *from consumption to waste*.

Thinking ahead, cycle corrections of this type cannot be enough. The challenge lays in the realisation of entirely *closed cycles*. Anything but closed cycles, which start from useful resources, and return to useful resources after their use, are unable to realise truly sustainable development: diffused, shared, and ideally endless for the entire human society.

The lesson learned from nature is simple and clear: only those organisms that succeeded in being part of closed cycles are able to survive, whereas those that are imprisoned in an open cycle disappear.

### *A New Definition of Sustainable Development*

Several statements are commonly quoted for defining sustainable development. Two among the most well-known are:

- "We do not inherit the Earth from our ancestors, we borrow it from our descendants".
- "Sustainable development is realised when the current society does not compromise with its development the possibility of development for future generations".

Many other statements apply, and most of them express similar concepts. What is missing to achieve a scientific theory of sustainability is a definition that goes beyond principles, indicating the pathway to follow and not just the objective to achieve. We have to give the Earth "back" to future generations in an appropriate manner—but how? How can we *measure* and evaluate whether a behaviour correctly leads towards sustainable development?

The proposal of *closed cycles of resources* provides enough elements to make a definition of sustainable development, together with stating the following principle:

"Sustainable development does not consume resources.
It uses and re-uses them, endlessly and without environmental consequences"

The trinomial "resource, use, re-generation", as opposed to the current model of "resource, consumption, waste" can achieve a type of development that, by definition, preserves the planet for future generations. The model of closed cycles, being a logical evolution of the current model based on open cycles, continuously provides the tools to measure the sustainability of human activities. Everything that consumes resources is, by definition, unsustainable. The more a process *preserves resources*, the less it *produces waste*, the closer it is to sustainability.

The challenge of sustainable development lays in stopping exploitation and using cycles with the re-formation of resources and with the elimination of waste generation. The closer our world gets to the ideal model of closed cycles, the closer it will be to sustainable development. The progress resides in the management of resources and waste, with the only exception of entropy increase: entropy increase is the only "waste" that it is possible to accept, and only its increase has to be considered as the thermodyanamic "limit towards infinity" of human development on Earth.

*The Application to Energy Systems*

The cycle of coal, oil and natural gas began several millions of years ago with organisms that fixed atmospheric carbon in their structures. They reached us through chemical-physical reactions that lasted millions of years and caused the formation of the most suitable natural energy resources that today's fast-growing and power-intensive society has found in nature ready to use.

The period of new formation of fossil carbons is several millions of years, and it is much greater than the use and consumption period, which is compressed by human activities into a number of years or decades. For this reason, the cycle of fossil fuels has to be considered as completely open from the human point of view. By using fossil fuels, our society quickly consumes incredible amounts of energy resources that the environment is absolutely unable to reproduce at the same pace [4–6]. Additionally, the consumption of energy resources entails chemical reactions that release substances harmful for the human health and for the environment: significant amounts of carbon monoxide, sulphur oxides, nitrogen oxides, hydrocarbons, particulate and traces of thousands of other compounds. Carbon dioxide represents a separate issue: it is released into the air through the combustion of fossil fuels in the form of "natural" re-coupling between carbon and oxygen. $CO_2$ is not a directly dangerous pollutant for human health, but its massive re-loading in terrestrial atmosphere in the time of a "geological instant" is a real threat to the climate.

The open cycle of fossil energy resources takes billions of tons of "potential" $CO_2$ securely stored underground in deep oil, coal and natural gas deposits, and releases them in the atmosphere. The scientific community is nearly unanimous in indicating an unsustainable increase in the Earth's temperature if this open cycle is not remarkably changed from its present trend:

$$\text{Hydrocarbons} \rightarrow \text{Carbon dioxide}$$

Paying attention to the small number of technicians and scientists who express doubts and scepticism about the projected temperature increase as a result of $CO_2$ and other greenhouse gas (GHG) emissions, it is evident that their scepticism is focused on the indicated effects of the GHG content variation in the atmosphere, and not on the presence of a tangible risk for the natural balance. Almost all of them oppose the calculations and projections on the effects of the change, and not the change itself. And most of them do not deny the presence of a risk.

Therefore, the vast majority of the scientific community identifies the cause and the effect as follows:

$$CO_2 \text{ increase} \rightarrow \text{Climate instability}$$

and a small minority does not agree on the effects, while agreeing on the cause. They state that anthropic carbon dioxide emissions only lead to a risk, and not to the certainty of a climate change:

$$CO_2 \text{ increase} \rightarrow \text{Unidentified Risk.}$$

The common element is at least the presence of a "risk", such as the risk of losing the environmental balance.

Even the chain of nuclear power is characterised by an open cycle:

$$\text{Radioactive materials} \rightarrow \text{Enriched materials} \rightarrow \text{Nuclear power production} \rightarrow \text{Radioactive waste disposal.}$$

Furthermore, in this case the key is the presence of a "risk" (during and) after the useful power production period:

$$\text{Radioactive waste} \rightarrow \text{Risk.}$$

Consequently, the risk of losing the environmental balance can only be avoided by the energy sector only if it abandons the scheme of open cycles of resources.

*Closed Cycles of Energy Resources*

Obtaining useful energy from *closed cycles* of resources is possible according to the Laws of Physics, although on a first analysis the integration of an energy vector in the energy system could be seen as an energy-wasting loop. The First Law states that to obtain an energy vector, at least the same amount of energy that it will return has to be spent. The Second Law states that to obtain an energy vector the amount of energy to be spent is always *higher* than stated by the First Law, and the energy it will return is always lower than the energy spent.

Therefore, the simple analysis of efficiencies would lead to conclude that the use of energy vectors has to be generally avoided and that it always brings to a loss in terms of use of primary resources. The energy spent has to be higher than the energy obtained.

The analysis is correct, as much as it is incomplete.

It is necessary to add to this well-known analysis framework that the "*use*" of a given amount of primary energy has a completely different meaning when the source of this energy is renewable, if compared to the "*consumption*" of a non-renewable one. In this case the concepts of "use" and "consumption" show a great difference in their meaning. And the difference between energy spent and energy obtained when passing through an energy vector has to be considered as an acceptable energy cost to be paid – when realising energy systems that do not "consume" resources that cannot be re-formed on Earth in the due time, and follow cycles that can theoretically reduce to zero their impact on the environment, both in terms of resource consumption and waste formation. So although the definitions of the term efficiency are correct, their consideration has to be reconsidered in the new "closed cycles" approach to energy systems for sustainable development.

The definition of sustainable development, based on the basic scheme of closed cycles, allows to objectively compare, for the first time, different development solutions from the point of view of sustainability. In fact, if development cannot be defined as sustainable until it consumes resources, the discrimination between sustainable and non-sustainable solutions becomes evident.

The application to the energy sector shows that developing human society without depleting the Earth's resources is possible by the use of incident solar energy. In this case, a closed cycle of resources can be acheived that respects the principle of "zero consumption and zero waste", also showing that it already has the potential for being technologically realised.

The practical implementation of closed cycles of energy resources can be achieved by exploiting renewable resources and structurally integrating energy vectors in the system. The inclusion of energy vectors (to be produced from appropriate primary resources) in the chain of energy systems becomes the key concept for the development model and, in a large-scale application, it would allow to meet all the energy end uses that human society needs for its social and economic development.

### 1.2.2 The Earth, as an "Open" Energy System, Allows the Realisation of Closed Cycles

The practical implementation of closed cycles of resources has long been deemed as impossible in the energy sector, since it was considered as non complying with the First and Second Law of Thermodynamics. Some confusion was generated between the concept of closed cycles of resources and the concept of closed energy cycles. The latter is evidently and clearly impossible to achieve, due to the qualitative degradation of the energy resource in any conversion, revealed by the increase of entropy, or related to the decrease of the energy content.

Fortunately, however, the particular conditions of the Earth characterise its energy status as this of *an open system rather than a closed one*; in this way, the real scenario in which actions are made in relation to the theoretical energy analysis changes completely.

The Earth constantly receives energy from the Sun. The planet-system, therefore, does not have an equalised situation in terms of energy balance, but rather, conversely, a situation of own energy balance largely positive that allows to hypothesise what would be considered as impossible without taking into account this simple but peculiar starting point. The constant and distributed inflow of additional energy in the Earth system allows it to be practically defined in energy terms in conditions of constant energy surplus.

This daily available – and thence usable – surplus is at the basis of the incredibly charming possibility of developing on Earth energy systems based on closed cycles of resources and on renewable forms of energy. What is renewable, in fact, is not the energy detected and exploited by a suitable energy system, but rather its cyclical and repeatable natural availability, largely linked to the new quantities of energy that from the Sun reach the surface of the Earth.

The continuous availability of an energy source, assimilable in human activities and practically inexhaustible, allows to face the energy spending necessary for the production of energy vectors, their exploitation, and the replenishment of the primary resources available.

## 1.3 Energy Systems

What is an energy system?

An unequivocal and standardised definition of energy system is necessary to perform any activity involving the development, analysis, planning and improvement in the energy sector as well as in the technologies suited for its exploitation and targeting to the carrying out of human activities.

Very often the definition of *energy system* coincides with the definition of *energy conversion system,* and in turn this is easily mistaken with the definition of energy conversion technology, which is traditionally considered as a technical

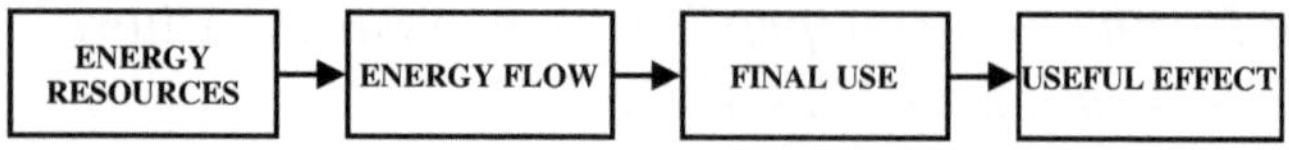


**Fig. 1.5** Elementary (or simple) energy system/theoretical

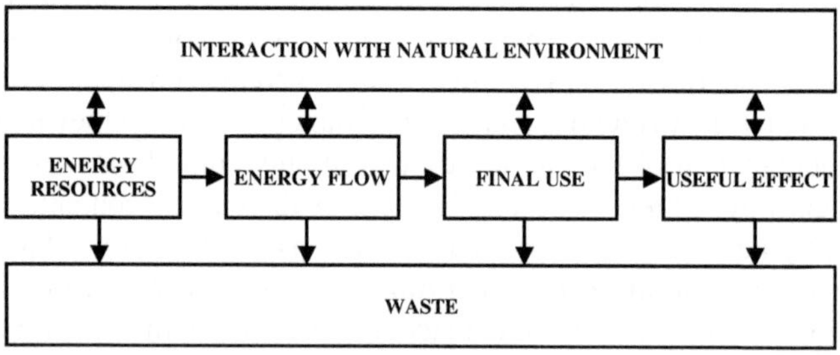


**Fig. 1.6** Simple energy system/real

solution aimed at leading a fluid to perform a thermodynamic cycle. It is possible to see how in these definitions the field of analysis is excessively limited and even targeted to specialists.

The modern needs for elasticity, flexibility, modularity and cross-sector solutions expressed by the human society are inappropriate for such sectorialisation.

Energy systems, therefore, have to be defined in a way that allows to exploit the choice, the comparison and the competition among different solutions in a scenario that takes their efficiency into consideration by referring to the entire energy cycle, from the primary resource exploited to the useful effect produced and with a clear indication of the impact caused on the environment in the different components of the energy chain that leads from the source to the useful effect.

An energy system is therefore a system that avails itself of energy resources to produce a useful effect (Fig. 1.5). The useful effect is obtained through an energy flow, and it may involve mechanic energy, lighting energy, thermal energy and electric/electronic energy.

The energy system must be therefore able to extract energy from the original resource (energy resources), to make it flow in the most appropriate form (energy flow) and to use it through suitable technologies able to make energy available in the form and with the appropriate characteristics for end use, in order to allow the realisation of the desired useful effect.

All this inevitably happens within a process of continuous interaction with the natural environment and also—at present in all the cases involved—through waste generation.

The scheme of a **simple energy system/real** is therefore more complex, and shown in Fig. 1.6.

According to this general definition, it is evident that an energy system can be more or less complex: in fact, it may encompass the entire world's energy system, or a single plant, or even a single household appliance (where a single equipment

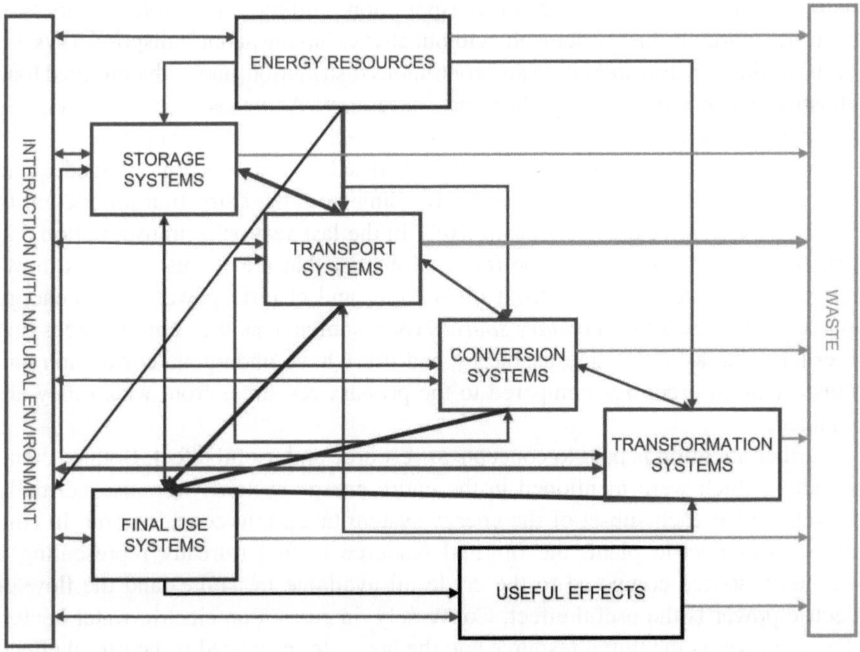


**Fig. 1.7** Complete energy system/real

includes the energy resource available in that case, the end use, and the achievement of the useful effect) passing through any possible intermediate "dimension". A complex energy system represents in this way the set of several simple energy systems interacting among themselves.

In its widest meaning, an energy system is made up of the set of primary energy resources, the processes and the relevant technologies for energy transformation, conversion, transportation, storage and end use, as well as the waste produced and the interactions with the environment generated in all the phases and processes mentioned above.

The basic structure of a **complete energy system** is therefore shown in Fig. 1.7.

In its most detailed representation, an energy system is made up of the set of primary energy resources, processes and technologies for conversion, transportation, storage and end use of energy, waste production and interactions with the environment in all the different phases.

The concepts implied in the definition of energy system (resources, processes and technologies for energy transformation, transportation, storage and end use) will be tackled more extensively afterwards; it looks however suitable and useful to immediately make a brief reference to them.

The "pathway" from the primary source to energy use, generally entails different needs.

First of all, a primary energy source is usually available in places other than the points of utilisation, which are also called end users. This poses – practically in the majority of cases – the problem of transportation. Furthermore, in order to meet at any time a variable energy demand, without always having at one's disposal a system for its production, transportation and continuous distribution, and with unlimited load adjustment capacities, storage becomes necessary. A means that allows energy transportation and/or storage is defined *energy vector* (or energy career).

Another need is that of having at one's disposal a form of energy with adequate characteristics at any step of the pathway going from the extraction-production to end users, which we define "*well-to-use*". In the last section of transportation and distribution, and for end use, the forms of energy that are extensively used nowadays include fuels deriving from oil refining and electric power. These energy sources are defined as *secondary sources* (or resources), as they are resources that in order to be available and usable by end users have undergone conversion and transformation processes compared to the primary resources from which they are obtained.

Further attention is paid to concepts as resource and useful effect. Both of them, in fact – which were mentioned in the entire energy system – are also normally considered for each subset of the energy system taken into consideration. In case of a thermoelectric plant, the original resource is fuel (already representing a secondary source compared to the crude oil available in nature) and the flow of electric power is the useful effect. Conversely, in case of an electric water heater, electric power is the initial resource and the hot water produced is the useful effect.

It is now suitable to provide a general definition – in case of a complete energy system (well to use) – of the beginning and the end of the process: in this case *energy resources* indicates the primary energy sources (namely, those available in nature) whereas *useful effect* means the effect desired and obtained by end users. In the case of room heating, the useful effect is the achievement and the keeping of the desired temperature in the room, and not the quantity of heat with given characteristics (that is to say a given flow of heat at a given temperature) necessary to air-condition the room under consideration. Similarly, the useful effect in case of mobility and transportation is the movement of people or goods from one point to another, and not the energy employed to perform such movement.

The energy necessary to obtain the useful effect, in fact, varies (both in terms of form and quantity) according to the technology used (air heating, thermal radiators; car, train, ship, and aircraft) and the modes carried out (central heating, individual heating; collective transportation, private transportation).

As concerns the definitions of primary and secondary source, and energy form for end use and energy vector, the following consideration is due: **definitions are not necessarily alternative**. *A primary energy resource may also be a vector*, as it can be the case of a secondary source, etc. Oil exists in nature, therefore it is a primary energy source, and has such characteristics that allow it to be transported and stored, therefore it is also an absolutely complete energy vector. The same goes for its derivatives (gasoil, petrol, fuel oil, LPG, etc.): they are the result of a process of transformation of a primary source; they can be transported and stored, *and* therefore they are secondary sources and energy vectors. Finally, electric energy is a form of energy that can be transported and not easily and economically stored, and it is as well a secondary resource and energy vector (however not "complete", meaning that it only owns one of the two characteristics applying for the definition of vectors: *yes* for the transportation capacity, *no* for the storage capacity).

## References

1. Ippolito F (1976) L'energia: fonti e produzione. Ed. Le Scienze, Milan
2. Dunn PD (1986) Renewable energies: sources, conversion and application. Peter Peregrinus Ltd., London
3. Makhijani A (1997) Economic and political weekly. Bombay
4. UN (2011) Energy Statistics Yearbook and Energy Statistics Newsletter, United Nations, New York, USA
5. BP Statistical Review of World Energy (2010) London
6. Key World Energy Statistics (KWES 2010) International Energy Agency, Paris

# Chapter 2
# Energy Resources

## 2.1 Definition of Energy Potential

A general assessment of the availability of energy resources[1] is essential for whichever long-term energy policy. As concerns renewable energy sources (RES), the difference vis-à-vis the most widespread fossil sources (RES are regularly-renewed energy flows), whereas fossils can be considered as "fixed" resources) induces to make some preliminary remarks about the definition of RES potential (Fig. 2.1).

The *theoretical potential* of a renewable energy source (like sun, wind, biomass, water, geothermal, etc.) is the annual energy flow without technical–economic references, estimated through a direct analysis. This analysis is made through the maps that take into consideration, respectively, the solar radiation equivalent (annual based), wind speed and distribution, biomass index, water capacity (difference between precipitations and evaporation), hydraulic heads (like differences of geodetic heights), geothermal fields, etc. Some authors add physical and social-economic considerations (slope of the land, eventual road access, presence of electrical grid, gas pipeline and aqueduct proximity to the power plant, etc.) and therefore determine the *theoretical available potential*. For fossil fuels alike, *global occurrences* take into account all the types of resources, either identified or not identified, both economic and not economic (also in this case, the resource available is considered without technical–economic limitations).

The *technical potential* of renewable energy sources is the fraction of the theoretical potential that current technologies allow to exploit. The equivalent for fossil resources is represented by resources (being economic and not economic, identified and not identified, excluding the resources that current technologies do not allow to exploit).

[1] World Energy Council (WEC) in 1998 defined *energy resource* a primary energy source with a demand and conversion and use technology.

F. Orecchini and V. Naso, *Energy Systems in the Era of Energy Vectors*,
Green Energy and Technology, DOI: 10.1007/978-0-85729-244-5_2,

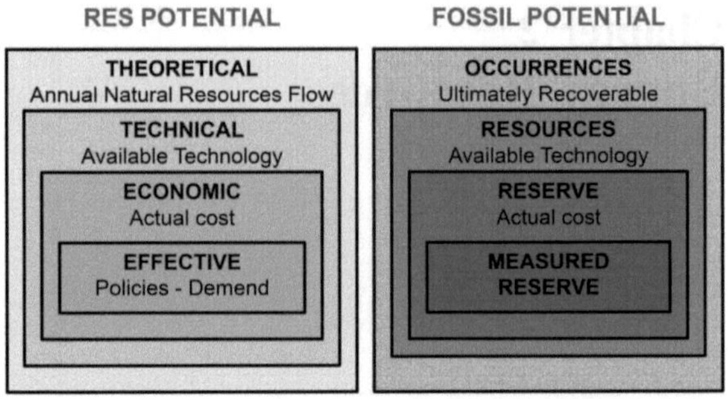


**Fig. 2.1** Potential of renewable and fossil resources

The *economic potential* is the fraction of the technical potential that is also cost-efficient to exploit. The equivalent for fossils resources is represented by the *reserves* (verifiable, possible and feasible). Finally, the *effective potential* of the RES is the fraction of economic potential exploited, taking into account the demand and the energy policies adopted in each country. As far as fossils are concerned, considering an existing demand and a "favourable" policy, the effective potential is analogous to the *verifiable reserves*, being the resources effectively measured, whose production is possible according to the economic and technological conditions.

Obviously potentials are influenced by several factors (environmental, technical, economic and social), therefore each numerical value changes dynamically according to environmental changes, technological development, market conditions and the relevant political situations.

## 2.2 The Earth's Energy Balance

The analysis of energy availability in nature starts from the global energy cycle of the system in which human beings operate: the Earth system. The Earth is not exchanging matter but only energy with the outside, mainly electromagnetic energy from the Sun and releases it in the space (so keeping its average temperature constant). Besides the direct *solarpower* (accounting for around 99.98% of the entire energy available on Earth) the other sources contributing to the energy balance of the Earth for the remaining 0.02% include the energy coming from *gravitational interaction* (practically reduced to the triad Earth, Moon and Sun), as well as the *thermal energy present inside of the Earth*, conveyed to surface through conduction and convection. A further contribution finally comes from *natural radioisotopes* (radioactive elements).

Solar radiation is radiant energy emitted by the Sun from a nuclear fusion reaction that generates electromagnetic energy. The spectrum of solar radiation is close to that of a black body with a temperature of about 5,800 K (according to Wien's Law, derived from Plank's Law, the Sun emits radiations with a peak wavelength of 500 nm). About half of the radiation is in the visible shortwave part

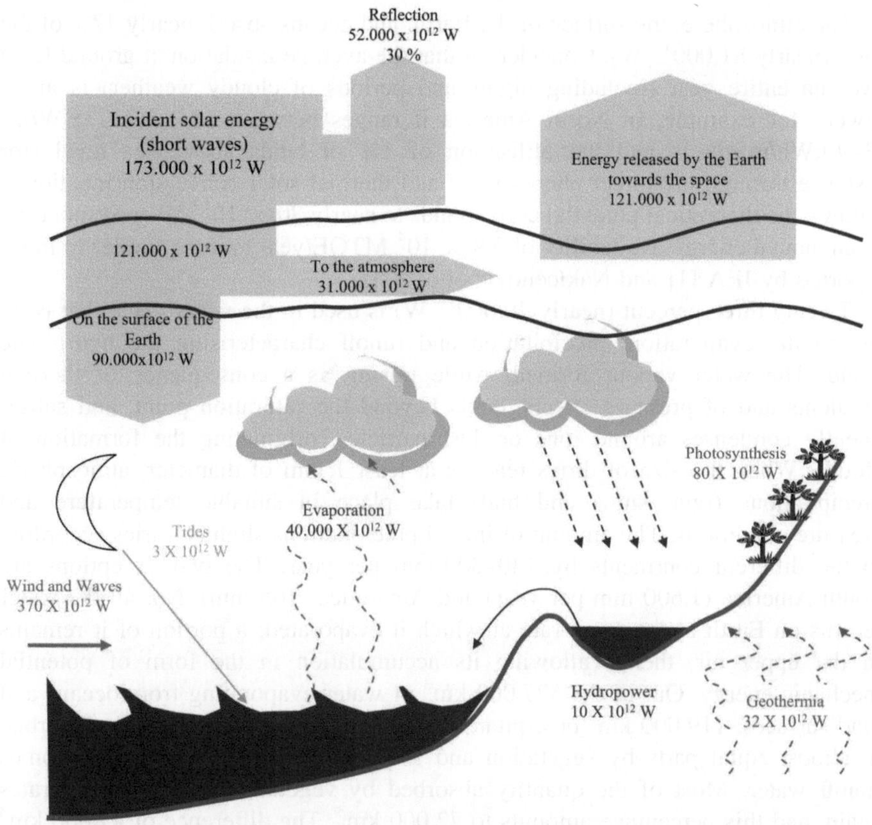


**Fig. 2.2** Instant energy balance of the Earth [3]

of the electromagnetic spectrum. The other half is mostly in the near-infrared part, with some in the ultraviolet part of the spectrum.

The average energy flow coming from the Sun is called *solar constant* and corresponds to the energy that, outside of the atmosphere (that is to say, without atmospheric absorption phenomena) and at the average distance of the Earth from the Sun, passes through a surface of 1 $m^2$, perpendicular to the direction of the flow, in the interval of 1 s. Measurements made through artificial satellites calculated for the solar constant a value of 1,353 $kW/m^2$, with a $\pm 3\%$ variability due to the ellipticity of the Earth's orbit. Considering the solar constant and the average Earth surface irradiated by the Sun, the estimated amount of the total average power intercepted by the Earth is nearly $173 \times 10^{15}$ W, to which corresponds an annual incident quantity of $5{,}404 \times 10^{21}$ J/year (that is to $1.3 \times 10^{8}$ MTOE/year).

As shown in Fig. 2.2, nearly 30% of the solar incident radiation is reflected at its arrival and dispersed in the external space under the form of short wavelength radiation (nearly $52{,}000^{12}$ W), whereas the remaining part proceeds towards the surface of the earth, through the atmosphere.

The atmosphere, the surface of the Earth, and oceans absorb nearly 47% of the total (nearly $81{,}000^{12}$ W). Considering that the average insulation at ground level over an entire year (including nights and periods of cloudy weather) is much lower—for example, in North America it ranges between 125 and 375 W/m$^2$ (3–9 kWh/m$^2$/day)—and the utilisation of 1% of lands above sea level (for instance through the direct photovoltaic and thermal solar conversion mentioned below), the theoretical potential corresponds to nearly $80 \times 10^{12}$ W corresponding to an annual energy availability of $5.9 \times 10^4$ MTOE/year (values similar to those reported by IEA [1] and Nakícenovíc et al. [2]).

Twenty three percent (nearly $40{,}000^{12}$ W) is used in the water cycle that is, to say in the evaporation, precipitation and runoff characterising the hydrologic cycle. The water vapour formed, while raising as a consequence of thermal gradients and of pressure effects, cools beyond the saturation point, and subsequently condenses around ions or dust particles, originating the formation of clouds. When the size of drops reaches at least 1 mm of diameter, atmospheric precipitations (rain, snow and hail) take place in suitable temperature and pressure conditions. The amount of inland precipitations slightly varies according to the different continents by 740–800 mm per year. The two exceptions are South America (1,600 mm per year) and Antarctica (165 mm). Not all the water returns on Earth at the same rate at which it evaporated; a portion of it remains in the upper air, thence allowing its accumulation in the form of potential mechanic energy. Out of the 577,000 km$^3$ of water evaporating from oceans and land surfaces, 119,000 km$^3$ precipitate on the soil. About two-thirds are absorbed in almost equal parts by vegetation and soil; the remaining one-third becomes runoff water. Most of the quantity absorbed by vegetation and soil evaporates again, and this percentage amounts to 72,000 km$^3$. The difference of 47,000 km$^3$ is, in principle, available for energy purposes. The convolution of runoff water allows to take into consideration the theoretical hydropower potential, whose amount (theoretical potential) is estimated as $5 \times 10^{12}$ W, with an annual energy amounting to $3.6 \times 10^3$ MTOE [13]. The largely prevailing energy use of this resource is electricity production in hydroelectric power plants; this is the reason for the word *hydroelectricity*.

A small fraction of the sun energy, amounting to 0.2% (nearly $370 \times 10^{12}$ W), is destined to winds, ocean circulation, and sea state, and is partly dispersed in the form of frictional heat. Wind is caused by heterogeneous air warming: the air above ground is warmer than air above the sea, and as a consequence of density variation, the warmest air rises and triggers the movement of air masses (AM). A region's mean wind speed and its frequency distribution have to be taken into account to calculate the amount of electricity that could be produced by wind turbines. Wind resources can be exploited in areas where wind power density is at least 400 W/m$^2$ at 30 m above the ground. If we consider the exploitable part of wind, then the technical potential is estimated around $15 \times 10^{12}$ W, to which corresponds an annual energy amounting to nearly $10 \times 10^3$ MTOE.

An even lower quantity of the sun energy, nearly $40 \times 10^{12}$ W (0.02%), is transformed from the chlorophyll contained in plants that fixes carbon in leaves and stores energy under the form of carbohydrates. Furthermore, this process releases oxygen and, through the decomposition processes, dissipates energy. However, a large quantity of organic matter is stored and accumulated in peat bogs or, in any case, in environments poor in oxygen that do not allow the return to balance conditions. The storage below the crust of the Earth of animal and vegetable remains is at the basis of the complex phenomenon relating to the formation of fossil fuels, which in this way end up storing the solar energy absorbed for millions of years by the Earth. According to estimates, the conversion from *biomasses* might provide $27 \times 10^{12}$ W ($2 \times 10^4$ MTOE per year).

The gravitational systems Earth–Moon and Earth–Sun are at the basis of the phenomenon of tides. The actions of the two systems, however, are combined in an extremely variable way. The movements of the Moon vis-à-vis the Earth take place, in fact, on an orbital level different from the Earth's equatorial plane and, on the other side, because of orbit eccentricity, the periods of the two celestial bodies differ, as well as the mutual distance and the distance to the Sun. Out of the two systems, the Earth–Moon system influences the phenomenon to a higher extend, due to the relative closeness of the Earth from its Satellite (mean distance $384.4 \times 10^6$ m), although the mass of the latter is 81.3 times smaller than the Earth's and over 2,500,000 times smaller than the Sun's. As regards the Earth–Sun system, given the higher distance between the two bodies (circa $149 \times 10^9$ m), it generates minor tidal forces, although not negligible, in consideration—this time—of the huge mass of the Sun (nearly 333,000 times as much as larger than the Earth's). The other celestial bodies, including those of the solar system, on the contrary, do not have any appreciable influence. Overall, the main gravitational action is the one that the Earth exchanges with the Moon, whereas the Sun only modifies the magnitude of tides. The tidal force peaks when the three celestial bodies are aligned and the lunar attraction adds up to the solar one, whereas it is at minimum levels when the Moon is at its first and last quarter (quadrature), namely when the line connecting the centre of the Earth and the centre of the Moon is perpendicular to the line connecting the centre of the Earth and the centre of the Sun. The level of springtides and of quadrature tides has a deviation of nearly 20%, compared to mean values. The power made instantly available by tides is evaluated as amounting to $3 \times 10^{12}$ W ($2 \times 10^3$ MTOE).

Geothermal power is a so-called energy source "of formation": in fact this Energy derives from star formation processes that originated the Earth itself. The heat flow coming from the internal part of the Earth was evaluated through geothermal gradient measurements as well as through the thermal conductivity of rocks. Vertically proceeding towards the interior, in the first kilometres a temperature mean increase of 3°C every 100 m is recorded; however, this increase is not constant, since the maximum temperature of the nucleus of the Earth amounts to nearly 4,500°C. By means of these measurements, a mean value of the energy flow was determined, amounting to 0.063 W/m$^2$; having considered that the Earth's surface amounts to nearly $510 \times 10^{12}$ m$^2$, an overall power of about

$32 \times 10^{12}$ W is obtained, that is $2{,}4 \times 10^4$ MTOE per year. Thermal sources and volcanoes, however, emit only 1% of total heat, thence the power actually available is reduced to nearly $0.3 \times 10^{12}$ W. Heat migrates from the hot nucleus of the planet towards the Earth's crust through two main processes: *conduction*, which contributes to a minor extent since rocks are poor conductors of heat, and *convection*, which conversely rapidly spreads thermal energy in fluid materials. Although heat propagation takes place towards the surface of the Earth, and from here towards the space, the temperature inside of the nucleus is rather constant thanks to the contribution of radiations and of the particles released by the radioisotopes present inside of the Earth's crust. Therefore, geothermal energy mainly derives from the decay of radioactive substances inside of the Earth and, having considered the quantities involved, it represents a practically inexhaustible source of energy.

The energy flows and figures making up the energy balance of the Earth are shown in Sankey diagram, where these flows are represented through bands whose width are propotional to the total contribution given by each item. In this framework, another interesting representation is shown, which highlights the water cycle (at the basis of the life on our planet and that, also from the energy viewpoint, has huge proportions and plays a major role).

It is important to consider the critical aspect of the energy issue: although, at any moment, the power available in nature is—as described afterwards—at least four times as much as higher than the needs of the entire world population, it is technically and/or economically impossible to meet energy needs at any time and for each user, through the only power that is directly made available in nature by the Sun or by the other sources mentioned above.

Meeting energy needs, therefore, is not a problem in terms of quantity but rather of availability of quantity in quality (suitable form), in the space, and at the time needed by users.

## 2.3 Renewable Sources

### *2.3.1 Solar Energy (Direct)*

If, on the one hand, solar energy has several advantages (such as the absence of polluting agents, the possibility of direct conversion into electric power, a practically unlimited duration as well as a global quantity available far higher than human kind's energy needs), on the other hand several problems must be overcome for its widespread use, including technical and economic ones, linked to the low energy density, and to its discontinuity (night/day alternation, seasons cycle, variation in weather conditions). It is clear, then, how big is the gap between potential availability and practical possibilities of use.

#### 2.3.1.1 Overview

As already mentioned in Sect. 2.2, outside of the Earth's atmosphere the incident power on an area perpendicular to the direction of Sun rays slightly exceeds 1,300 W/m$^2$ and this value is named *solar constant*. In its passage through the atmosphere, however, the spectrum of solar radiation changes both quantitatively and qualitatively. More in detail, the interaction of solar radiation with the earth's atmosphere and surface is mainly determined by 3 groups of factors:

1. The Earth's geometry, revolution and rotation (declination, latitude, solar angle);
2. Terrain (elevation, surface inclination and orientation, shadows).
3. Atmospheric attenuation (scattering, absorption) by:

    a Gases (air molecules, ozone, $CO_2$ and $O_2$).
    b Solid and liquid particles (aerosols, including non-condensed water).
    c Clouds (condensed water).

The first group of factors determines the extraterrestrial radiation available according to the solar position above the horizon, and it can be precisely calculated through astronomic formulas.

The radiation input to the Earth's surface is then modified by the topography of its terrain—namely slope, inclination and aspect—as well as by the shadowing effects of neighbouring terrains. This group of factors can also be modelled at a high degree of accuracy. The elevation above the sea level determines the attenuation of radiation by the thickness of the atmosphere.

The intensity of the extraterrestrial solar radiation passing through the earth's atmosphere is attenuated by various atmospheric constituents, namely gases, liquid and solid particles and clouds. The path length through the atmosphere is also critical. Because of its dynamic nature and complex interactions, the atmospheric attenuation can be modelled only at a certain level of accuracy.

The quality of radiation, that is to say the distribution of wavelengths in the solar spectrum, undergoes a variation due to the scattering phenomenon, based on which the photons of different wavelengths are spread according to different angles. The radiation that reaches the ground has then a different composition from the radiation originally released by the sun.

To quantify the spectrum of ground radiation, reference is often made to the concept of AM: it is defined as the ratio between the length of the real path of Sun rays and the length of their shortest possible path, that is to say with the Sun at the zenith. By convention, it is agreed that outside of the atmosphere the solar spectrum has an AM equal to 0. At the sea level, on a clear summer day, the radiation coming from the Sun at the zenith corresponds to AM1, whereas other positions correspond to an AM of 1/sin h, where h is the zenith angle (Fig. 2.3). As a reference magnitude for the measurement of cells and of photovoltaic modules, an AM of 1.5 is used.

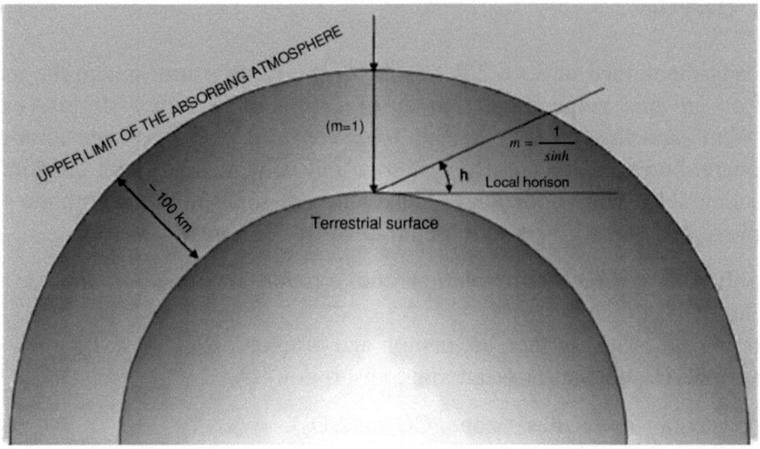


**Fig. 2.3** Air mass (AM) and schematization of the solar radiation incidence angle

**Fig. 2.4** Comparison between the extra-atmospheric solar spectrum (AM0) and a ground solar spectrum at ground level (AM1)

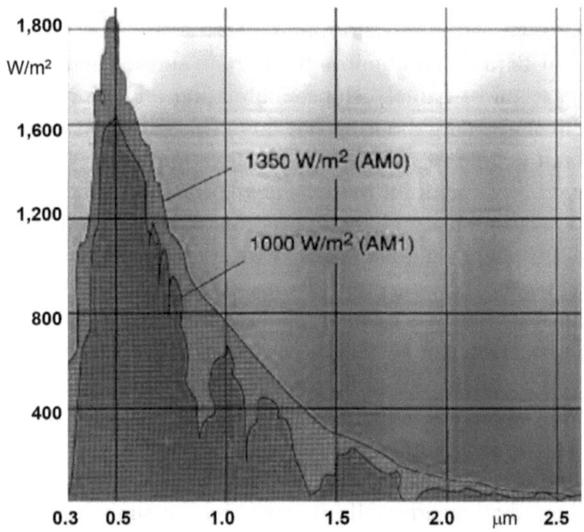


An interesting comparison concerns (Fig. 2.4) the extra-atmospheric solar spectrum (AM0) and an atmospheric solar spectrum (AM1), where the already mentioned qualitative and quantitative differences between the two radiations are highlighted.

The radiation impacting on a given surface, called *global solar radiation*, is the sum of three components (Fig. 2.5). The component that is neither absorbed nor spread by atmospheric particles, and that directly reaches the surface in question with a single and well-defined angle, is called *direct* radiation. The component

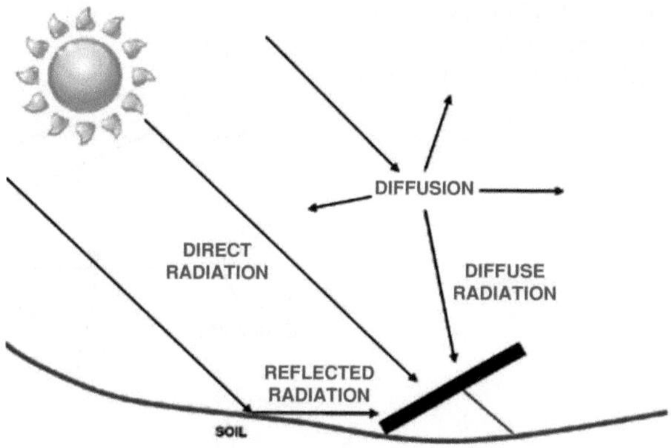


**Fig. 2.5** The components of solar radiation

that, through the atmosphere, is deflected in random directions (scattering), mainly due to the particles of water vapour in the air, but that in any case succeeds in impacting the surface, is named *diffuse* radiation. The grey areas, that is to say the areas of surfaces not reached by direct radiation due to the presence of hindrances, are not completely darkened, also thanks to the contribution of diffuse radiation. The radiation that reaches the surface after being reflected by the ground or by other surfaces is called *reflected* radiation. As concerns the tendency to reflect incident energy, the major contribution is given by light surfaces (for instance, areas covered with snow in winter) whereas dark surfaces (as ploughed fields or grasslands) are characterised by a high absorption coefficient.

The relative importance of these three components varies depending on:

- *Weather conditions* In a cloudy day, the diffuse component prevails over the direct one.
- *Inclination of the collector surface* During summer months, a surface with a minor inclination on the horizontal plane shows a higher contribution of direct radiation, and the opposite happens in winter months.
- *Period of the year* In a collector surface placed vertically, the contribution of direct radiation will be lower in summer months, and the opposite applies in surfaces with a minor inclination.
- The global value of the radiation reaching the ground is subject to:
  - Regular variations (daily and annual) due to the apparent motion of the Sun in the sky.
  - Irregular variations due to weather conditions (cloudiness, humidity, etc.).

The position of a surface that allows to maximise the quantity of solar radiation collected is the one in which this surface is oriented southward,[2] with a tilt angle

[2] This condition is true for the regions of the Northern Hemisphere placed north of the Tropic of Cancer (lat. 23°27'N); in the regions of the Southern Hemisphere placed south of the Tropic of Capricorn (lat. 23°27'S) the suitable orientation is northward; finally, in case of the regions of the tropical zone, it depends on the period of the year.

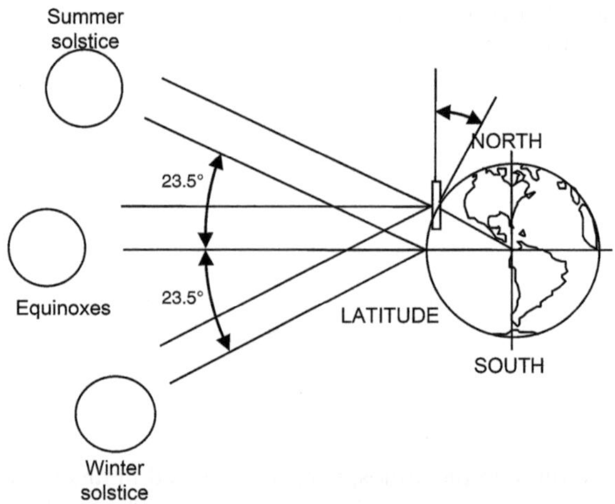


**Fig. 2.6** Radiation variations depending on the annual oscillation of the earth's axis

equalling the latitude of the site. Orientation optimises reception during the entire day, whereas inclination minimises the radiation variations due to the annual oscillation of the earth's axis that determines a variation in the direction of Sun rays compared to the vertical axis (Fig. 2.6).

#### 2.3.1.2 Resource Evaluation

Let us turn now to the calculation of global solar incident radiation on a surface placed in any position. Clearly, this is the most interesting calculation for the evaluation of electric power production potential by a photovoltaic system installed in a given site.

The main sources to estimate the incident solar radiation in a given period on a surface are the Solar Radiation Atlas. They collect data mostly provided by national meteorological institutes or, in exceptional cases, by universities. This set of data is collected in maps and tables, and usually derives from 10-yearly measurements. In the case of Europe, the European Solar Radiation Atlas is divided into two volumes: the first one relates to horizontal surfaces; the second volume concerns tilt surfaces. In the first volume of the Atlas tables show the values of mean daily radiation for each place considered, expressed in Wh/m$^2$ (o kWh/m$^2$), that is to say in *energy density*.

Each place is characterised by its coordinates (latitude and longitude), and by its altitude above the sea level. The most important value is the one appearing in the first row of each table, indicated with letter G. It expresses the mean daily radiation and is shown for each month and then as annual mean value.

The values of incident global radiation on a horizontal surface; however, have a limited interest for the design and development of photovoltaic systems, since, in order to optimise their electric output, these systems are always arranged with a certain

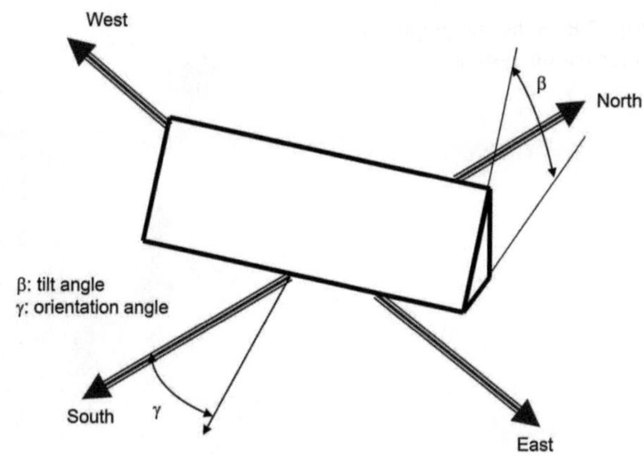


**Fig. 2.7** Angles characterising the position of a surface

inclination on the horizontal plane. Therefore, the determination of the incident solar energy on a tilt surface with a given angle becomes of paramount importance.

This problem can be solved through two different approaches: the consultation of the second volume of the European Solar Radiation Atlas, or the mathematical calculations of the radiation on a tilt surface starting from the corresponding data for a horizontal surface.

The first approach resorts to the second volume of the Atlas, which refers to tilt surfaces. For each place, tables show the mean daily radiation (both global and diffuse) per each month and on an annual basis, corresponding to different positions of the collector surface. This position is characterised by two angles (Fig. 2.7):

- The tilt angle (shown in the figure with letter $\beta$) that measures the tilt of the surface considered, as compared to a horizontal plane.
- The orientation angle (or azimuth, shown in the figure with letter $\gamma$), which indicates the orientation of the surface vis-à-vis the cardinal points (the orientation southward corresponds to a null azimuth); more precisely, it is the angle that the projection on the horizontal plane of the normal surface forms with the southward direction in the northern hemisphere (or northward in the southern hemisphere).

In order to know the values of solar radiation according to the reference geographical site, suitable maps, called *iso-radiation* maps, are used. Radiation values mainly depend on orographic characteristics and, to a lower extent, on latitude variations.

It is necessary to specify that, in most of data available, no distinction can be made between the component of direct radiation of diffuse radiation.

Maps show the distribution of daily mean radiation for the different months and on an annual basis (Fig. 2.8). The black lines in the maps are the isoradiative

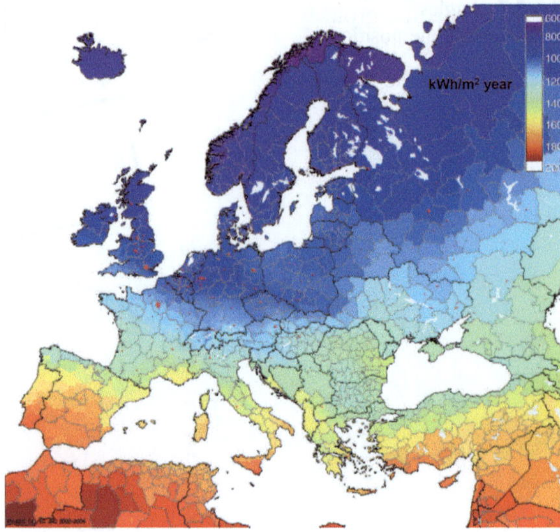


**Fig. 2.8** Solar geographical information system

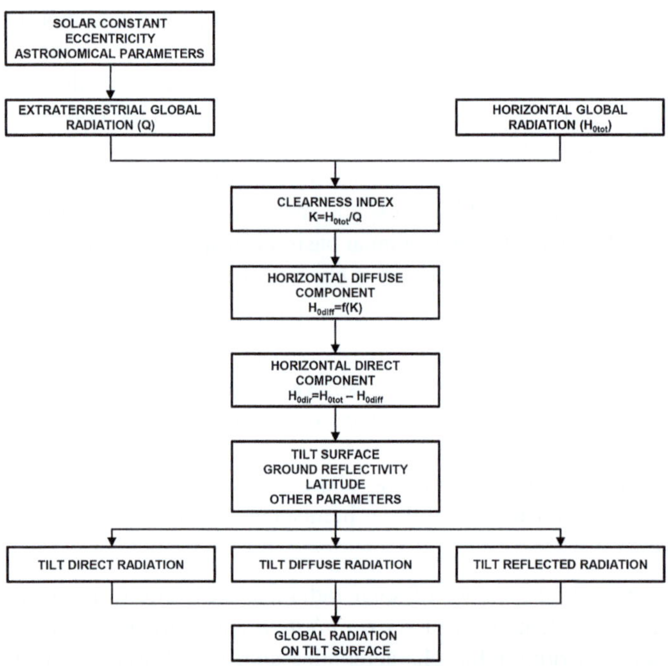


**Fig. 2.9** Procedure for the calculation of the global solar radiation on a tilt surface

curves that is, to say lines indicating the constant value of the mean daily radiation, measured in kWh/m$^2$.

Note that the values shown in the second volume of the European Atlas were calculated by hypothesizing a fixed value for ground reflectivity, amounting to 0.2

(meaning that the ground reflects 20% of incident radiation; a value corresponding to a grassland). If the site in question shows a different reflectivity, corrective formulas shall be used to determine the exact value of the reflected component. However usually the reflected component provides global radiation with a rather limited contribution, amounting to a few percentage points.

The second approach for the determination of solar radiation on a tilt surface is the calculation starting from the data on a horizontal surface, which in turn can be derived from the Solar Radiation Atlas.

Without lingering on the detail of the calculation procedure, here it follows a short description of the fundamental steps at the basis of this approach (Fig. 2.9).

The first step consists in determining the extra-atmospheric radiation starting from the evaluation of a few astronomic parameters as the eccentricity of the earth's orbit. The comparison between the global values of extraterrestrial radiation and the incident radiation on a horizontal surface shows the so-called "*clearness index*".

The diffuse component of radiation (always on a horizontal surface) is then calculated as a complex function of this newly-determined index. The direct component is found as a difference between the global and the diffuse radiation. At this point of the procedure, the three components of solar radiation on a horizontal surface are taken into consideration. It is then possible to calculate, through rather complex formulas, the values of these three components for a given surface arranged in space. Different input parameters are necessary in this calculation phase, and in particular surface tilt, ground reflectivity (for the determination of the levels of the reflected component) and the latitude of the installation site.

### *2.3.2 Hydropower*

The rain and the snow give origin to the water flowing on our planet's lands. The quantity of precipitations, the configuration of the territory and its type of rock or ground, are the factors that determine the characteristics of a watercourse. It is therefore important to know "how much water falls" on a territory to imagine and plan the use of this resource that is so important for all the human kind. The measurement of precipitations is indicated in millimeters of water, and is carried out in the "Precipitation Stations" suitably installed throughout the territory.

As mentioned in Sect. 2.2, hydropower directly depends on the portion of solar radiation that originates the *water cycle on Earth*. This resource is, to all intents and purposes, a *renewable source*; but, due to its intensive use in electricity production, it is traditionally included among *conventional* energy sources.

In the last few years, a renewed interest was expressed vis-à-vis the use of water basins that can fuel small plants with a maximum power of 3,000 kW. *Mini-hydraulic energy* (or minihydro is the name of the branch dealing with these plants) can, in fact, remarkably contribute to the electrification of developing countries, where small-sized isolated users are present, although almost lacking in a distribution grid linking them to large plants. Finally, particular importance is ascribed to water basins or watercourses that can fuel very small plants of tens of kW, *micro-hydraulic plants*.

#### 2.3.2.1 Overview

Drainage Basins

River courses always develop at valley bottoms, having considered that water flows due to the effect of gravitational forces; the valley where the river flows, and the mountains surrounding it, up to the watershed peak, is called *catchment area*.

Through the analysis of the basin ground and the measurement of precipitations in that given area, it is possible to define the behaviour of the river and the quantity of water that can be used.

A watercourse can be named in different ways: river, torrent, brook etc.; each of these names indicates its different characteristics dealing with the continuity of the flow rate, its volume and the size of the watercourse itself.

#### 2.3.2.2 Resource Evaluation

The study of water resources is fundamental, and involves the science of hydrology, the study of rainfall and stream flows, the measurement of drainage basins, catchments, evapotranspiration and surface geology. While flowing from point A to point B of a watercourse, water looses potential energy according to (2.1). This available power can be converted, by limiting friction, into useful kinetic energy.

$$P = Q \cdot \rho \cdot g \cdot H \tag{2.1}$$

In (2.1), $P$ is the power [W], $Q$ the *flow* in [m$^3$/s], $\rho$ the water density (1,000 kg/m$^3$), g the acceleration of gravity (9.81 m/s$^2$) and $H$ the gross *head* [m] (the vertical useful difference in level). Therefore, to estimate the water potential, it is necessary to know the variation of the discharge throughout the year as well as level of the gross available head. A United Nations Organisation was established—the World Meteorological Organisation—providing hydrologic information service besides National Hydrological Services, River Basin Authorities and other institutions, but it is unusual to make regular gauging in the stretch of river where the development of a small hydro scheme is proposed. Thus, normally, through the models of local data in terms of *precipitation* (monthly rainfall), *evaporation* discharge (linked to temperature values, etc.), *stage* (water level), *sediment and water quality* and *groundwater of the areas*, it is possible to calculate the variation of the discharge throughout the year and the available head. Generally, an area with mountains and high precipitation has several good hydropower sites (Fig. 2.10).

To estimate the hydroelectric potential of a site, the annual variation of the flow rate during the year must be known, as well as the gross head available. In the best

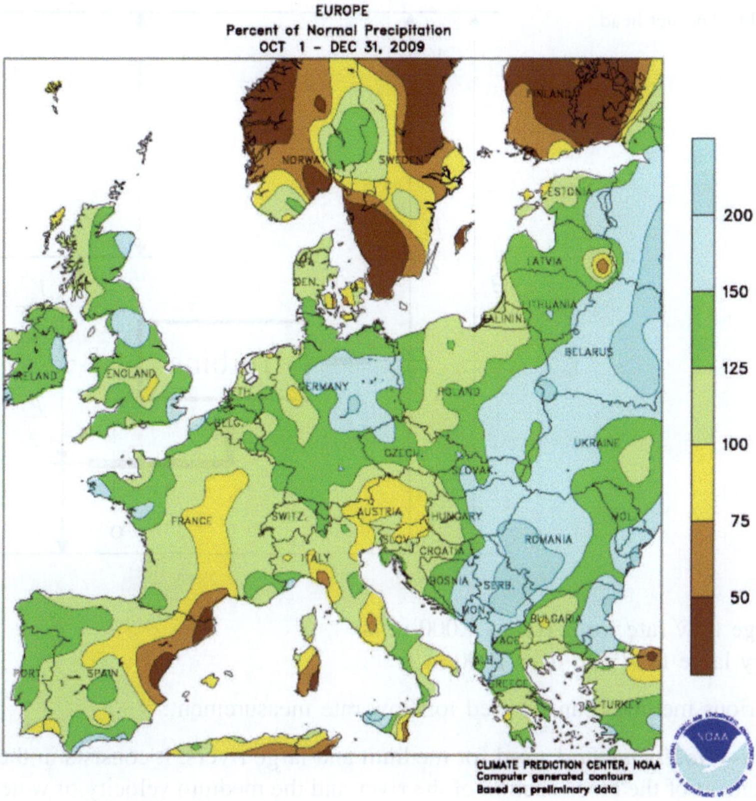


**Fig. 2.10** World hydrological cycle observing system

of cases, the drainage services have installed a measurement station, and the historical series of effluent flow rates will be collected. In case of unknown hydrological data, the flow rate for at least one year will have to be measured.

Equation 2.1 clearly shows the power (and thence the energy) obtainable as determined by the *head* ($H$) and by the *flow rate* ($Q$) of the watercourse considered.

The head is the difference in level between the point of water inlet and the point of water outlet [m].

Hydroelectric power plants are divided according to the value of this size into:

- High head over 100 m
- Medium head 30–100 m
- Low head 2–30 m

The flow rate taken from a hydroelectric plant is the water volume taken in a time unit [m$^3$/s]. According to the flow rate, reference is made to

- Small flow rate up to 10 m$^3$/s
- Medium flow rate from 10 to 100 m$^3$/s

Fig. 2.11 The net head

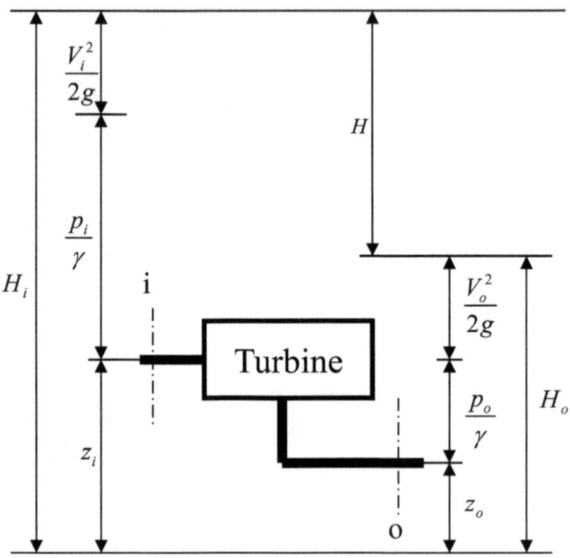


- Large flow rate from 100 to 1.000 $m^3/s$
- Very large flow rate over 1000 $m^3/s$

Various methods can be used for flow rate measurement:

- *Area–velocity method*, used for medium and large rivers, it consists in the measurement of the cross-section of the river and the medium velocity of water. The section is detected through topographic methods, whereas the mean speed can be measured through different tools: float, swivel and electromagnetic meters.
- *Direct measurement of flow rates through dilution of a solute in the current.* Suitable for small watercourses characterised by turbulent flow; the simplest method consists in the dilution of common salt in water and in the subsequent measurement downstream of the electrical conductivity of the solution formed;
- *Measurement through overflow*, if the river is small enough, it is possible to make a provisional weir and allow the water to flow through a known section (the overflow); through measurements on the geometry of the system, it is possible to trace back the flow rate;
- *Slope–area measurement*, this method is suitable for high flow rate values and is based on the principles of hydraulics.

#### 2.3.2.3 Net Head

The net head of a hydraulic turbine is the difference of hydraulic loads, namely of fluid mechanic energy, between two conventional sections taken before and after the machine.

In relation to Fig. 2.11, the net head is obtained as follows:

$$\frac{V_i^2}{2g} + \frac{p_i}{\gamma} + z_i = \frac{V_o^2}{2g} + \frac{p_o}{\gamma} + z_o$$

### 2.3.3 Wind

The power owned by the wind, so called *wind energy*, has been used by men since ancient times for purposes ranging from sailing (Egyptian images of 2,500 BC.) to the winnowing of cereals, to the drying of farming and fishery products.

Windmills in Europe first appeared in the middle ages, at the time of crusades. Completely different from the conceptual viewpoint, with horizontal-axis wheels, these windmills were larger in size, technologically more complex, with a higher yield. They were used for the grinding of cereals, olive pressing, water pumping, setting in motion and operation of sawmills, paper mills, dye-works and tobacco industries.

In the last century, multi-blade rotor ram-air turbines became widespread, subsequently used in large numbers in the farms of newly colonized territories.

A further step forward in the transition from windmills to modern aero-generators was made in Denmark by Paul La Cour, between 1891 and 1907. His rotor had four rectangular warped blades, typical of a European wind mill.

During World War II, some studies were launched in the USA, in the Netherlands and in Denmark that—boosted by the problems of a war economy—led to the design and development of machines (at that point *aerogenerators*) with two-bladed and three-bladed rotors.

In 1924, the Finnish engineer Savonius tested a new and innovative type of *vertical-axis aerogenerator*.

Another type of rotor, Darrieus, derives from an original plan dating back to the beginning of the '30 s by Darrieus. At the moment, it represents the only vertical-axis machine able to really counter the supremacy of the HAWT (Horizontal Axe Wind Turbine).

The possibility of exploiting the wind source with directly competitive costs, or at least costs comparable to conventional sources, is heavily limited by its *low energy density* (in the windiest areas, it is possible to reach the maximum[3] value of 400–600 W/m$^2$), as well as by its marked *changeableness and uncertainty*, also while considering seasonal and annual mean values. Furthermore, there is also the problem of *windblasts* that is, to say winds whose high and suddenly variable intensity generate—in wind power plants—mechanical resistance problems and may also cause breakings.

[3] These are annual mean values of power per surface unit affected by the blades of a wind machine.

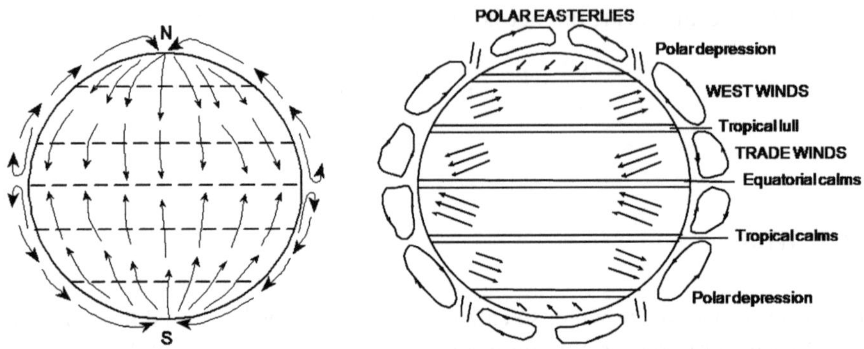


**Fig. 2.12** Wind circulation linked to phenomena operating at the Earth level (*Source*: **ENEL SpA**)

From a merely operational viewpoint, a fundamental step to usefully exploit the wind source is the choice of the suitable place where installing conversion plants, that is to say the determination of a suitable choice of the site.

#### 2.3.3.1 Overview

*Wind energy* means the kinetic energy owned by AM moving in the Earth's atmosphere. In general, their movement is linked to the action of atmospheric pressure, or better, pressure gradient, and therefore of temperature, in the sense that air tends to move from high-pressure areas to low-pressure adjacent areas, with a speed related to such gradient.

The wind and its main characteristics, namely *intensity* and *direction*, depend upon several factors acting at a local and global scale.

The most important factor is *solar radiation and its different intensity*, higher in the equatorial zone and lower in the areas close to the poles. The Earth's surface releases to the atmosphere part of the heat received from the Sun; the equatorial air, warmer thus with a lower density, tends in any case to rise, generating a depression that attracts other AM from the poles, and triggering—at the planet level—a *circulation of cold air from polar zones to the equator*.

Among the other factors influencing the circulation of winds, from a global viewpoint, there are the *inclination of the Earth axis* and the movement of *revolution around the Sun*, which seasonally move the most insulated areas inside of the tropical zone. On the contrary, the movement of *rotation of the Earth* determines a daily alternation of solar radiation, and induces the *Corioliscomplementary acceleration* on the moving AM, producing the typical *spiral* movement of wind currents; the *lack of uniformity and homogeneity of the earth's surface* is finally responsible for the different absorption capacity and for the exchange of solar heat between the Earth and the atmosphere above it.

**Fig. 2.13** Wind speed variation according to the distance from the ground

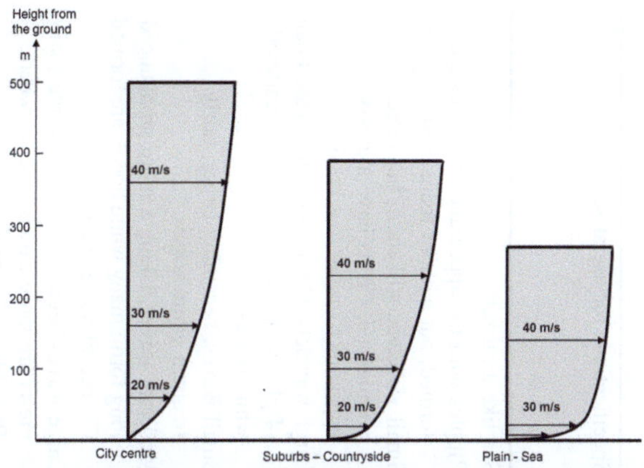

**Fig. 2.14** Anemometers

Figure 2.12 shows a global representation of wind circulation due to high and low pressure flows,[4] which develop at a regional level as a consequence of the phenomena mentioned above.

[4] As regards Fig. 2.12, there are in particular one zone of low pressure at the equator; two tropical areas of high pressure; two areas of low pressure in sub-polar zones; two high pressure zones on the polar caps.

**Table 2.1** Beaufort scale

| Force | Equivalent speed 10 m above ground | | Description | Specifications for use on land | Specifications for use at sea |
|---|---|---|---|---|---|
| | miles/hour | knots | | | |
| 0 | 0–1 | 0–1 | Calm | Calm; smoke rises vertically | Sea like a mirror |
| 1 | 1–3 | 1–3 | Light air | Direction of wind shown by smoke drift, but not by wind vanes. | Ripples with the appearance of scales are formed, but without foam crests. |
| 2 | 4–7 | 4–6 | Light breeze | Wind felt on face; leaves rustle; ordinary vanes moved by wind. | Small wavelets, still short, but more pronounced. Crests have a glassy appearance and do not break. |
| 3 | 8–12 | 7–10 | Gentle breeze | Leaves and small twigs in constant motion; wind extends light flag. | Large wavelets. Crests begin to break. Foam of glassy appearance. Perhaps scattered white horses. |
| 4 | 13–18 | 11–16 | Moderate breeze | Raises dust and loose paper; small branches are moved. | Small waves, becoming larger; fairly frequent white horses. |
| 5 | 19–24 | 17–21 | Fresh breeze | Small trees in leaf begin to sway; crested wavelets form on inland waters. | Moderate waves, taking a more pronounced long form; many white horses are formed. Chance of some spray. |
| 6 | 25–31 | 22–27 | Strong breeze | Large branches in motion; whistling heard in telegraph wires; umbrellas used with difficulty. | Large waves begin to form; the white foam crests are more extensive everywhere. Probably some spray. |
| 7 | 32–38 | 28–33 | Near gale | Whole trees in motion; inconvenience felt when walking against the wind. | Sea heaps up and white foam from breaking waves begins to be blown in streaks along the direction of the wind. |
| 8 | 39–46 | 34–40 | Gale | Breaks twigs off trees; generally impedes progress. | Moderately high waves of greater length; edges of crests begin to break into spindrift. The foam is blown in well-marked streaks along the direction of the wind. |

(continued)

**Table 2.1** (continued)

| Force | Equivalent speed 10 m above ground | | Description | Specifications for use on land | Specifications for use at sea |
|---|---|---|---|---|---|
| | miles/hour | knots | | | |
| 9 | 47–54 | 41–47 | Severe gale | Slight structural damage occurs (chimney-pots and slates removed). | High waves. Dense streaks of foam along the direction of the wind. Crests of waves begin to topple, tumble and roll over. Spray may affect visibility. |
| 10 | 55–63 | 48–55 | Storm | Seldom experienced inland; trees uprooted; considerable structural damage occurs. | Very high waves with long over-hanging crests. The resulting foam, in great patches, is blown in dense white streaks along the direction of the wind. On the whole the surface of the sea takes on a white appearance. The 'tumbling' of the sea becomes heavy and shock-like. Visibility affected. |
| 11 | 64–72 | 56–63 | Violent storm | Very rarely experienced; accompanied by wide-spread damage. | Exceptionally high waves (small and medium-size ships might be for a time lost to view behind the waves). The sea is completely covered with long white patches of foam lying along the direction of the wind. Everywhere the edges of the wave crests are blown into froth. Visibility affected. |
| 12 | 73–83 | 64–71 | Hurricane | - - | The air is filled with foam and spray. Sea completely white with driving spray; visibility very seriously affected. |

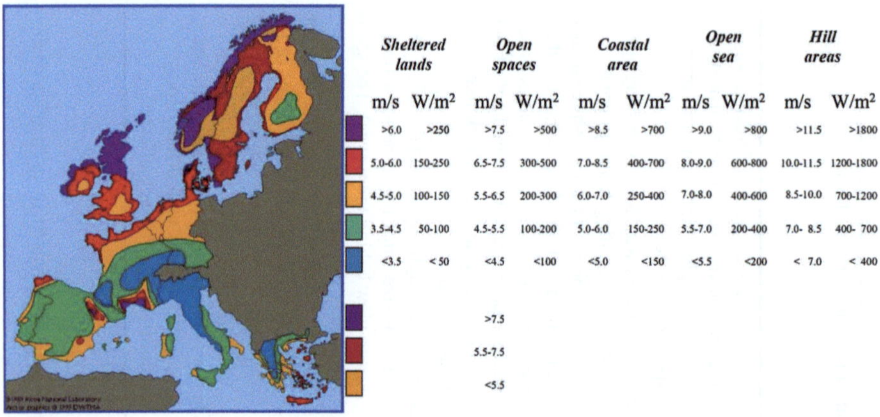


**Fig. 2.15** Example of a wind map

#### 2.3.3.2 Resource Evaluation

The last factor recalled that is, to say the varied and uneven type of the Earth's crust (oceans, lands above sea level with mountain chains, valleys, plains and depressions, glaciers, etc.) also acts on a local scale, together with other elements as *frictional force* and *orography*. The frictional force, due to the *roughness* of the earth's surface, determines a dissipation of the energy of wind currents, as well as the development of a border layer of the atmosphere (1–1.5 km of thickness), inside of which wind speed increases with height and its trend can be represented through a logarithmic function (Fig. 2.13).

The most immediate way to quantify the wind consists in measuring its speed. To this end, devices called *anemometers* were developed, the most common types of which are shown in Fig. 2.14.

Approximate expeditious measurements of wind speed can be made through the Beaufort scale (Table 2.1).

Wind up to 5 m/s (breeze) is a low wind (scarcely perceivable: thus class 1 areas are unsuitable for wind energy development); from 5 to 15 m/s (around 54 km/h) is a "normal" wind (but only areas designated as class 4 or greater are suitable with the actual wind turbine technology; class 2 and 3 areas may be suitable for future technologies, and over is the storm range (35 m/s, i.e., 126 km/h is normal speed of a hurricane; these values are very rare).

On the basis of the measurements made through anemometers, the *isovelocity curves* are defined. The measurements made in a place are extended to the neighbouring areas under suitable hypotheses and through the use of suitable models (Fig. 2.15).

To evaluate the energy provided by the wind, some quantitative analyses have to be made, starting from the physical and geographical values involved. In this connection, the two following theorems are referred to:

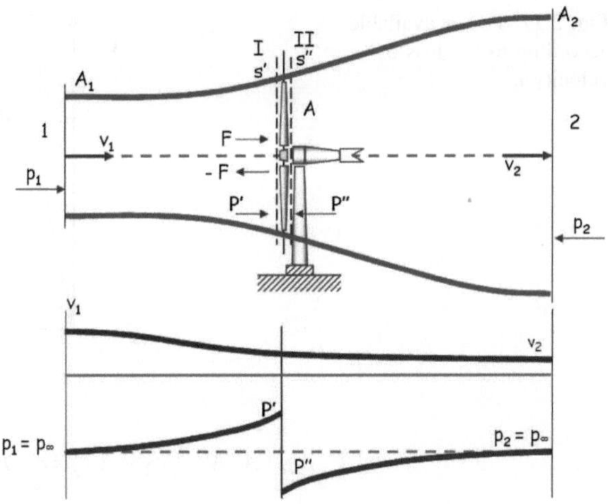


**Fig. 2.16** Stream tube, velocity and pressures in an aerogenerator

- *Froude's theorem*: the speed of fluid vein at the level of the propeller disk equals the arithmetic mean of the speed that the fluid vein has upstream and downstream the blade itself.
- *Betz theory*: the maximum possible energy derived from the wind for a windmill is 0.59 of the power available.

Froude's theorem can be demonstrated by assuming the following hypotheses:

- Fluid density, due to the limited velocity variations, is considered as constant.
- The flow rate inside of the stream tube is constant.
- Energy exchanges are only axial that is, to say the variations in momentum are only axial.
- The propeller has infinite blades.
- There are no rotational motions.

The analysis of pressure distribution shows that:

- The fluid vein stresses the windmill with a strength amounting to F
- The windmill stresses the fluid vein with a strength amounting to -F
- The fluid vein holds back both upstream and downstream.

All this determines that:

- The stream tube takes the shape of a bottleneck (for continuity).
- Velocity variation is continuous along the stream tube.

As regards Fig. 2.16 the pulse theory is applied to the windmill. According to the law of continuity, it follows that:

$$A_1 \cdot V_1 = A_2 \cdot V_2 = A \cdot V = \cos t \tag{2.2}$$

By applying the pulse theory between sections 1 and 2 and sections I and II

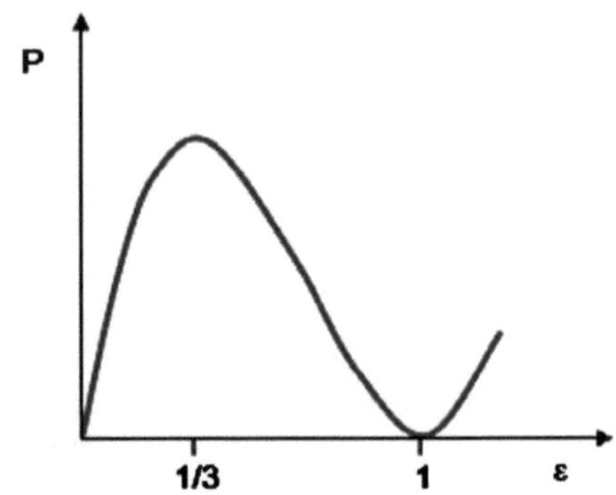


**Fig. 2.17** Power available according to the loss of velocity ε

$$\rho \cdot A \cdot V \cdot (V_2 - V_1) = -F \tag{2.3}$$

$$\rho \cdot A \cdot V \cdot (V - V) = 0 = A \cdot p' - A \cdot p'' - F \tag{2.4}$$

From (2.4) it can be easily derived that:

$$F = A \cdot (p' - p'') \tag{2.5}$$

By applying the Bernouilli's theorem between a section upstream the blade (undisturbed flow) and section s', and between section s'' and a section downstream the blade (undisturbed flow), it follows:

$$p_1 + \frac{1}{2}\rho V_1^2 = p' + \frac{1}{2}\rho V^2 \tag{2.6}$$

$$p_2 + \frac{1}{2}\rho V_2^2 = p'' + \frac{1}{2}\rho V^2 \tag{2.7}$$

Subtracting (2.7) from (2.6), therefore:

$$\frac{1}{2}\rho \cdot \left(V_1^2 - V_2^2\right) = p' - p'' \tag{2.8}$$

By replacing (2.8) in (2.5) it is obtained:

$$F = \rho \cdot A \cdot V \cdot (V_1 - V_2) = \frac{1}{2}\rho \cdot A \cdot \left(V_1^2 - V_2^2\right) \tag{2.9}$$

From which it is possible to easily derive the $V$ expression, amounting to the mean of the velocity that the fluid vein owns upstream and downstream the blade:

$$V = \frac{1}{2} \cdot (V_1 + V_2) \tag{2.10}$$

To demonstrate the Betz theory, please consider, while recalling (2.9), the expression of the power available to the blade:

$$P = F \cdot V = \frac{1}{2}\rho \cdot A \cdot V \cdot \left(V_1^2 - V_2^2\right) \tag{2.11}$$

The velocity of the fluid vein decreases along the stream tube. Indicating the percentage of loss (as compared to the value of the velocity of the undisturbed flow upstream the blade $V_1$) with ε, it is possible to express $V$ and $V_2$ according to $V_1$:

$$V = V_1 - V_1\varepsilon = V_1 \cdot (1 - \varepsilon) \tag{2.12}$$

$$V_2 = V - V_1\varepsilon = V_1 \cdot (1 - 2\varepsilon) \tag{2.13}$$

By replacing (2.12) and (2.13) in (2.11) it follows:

$$P = 2 \cdot \rho \cdot A \cdot V_1^3 \cdot \varepsilon \cdot (1 - \varepsilon)^2 \tag{2.14}$$

The trend of power available according to the loss of velocity ε is shown in Fig. 2.17. Deriving (2.14) compared to ε and equating to zero, the values of ε are obtained to which correspond the minimum and the maximum of the power available.

$$\frac{\partial P}{\partial \varepsilon} = 2 \cdot \rho \cdot A \cdot V_1^3 \cdot (1 - \varepsilon) \cdot (1 - 3\varepsilon) \tag{2.15}$$

This derivative is null for

$$1 - \varepsilon = 0 \Rightarrow \varepsilon = 1 \tag{2.16}$$

$$1 - 3\varepsilon = 0 \Rightarrow \varepsilon = \frac{1}{3} \tag{2.17}$$

As shown in the chart of Fig. 2.17, the value $\varepsilon = 1$, corresponds to a minimum value of the power available ($P = 0$), whereas for $\varepsilon = 1/3$, there is a maximum of the power that can be supplied. It is evident that (2.14) is null for $\varepsilon = 1$ and for $\varepsilon = 0$. By recalling (2.12) and (2.13), for $\varepsilon = 0$ it follows:

$$V = V_1 = V_2 \tag{2.18}$$

The velocity and therefore the kinetic energy of the air remains constant: in this case the power is null since there is no energy exchange between the AM and the blade. For $\varepsilon = 1$, on the contrary:

$$V_2 = 0 \tag{2.19}$$

This means that in this case the power is null since the flow rate is null.

Assuming $\varepsilon = 1/3$ in (2.14) the maximum value of the power available is obtained. This value represents the limit of Betz.

$$P = \frac{1}{2} \cdot \rho \cdot A \cdot V_1^3 \cdot \frac{16}{27} = 0,59 \cdot \left(\frac{1}{2} \cdot \rho \cdot A \cdot V_1^3\right) \tag{2.20}$$

This shows that the maximum power that a windmill can take from the wind accounts for 59% of wind power.

Equation 2.20 shows that power increases by a factor of three as wind speed increases, and it is proportional to air density and to the swept area (for a HAWT the area through which the rotor blades pass by the rotors). Because of the low air density (1.25 $kg/m^3$), *wind power density is much lower than water power* (1.000 $kg/m^3$). If the *diameter* of the rotor blades is *doubled*, the power increases by a factor of*four*. If the wind *speed* then *doubles*, power increases by a factor of *eight*.

The wind source is an intermittent energy source, with characteristics of space–time uncertainty, and therefore characterised by its own availability in time that can be very much variable although within the statistical limits of the phenomenological. Therefore, the following definitions apply:

- *Maximum resource available*: fifty nine percent of the energy actually contained in the incident flow.
- *Resource actually available*: the resource that, following the meteorological phenomena related to local and global climate, is present on the site.
- *Accessible resource*: the resource obtained by applying to the maximum resource the maximum efficiency of present machines.
- *Actual accessible resource*: the one obtained by applying to the resource actually available the maximum efficiency of present machines.

## 2.3.4 Biomass

The word *biomass* indicates each and every organic matter, either vegetal or animal, in a non-fossil form. Biomass represents the most ancient energy form ever known by human beings. By storing solar energy in chemical bonds, it also represents the most sophisticated form of energy storage in nature. In fact, through photosynthesis, plants containing chlorophyll convert water and $CO_2$ into organic matter during their growth, that is to say they use solar energy to create long chains mainly made up of carbon, oxygen and hydrogen. However, only one part of the solar spectrum (45%) intervenes in photosynthesis and a further 20% is lost due to reflection or poor absorption phenomena. Therefore, the yield of this process is low (between 0.1% (world average) and 1% (crops)); however, due to the large spreading of biomasses, energy values are considerable. In this way, in fact, nearly $2 \times 10^{11}$ tons of carbon are "fixed" overall every year, and 120 Gt of biomass are produced with an energy content of $70 \times 10^9$ TOE distributed as follows:

- 50% in water
- 35% in forests
- 10% in savannahs and grasslands
- 5% in farmlands.

In consideration on its distribution in the planet, biomass is the most common form of renewable energy, and the main reservoir of solar energy. Although the real ratio of compounds varies from species to species, on average biomass is made up 75% by carbohydrates or sugars and 25% by lignin.

Since ancient times, the residues of organic matters derived from the photosynthesis activity—either direct (plants) or indirect (animals)—have been used for heat generation and for the production of derived solid, liquid and gaseous fuels. They represent the most sophisticated form of storage of solar energy and, therefore, renewable primary energy sources.

Also wood, considered as an energy source, falls within the most general category of biomass. As it happens with the other *non-commercial* sources, even in case of wood it is difficult to determine its market value and include it in official energy statistics. One of the main reasons shall be sought in its energy characteristics and, most of all, in its limited calorific value,[5] which among other things varies a lot according to humidity, to wood species and to its origin. Furthermore, its large-scale trade is hindered by the high incidence in the transportation cost.[6] The level of energy dependency of a Country from this type of fuel varies on average from below 3%, in industrialised Countries, to over 40% in developing Countries (with peaks of 70%).

In evaluating the utilisation opportunities and the energy purposes, it is necessary to carefully consider, on a case-by-case basis, the time needed to restore the timber taken from woodlands. In fact, everyone is aware of the *deforestation* and of subsequent *desertification* problems, which are reaching alarming levels globally; however, they are rarely ascribable to energy needs, although they are frequently linked or ascribed to them. It is sufficient to think that the thinning out of plants and the removal of coppices from a wood, provided that it is not excessive, is even beneficial for the wood, since it represents a form a "maintenance" able to favour the renewal of plants and their growth.

Also for biomass in general it is very difficult to exactly quantify their availability and consumption for energy purposes. For instance, it is not possible to estimate precisely and in detail the direct consumption in rural areas (especially in developing Countries) and industrial waste destined to combustion.

Biomass can also originate from the residues of crops expressly destined to the food sector, or from farming activities destined, either totally or partly, to energy production (sugar cane plants, plants rich in starch, etc.), which for this reason are called *energy crops*. High quantities of organic matter can also be derived from self-sown vegetation (wooden material) and from the zootechnical waste of the farms.

Within the European framework of energy use of biomasses, Italy is in a condition of scarce development, despite the high potential at its disposal. In fact,

---

[5] Nearly 19,000 kJ/kg for high-quality wood, with 15% in weight of humidity.

[6] Generally, for distances exceeding 80–100 km the cost of wood becomes higher, given the same energy content, than other conventional sources.

the availability of residual biomasses (wood, farming and agri-food industry residues, urban waste and zootechnical industry) in Italy amounts to nearly 66 million tons of dry matter every year, equivalent to 27 MTOE.

Through a suitable reforestation and forest maintenance programmes, in fact, new biomasses could be available for an annual amount of nearly 2 MTOE. Besides the existing forests, new conventional wood surfaces could be obtained by exploiting part of the over 2,000,000 ha that are not destined to farming due to their low productivity. Furthermore, coppice stands and ground crops mainly destined to energy use could be planted, hence reconverting part of the 250,000 ha that are presently left uncultivated, in compliance with the EU directives enacted in relation to the problem of farm surpluses: the present surface destined to energy crops—extremely limited indeed—should be extended to 3,500–5,000 ha of short-cycle cultivations, but the potential surface amounts to nearly one million ha.

The use of this resource varies according to the origin, the chemical composition (in particular the C/N ratio) and the humidity content. Biomass can be burned directly to obtain thermal energy, or transformed into various types of fuels (biogas or bio-fuels), which improve their quality and usability but lower the level of energy content. In particular, transformation processes aim at producing methane gas compounds with a low or medium calorific value, or other fuel substances (methanol, ethanol, carbon substances, etc.).

As regards the possibility of producing fuels from human waste, it is necessary to consider the potential interest of energy enhancement from municipal solid waste (MSW), a percentage of which is made up of organic matter and which can therefore also be considered as biomass.

#### 2.3.4.1 Overview

Being *every renewable organic matter biomass* is integral to life. This means that biomass is the "fuel" of life, a source of energy and materials.

In practice, organic materials (carbon compounds) can be used either "directly" by organisms as their own source of energy and nutrients (in this case, reference is made to *food*) or "indirectly" as a source of external energy or *materials* (clothing, furniture, building, chemicals, etc.).

Thus *food, biomass and renewable materials are strictly connected.* From these preliminary definitions, it is possible to understand the usefulness of *energy-farm* and *bio-refinery*: they propose, respectively, the integrated production of power-food and power-chemicals from biomass.

Biomass has been (and in many cases, in Africa in particular, remains) the first "fuel" of mankind. Today about 11% (around 1,200 MTOE) of total world primary energy supply (nearly 13,200 MTOE) comes from biomass [4].

Biomass is the *fourth primary energy resource worldwide* (following oil, coal and natural gas) although in industrialised Countries it only accounts for 3% of the total, whereas in developing Countries it accounts for 35%, being in many cases the main energy resource. As a matter of fact, a change is occurring in the use of

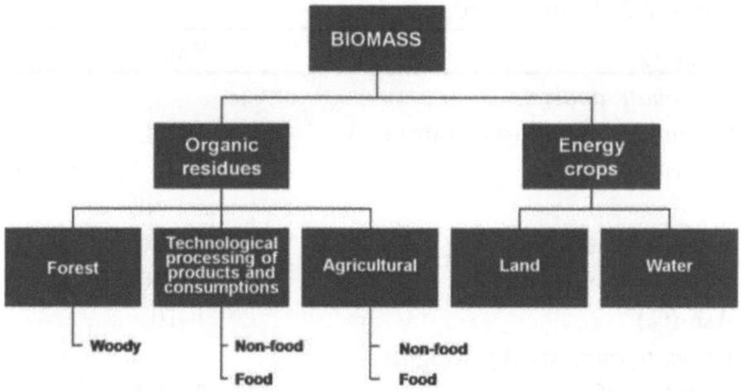


**Fig. 2.18** Biomass typologies

*biomass energy* from traditional and non-commercial to modern use (high *efficiency*, high-value energy vectors produced, such as *electricity and bio-fuels*, and *integration* with industries operating in the food sector or in the field of renewable materials). The transformation as well as the increase in the use of biomass energy are due to the trends explained in the previous chapter on renewable energy sources, and to the potential of energy biomass which is not fully exploited. Therefore, not only the theoretical potential but also the technical and economic ones (that take into account actual technologies and costs) are sometimes higher than actual energy consumption [5].

In brief, biomass is one of the main energy sources available: *the challenge, therefore, resides in its sustainable use and not in its availability.*

As it happens with other energy sources, the use of biomass depends on the global analysis of the production-use system. Different from the other renewable energy sources, biomass—being a fuel (and not a directly convertible energy source, like solar, wind, hydropower and geothermal energy)—is *subject to the traditional energy chain* (and relevant impacts) that provides a "natural" resource to our daily life.

Current energy chains deplete non-renewable resources and release harmful substances, producing waste as well as local and global pollution.

Given that long-lasting development can only exist through the realisation of energy cycles by using renewable energy sources; in this way, the main aim of the energy sector shifts to the so called closed-cycles, and aimed at achieving a "zero-impact" on the environment.

In the case of biomass, the first part of the entire energy chain (extraction and processing) is replaced by growing, harvesting and treatment. Therefore, the entire energy chain encompasses the following steps:

1. *Production* (growing, harvesting and treatment)
2. *Transportation*
3. *Conversion* (and relevant initial and final treatment)
4. *Distribution*
5. *End use.*

Different from the other renewable energy sources, biomass (owing to the large variety of renewable multi-carbon compounds) can be converted through several

**Table 2.2** Biomass properties

| Property | Range |
|---|---|
| *Proximate analysis* | |
| Humidity content (moisture) (%) | 10–70 |
| Calorific value[a] | 2–40 MJ/kg (from very wet biomass to very dense oil) |
| Volatile matter (%) | 30–80 |
| Fixed carbon (%) | 15–30 |
| Ash (%) | 1–10 |
| *Ultimate analysis* (by weight) | |
| Cellulose (%) | 30–50 |
| Hemi-cellulose (polysaccharides) (%) | 20–40 |
| Lignin (%) | 5–30 |
| Carbon (%) | 40–50 |
| Oxygen (%) | 38–43 |
| Hydrogen (%) | 5–7 |
| Alkali metal and inorganic element (%) | 1–15 |
| *Bulk volume (density)* | 1–50 $m^3$/t (daf) |
| *Times of cultivation* | 6–24 months |
| *Yield (potential annual production capacity)* | 1–100 (dmt[b]/ha)/yr |
| *Characteristics of cultivation* | Depends on climates, land, water, pesticides and fertilisers demand |
| *Production cost* | Negative (for waste) to 1 (DC)–5 (IC) €/GJ |
| *Transportation cost* | Function of the distance travelled and the energy density (from 0 to 1 MJ/km) |
| *Conditions of suppliers* | Depends on local conditions |

[a] NCV on daf (dry ash-free) basis
[b] *dmt* Dry matter tons

processes in order to obtain various products and by-products. From a *chain perspective*, biomass can be divided, as showed in Fig. 2.18, into:

- *Energy crops*: biomass especially planted for energy purposes in marginal or productive land or water.
- *Organic residues*: from woodlands, public parks and gardens, human or animal cultivations, industry (food, wood, textile, …) or Municipal Solid Waste (MSW).

The former type is subject to the entire chain of biomass, whereas the latter only includes the last three processes (conversion, distribution and end use), if is used in situ.

From a quality point of view, biomass can be divided into

- *Lignocelluloses herbaceous and arboreal plants,* like Miscanthus, poplar, etc.
- *Sugar*, like sugarcane, beetroot, etc.
- *Starch*, such as corn, wheat, etc.
- *Oil* from sunflower, rape, palm, soya, etc.

- *Moisture:* manures, aquatic plants, etc.
- Others.

#### 2.3.4.2 Resource Evaluation

Biomass is characterised by many properties which vary within a wide range (from countries to countries, land, season, management, etc.). In order to use biomass for energy production purposes, all the properties shown in Table 2.2 must be taken into account [5, 6]. Obviously, owing to the very large range of biomass typologies, the value range has to be considered as example (Table 2.2 mainly refers to crops and vegetable residues, the most frequently used biomass).

The first two categories (proximate and ultimate analysis) and the bulk volume show the main chemical and physical biomass proprieties. The moisture content and the cellulose/lignin ratio are of primary concern in the biochemical conversion processes while the moisture, fixed carbon, ash, alkali content and the calorific value are of prime concern in the thermo-chemical conversion processes. The bulk volume is relevant in the transportation and feeding systems (together with the friction and others particular properties).

Times and requirements of cultivation and yield are *general characteristics* of crops. The energy yield (yield and CV multiplied) is the main quantity to consider.

Production and transportation costs are the main *economic characteristics* to take into account. Owing to the large amount needed (the biomass, as shown, has a low energy density) and to the great variability, the *supply conditions* are paramount.

After all, since biomass is a fuel with a large property variety, the possible *variation* of all the related parameters must be analysed. Thus, the biomass energy use implies a careful study of:

- the territory and biomass properties and characteristics
- the *place* where the power plant is setup (not in distant from the biomasses to be used)
- the *reliability* of supplies, conversion technologies, etc.

### *2.3.5 Geothermal*

Geothermal energy (from the Greek "gê" and "thermòs", "heat of the Earth") is the energy contained inside of the Earth in the form of "heat".

The origin of this heat relates to the inner nature of our planet and to the physical processes taking place in it.

This heat, although available in huge and practically inexhaustible quantity, is very much dispersed and only rarely concentrated.

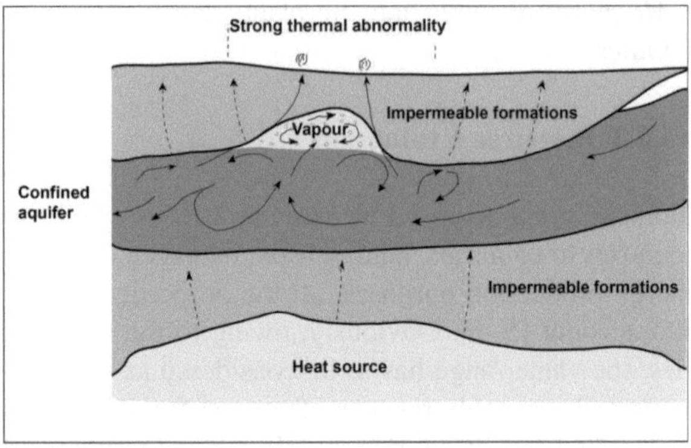


**Fig. 2.19** Outline of a hydrothermal system

The internal heat regularly dissipates towards the surface of the Earth; its existence can be perceived through the progressive increase of rock temperature as depth increases; the geothermal gradient is, on average, 3°C every 100 m of depth, namely 30°C every km.

*Geothermal energy* here indicates the exploitable part of natural heat present inside of the Earth's crust. The subsoil of our planet, in fact, shows a thermal gradient linked to the hot nucleus at the core of the Earth according to which temperature increases as depth increases, but mostly to spontaneous radioactivity phenomena, which develop thermal energy locally. The main radioactive elements include Uranium $^{235}U$ and $^{238}U$, Thorium $^{232}Th$ and Potassium $^{40}K$.

Most of the endogenous heat is available at too high depths or in conditions that make it technically impossible to allow its exploitation as energy source. However, there are zones on the surface of our planet whose thermal energy concentration is very high locally (*high thermality areas*). Heat concentration in less deep layers of the Earth's crust, as volcanoes, geysers and hot-water springs, originates the so-called *geothermal systems*; they are effects of natural radioactivity, and not of the gradient.

#### 2.3.5.1 Overview

The only systems presently used on an industrial scale are the so-called *hydrothermal systems*. They are made up of permeable rocks containing hot fluid (water or steam); named *aquifers*, they are delimited by layers of impermeable rock (Fig. 2.19).

The feeding of aquifers, which can be considered as natural reservoirs, is due to the introduction of meteoric water—that is to say coming from atmospheric precipitations—through the recharging surfaces, formed by zones where the reservoir itself comes to surface. *The heat that warms the fluid is provided by the natural decay of isotopes within the Earth*; the particles emitted by the basins of radioactive material are, in fact, slackened by the surrounding masses, converting their kinetic energy into heat.

Some necessary conditions (although not sufficient) must be assessed for the existence of hydrothermal systems. They include the *presence of a heat source at a suitable temperature* (at least 600°C) at a not too high depth; the *presence of permeable rocks*, making up the aquifer, and *impermeable rocks*; and the suitable renewal, in the subsoil, of *meteoric water*. In general, the renewal is guaranteed by the fact that meteoric waters, while warming up, undergo a pressure increase and often turn into vapour, hence tending to raise through either natural or artificial shafts. The hollow deriving from this process is filled with other meteoric water that, through its infiltration in the ground, feeds the cycle. The upper impermeable layer guarantees a suitable pressure and reduces the thermal flow towards the surface.

According to geological and thermodynamic conditions, the endogenous fluid can appear at the shaft outlet in three different ways: *in liquid phase*, as *saturated vapour* (low enthalpy), or *superheated vapour* (high enthalpy). In case of presence of high quantity of vapour, the hydrothermal system is defined as *vapour-dominated*; in case of presence of liquid and low quantity of vapour, a *water-dominated* system takes place.

All the reservoirs contain, in variable percentages, incondensable minerals and gases, mainly formed by carbon dioxide.

Presently, most of the electric energy produced from geothermal energy is based upon high-enthalpy sources, although there are already several applications using saturated vapour and even hot water.

*Non-hydrothermal systems* are represented by the so-called *hot dry rocks* and by *magmatic systems*.

The *former* are highly-thermal zones of the Earth's crust, although free from aquifers, therefore lacking in a vector transporting heat to the surface. In order to exploit this energy source, it is possible to cause, through artificial fracturing, a forced circulation of endogenous fluid.

*Magmatic systems* are made up of high-temperature fused rocks (magma) and entail huge difficulties in terms of energy extraction; the possibility of their exploitation is remote even if recently innovative energy systems have been proposed.

The *disadvantages* relating to the use of geothermal energy are mostly linked to the *uncertainty of its finding and of the consistency* of the source, as well as the *high extraction costs*. Huge investments are needed for the more and more difficult determination of exploitable zones and for shaft drilling. Before its use, the fluid coming from the subsoil always has to be processed since it contains, as already said, gases, minerals and dissolved water (in case of vapour-dominated systems), which are harmful both for the conversion plants and for human health.

#### 2.3.5.2 Resource Evaluation

The world distribution of geothermal sources sees a concentration in the zones of contact between tectonic plates, where the earth's thermal gradient is higher due to

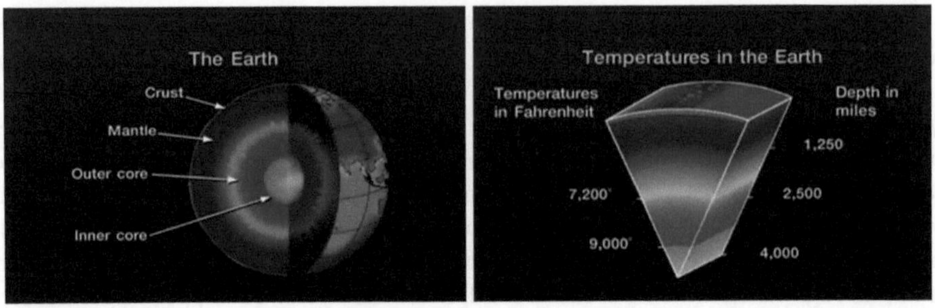


**Fig. 2.20** Earth layers and relevant temperatures

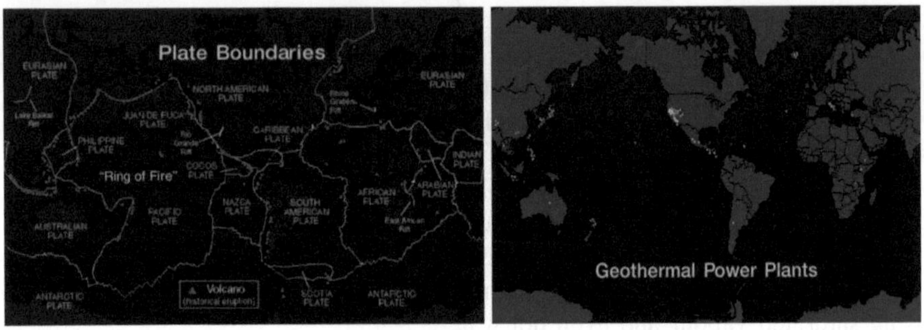


**Fig. 2.21** Plate boundaries and geothermal power plants

**Fig. 2.22** Geothermal phenomena

the fractures of superficial layers; some of these zones are traditionally called "volcanic zones" (Fig. 2.21).

Globally, the first experiments entailing the industrial exploitation of geothermics were performed in Larderello (Tuscany, Italy) in 1904, whereas the first geothermal power station (with a power of 250 kW) was established in the same area in 1913.

Having considered that the word *geothermal* comes from the Greek *gea* (Earth) and *thermos* (heat), geothermal energy is usually defined as the heat stored within the Earth. Heat originates from the Earth's molten interior and from the decay of radioactive materials. Its quantity is larger than the solar energy theoretical potential (the annual solar radiation). In fact, not only the total geothermal theoretical potential is more than 100,000 times higher than the world's energy consumption, but also its smaller total technical potential (about $5 \times 10^{12}$ GJ) is higher than the world's energy consumption (about $45 \times 10^{9}$ GJ), despite an actual use of about $2 \times 10^{9}$ GJ. However, like other renewable sources (solar energy, wind energy), geothermal energy is widely dispersed. Consequently, the technological ability to use geothermal energy will determine its future spreading.

The Earth has four main layers, and each layer has different compositions, functions and temperatures, as is shown in Fig. 2.20.

The crust is the outermost layer of the Earth, the land that forms continents and ocean floors. It can be from 3 to 8 km thick under the oceans and from 25 to 55 km thick on the continents.

As a matter of fact, the heat of the Earth dissipates towards its surface in a way that leads the Earth's temperature to increase by nearly 3°C every 100 m of depth, being therefore a difference in temperature too low to generate a great local power. However, the Earth's crust is broken into pieces called *plates*; magma therefore comes close to the Earth's surface, near the edges of these plates. Consequently, geothermal energy is generally greater in areas where tectonic plates collide and generate volcanic activity, as it can be seen in Fig. 2.21, which shows where plate boundaries and geothermal power plants are located. For example the Pacific ring (ring of fire) is a prime spot for the harnessing of geothermal activity because it is an area where tectonic processes are always taking place.

Even if most geothermal reservoirs are deep underground with no visible clues showing above ground, the geothermal energy can sometimes find its way to the surface in the form of *volcanoes* and *fumaroles* (holes where volcanic gases and vapour are released), *hot springs* and *geysers* as shown in Fig. 2.22.

When magma comes close to the surface, it heats the ground water trapped in porous rock or water running along fractured rock surfaces, and faults. These features are called hydrothermal and have two common ingredients: water (hydro) and heat (thermal). Thus, water at high temperatures (300°C) can also be available for many applications. Large areas with hydrothermal resources are called *geothermal reservoirs*. Therefore, four types of geothermal energy are usually identified:

- *Hydrothermal* hot water or steam at moderate depths (0.1–4.5 km).
- *Geopressed* hot-water aquifers containing dissolved methane under high pressure, 3–6 km depth.
- *Hot dry rock* abnormally hot geologic formations with little or no water.
- *Magma* molten rock at temperatures of 700–1,200°C.

Today only hydrothermal resources are used on a commercial scale for electricity generation and as a direct heat source. These four types of geothermal energy located in particular areas have to be added to the use of Earth's stable ground or water temperatures near the Earth's surface; therefore, it is possible to take advantage of the relatively constant temperature of the core of the Earth, using it for both heating and cooling purposes. Moreover this source of energy is available everywhere.

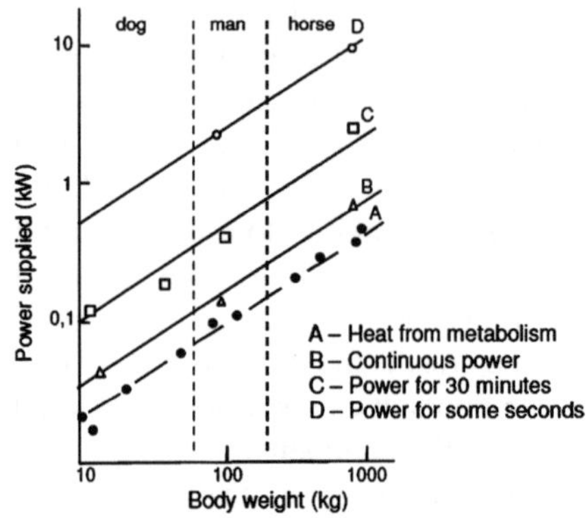


**Fig. 2.23** Total power provided by different living beings according to their weight

Since some renewable energy sources can only operate under favourable weather conditions, they are often considered as limited in their ability to meet the looming large-scale power needs of the 21st century. However, geothermal energy, has the potential to provide reliable electricity sources while still offering significantly lower emission levels than fossil fuels, as well as avoiding problems in terms of radioactive waste disposal. Geothermal relies on a readily available, constant source of heat for generation, and is therefore considered a base-load resource. A base-load resource operates most efficiently at a relatively constant level of generation, and is not limited by changes in weather conditions or other factors. The *availability factor* is measured as the number of hours that a power plant is available to produce power divided by the total hours in a given period of time, usually 1 year. Geothermal energy *availability factor* is about 95%. This means that geothermal power plants are available for generation 95 % of any given time, based on decades of observations by plant operators. While availability factor measures a plant potential for use, the *capacity factor* measures the amount of real time during which a facility is used. A Geothermal plant also has a great capacity ranging from 89 to 97%, depending upon the type of geothermal system in place.

## *2.3.6 Other*

### 2.3.6.1 Human and Animal Power

Since the prehistory, human beings learned how to use animal energy, besides human energy, for the individual and social needs of their community. Still nowadays, in many developing countries, this energy resource is among the most common and widespread energy sources, also because very often it is the only one easily and conveniently available.

The size of this source varies according to the animal providing energy and, different from almost all the other sources, from the duration of the period in which the effort is provided; in this case, then, it is more suitable to refer to the available power. Figure 2.23 shows some power values according to the weight of the animal involved.

The analysis of human and animal energy lies outside of the purposes of this research; however, it has to be taken into consideration, not only in relation to the situation of some Countries that largely use it still nowadays, but also as a basis of comparison of any energy system.

#### 2.3.6.2 OTEC

The *Ocean Thermal Energy Conversion* (OTEC) is the technology that exploits the heat stored in oceans to generate electricity. In order to allow an OTEC plant to produce a suitable quantity of energy, it has to be installed where the difference in temperature between hot surface waters and cold deep waters amounts to nearly 20°C. These conditions are identified in the coastal areas of the tropical seas between the Tropic of Capricorn and the Tropic of Cancer.

#### 2.3.6.3 Tidal Power

Cyclic phenomenon extended to the entire hydrosphere and generated by the pull of the moon and of the Sun on bodies of water. *Tidal excursion* is the tidal range between the two limits of high and low water (*intertidal height*).

Depending on the positions of the moon and of the sun, the higher high water occurs after the phases of full moon and new moon, and is named *springtide*, whereas the lower low water occurs after moon quarters and is called *neap tide*. According to the size and the geometry of the sea and ocean basins involved in the passage of the tide, the tidal stream can reach remarkable levels of flow and velocity. A tide gauge is installed in water for the measurement and recording of tidal range.

Energy can be derived from the regular lowering and increasing of water bodies. To build a tidal plant, an estuary must be dammed towards the sea with an artificial dam. The energy technique exploits tidal excursion (the high-water and low-water range). A sufficient tidal excursion, as it happens in Northern France, at St. Mâlo, where the difference between the lower and higher water is around 12–13 m.

Tidal power plants; however, have the disadvantage that their maximum yield daily shifts by nearly 50 min with the rhythm of the tides. The optimal conditions for tidal power plants are rarely met. Besides St. Mâlo, small devices are operating on the banks of the White Sea in Russia, in Scotland and in China.

#### 2.3.6.4 Wave Energy

Wave energy can be exploited in different ways for the production of electric energy. The main ones are the following:

**Table 2.3** Summary of the RES potential in EJ/y [7]

| Resource | Current | Share (%) | Technical Potential | Share (%) | Theoretical potential | Share (%) |
|---|---|---|---|---|---|---|
| Biomass | 48.3 | 79.3103 | 276 | 3.6600 | 2,900 | 0.0002 |
| Hydro | 10.0 | 16.4204 | 50 | 0.6630 | 147 | 0.0001 |
| Solar power | 0.2 | 0.3284 | 1,575 | 20.8858 | 3,900,000 | 2.7100 |
| Wind power | 0.3 | 0.4926 | 640 | 8.4869 | 6,000 | 0.0042 |
| Geothermal power | 2.1 | 3.4483 | 5,000 | 66.3042 | 140,000,000 | 97.2837 |
| Total | 60.9 | 100.0000 | 7,541 | 100.0000 | 143,909,047 | 100.0000 |

- *Waves are directed towards a concrete chamber* the increase in water level compresses the air inside of the chamber that activates a turbine. Also the depression during wave return is exploited to activate the turbine.
- *The vertical motion of waves moves a piston* designed as a floating body.
- *A collection tunnel lifts waves for several metres* and lets them fall on a turbine.

#### 2.3.6.5 Sea Currents

The kinetic energy of sea currents can be successfully exploited. Indicating with $Q$ the volume flow rate ($m^3/s$), with $\rho$ seawater density ($kg/m^3$), with A the passage section of sea current ($m^2$) and with V current velocity (m/s) the power available (W) is:

$$P = \frac{1}{2}\rho \cdot Q \cdot V^2 = \frac{1}{2}A \cdot \rho \cdot V^3 \tag{2.21}$$

Considering a unit section ($A = 1$ $m^2$) and assuming $\rho = 1.000$ $kg/m^3$, (2.21) becomes:

$$P = 500 \cdot V^3 \tag{2.22}$$

From the economic point of view, this resource is exploitable if the power available is not lower than 500 $W/m^2$. Equation 2.22 shows therefore a minimum current velocity of 1 m/s.

### 2.3.7 *Current Renewable Energy Use*

The current use of RES is about 10% of world energy consumption, but the theoretical and technical potential of renewable energy is huge and it is mainly due to solar and geothermal energy (Table 2.3). In fact, only about 0.02% of global contribution to the Earth energy balance[7] does not come from solar energy, but rather from geothermal, gravitational and nuclear energy. Moreover, the technical potential of solar indirect power (wind, hydro, biomass, wave) are about half or one-third (depending on individual evaluations) of solar direct technical potential. Geothermal energy theoretical potential originates from the Earth's molten interior and from the decay of radioactive materials; its quantity is larger than the solar energy theoretical potential (the annual solar radiation). In short, not only the total theoretical potential is 100,000 times higher than world energy consumption (geothermal 100,000, solar 10,000, and the rest about 10),[8] but also the total technical potential of RES is 20 times higher. Therefore, its implementation mainly depends on technological capacity and costs, not on availability. Therefore, in consideration of the huge unexploited potential, perspectives are very interesting; it is necessary to take into account the expected capacity, efficiency, as well as economic, social and environmental impacts.

If for large hydropower and geothermal plants it is difficult to foresee technological improvements, thermal solar power plants can only be considered in a future perspective (in fact, exception made for some pilot plants and projects, there are no commercial power plant). For mini-hydropower, wind and photovoltaic power, an efficiency increase can be foreseen (and a cost decrease of 10 and 50%, respectively). The same considerations apply for bio-energy systems, but in this case improvements are linked to an increase in annual hours of operation (i.e., from 5,000 to 7,000).

## 2.4 Non-renewable Energy Sources

Energy sources with a well-established position in world trade, and underlying economic relations and balances, as well as energy trade between different Countries, are generally referred to as *conventional* (or *commercial*) energy sources: these are mainly non-renewable sources. Their real availability, the long-term predictability of price levels and the impact on the environment produced by the operating cycle (finding, transportation, utilisation and disposal of residues or waste) evidently are critical problems for the political, socio-economic and environmental sustainability of the use of these sources.

Availability depends on several factors of physical, technological, economic-financial and geopolitical nature. Once a deposit—or a primary energy source—is

[7] The Earth is a physical open system (exchange only energy no matter) so, normally, receives energy from the Sun and returns almost the same energy quantity to maintain its average temperature.

[8] From the mean average value of *annual* solar constant (1.37 $kW/m^2$) and the mean radiate terrestrial surface, it is possible to calculate a terrestrial annual solar irradiation of about $5.44 \times 10^{24}$ J (1.370 $W/m^2 \times 1.27 \times 10^{14}$ $m^2 \times 3.600$ J/W $\times 24 \times 365$) against an annual primary energy consumption of about $550 \times 10^{18}$ J = 550 EJ (1 oil kg has 10,000 kcal = $4.1868 \times 10^7$ J; thus 1 Gtoe = $4.1868 \times 10^{19}$ J = 41.868 EJ).

**Table 2.4** Main features of the different types of coal

| | Humidity (%) | Ashes (%) | HCV (kcal/kg)[a] | Volatile substances (%) |
|---|---|---|---|---|
| Peat | – | – | 5,500 | – |
| Lignite | 20–35 | 15–30 | 7,200 | – |
| Bituminous coal | 5–15 | 10–20 | 7,200 | 12 |
| Anthracite | 5–8 | 10–35 | 7,700 | 12 |

[a] The calorific value is referred to pure fuel, free from ashes and humidity

identified, the suitable technological devices to stock up at the minimum possible costs are needed; subsequently, the primary energy obtained has to be eventually processed by transforming or converting it into an adequate energy vector, which finally has to be transported to end users. All this requires know-how, investments and organisation.

The "energy issue" overwhelmingly came to the fore at the beginning of the 1970s, mainly as an international political issue: it became evident, in fact, that the business chain had a weak link in the relations between oil consuming Countries (mainly industrialised Countries), and oil exporting Countries (mainly developing Countries, mostly in the Arab Middle East).

However, there are also objective physical factors that influence the availability of conventional energy and, in particular, fossil energy. Being "non-renewable", these sources are destined to deplete, although in the medium-long term, since the time needed for their replenishment is largely higher (even millions of years) than the time of their consumption, amounting to tens or hundreds of years.

Additionally, availability also depends on the good value for money relating to the exploitation of each deposit. Very often, the so-called "*available resources*" are not used since their marketing and trade is not cost-effective: many deposits, although localised long time ago, are still unexploited since their extraction costs and transportation charges are too high. Therefore, depending on market prices and available technologies, a distinction is made between *economically extractable reserves* and *potentially extractable reserves*, being clear what these definitions refer to.

Among the causes for additional costs in the exploitation of energy sources, it is also necessary to consider the presence of technological and environmental constraints. The environmental impact, in fact, is the other critical factor for the use of conventional sources.

The case of coal is emblematic: in its entire cycle—from its production to its energy use—it is responsible for various forms of pollution. Thereby, mining activities, either underground or open-cast, cause different types of environmental problems with complex solutions, also linked—among other things—to the safeguard of workers' health and of the environment in general. On the other hand, open-cast mines—that are much more productive—very often have a non-sustainable environmental impact on the areas involved, causing the disruption of the social, agricultural, hydrological and economic fabric. Additionally, the problems related to transportation, combustion and, last but not least, ash disposal, must be considered.

In general, each commercial energy source is indissolubly linked to an environmental impact cycle, more or less critical and dangerous.

## 2.4.1 Coal

### 2.4.1.1 Definition and Structure

*Coal*, or better, *hard coals* are solid mineral substances and, from the energy viewpoint, represent the most important and abundant class of natural solid fuels.

Traditionally, coal is divided into four categories:

- *Peat*
- *Lignite*
- *Bituminous coal*
- *Anthracite*

Out of them, only *bituminous coal* and *anthracite* represent proper coal. Table 2.4 shows some pieces of data on the main features of coal according to this subdivision.

These values shall be considered as indicative of different categories that are not however strictly differentiated.

As to its structural features, coal is made up of a large variety of macromolecules,[9] very much different in terms of weight, although; however, sharing a common lamellar morphology.

In each macromolecule two areas can be identified; the *condensation nucleus* of aromatic rings, whose size characterises the *carbonisation degree,*[10] and one *peripheral area*, made up of different types of functional groups, which characterise the molecule itself.

The total percentage of carbon depends on the degree of transformation achieved, and ranges from 55 to 95% (on the dry substance); the uncombined portion (aromatic nuclei) ranges from 20 to 90%.

### 2.4.1.2 Genesis of Coal

The processes that led to the formation of fossil fuels are still analysed nowadays, although it is considered as clearly determined that the different types of coal share a sedimentary and phytogenic origin. In fact, it is certain that, starting from the paleozoic age,[11] huge quantities of organic matter, mainly of vegetable origin,

---

[9] These are mainly high-molecular mass carbon compounds with hydrogen and oxygen, containing, in minor percentages, nitrogen, phosphorus and sulphur.

[10] From the chemical viewpoint, *carbonisation* consists of a process that increases the size of the aromatic nucleus and increases its condensation; the latter inversely depends from the quantity of non-aromatic bridges that connect aromatic rings.

[11] More precisely, one of the periods of the Paleozoic era, which goes from 570 to 225 million years ago. It is the *Carboniferous* period, which began nearly 350 million years ago and takes its name from this important phenomenon.

were accumulated in wet environments, originating a process of carbon enrichment (*carbonisation*). This circumstance is also demonstrated, among other things, by the finds of vegetable organic remains[12] in less evolved coal-beds.

To clarify the topic of coal genesis, it is necessary to take into consideration at least the main factors that, more or less directly, have caused it; among them, the types of vegetables transformed, the geological and environmental conditions and chemical–physical phenomena.

The paleobotanic studies carried out on hard coals demonstrated how the biological nature of vegetable combinations, which underwent fossilisation in the carboniferous Age is rather similar to the nature of the plants that nowadays grow in marshes and in the lagoons of equatorial regions with hot-wet climate. The reason why the main coal basins are, on the contrary, almost all included in the temperate climate area has to be ascribed to the fact that the territories in the intertropical area underwent, in the areas following the Paleozoic Age, a series of slips on the Earth surface,[13] that determined its present position.

Coal deposits typically have a configuration with repeated layers or seams,[14] each one with a constant or gradually variable thickness, whereas the rocks associated and alternate to coal-beds are generally sedimentary (calcareous, sandstone, etc.), sometimes of sea origin. Furthermore, it is possible to observe that the different layers have a marked parallelism: it is possible to infer that the arboreal organisms growing in marshy areas, or in any case in flat lands, were periodically flooded and covered by the sea, due to local bradyseisms, originated in this way vegetable organic sedimentations. Each time they were covered by inorganic sedimentations, originated in this way the stratigrafic alternation between hard coal and sedimentary rock. This also explains the existence of thick coal-beds,[15] which probably represent the result of the deposit of vegetable substances in subsequent stages, each time on pre-existing material of the same nature but already partly transformed.

Apart from the indications suggested by the growing percentage of carbon, the fact that neither peat-beds were ever found below lignite deposits, nor layers of lignite below bituminous coal and anthracite beds, shows that the process occurred in this sequence: peat, lignite, bituminous coal and anthracite.

As to the chemical-physical processes involved in coal formation, two phases can be identified, although there is no sharp separation between them.

At the beginning, *biochemical factors* are involved. Fermentative and putrefactive changes of the organic matter take place in an acid environment ($pH < 7$); they are caused by *aerobic bacteria* and fungi, which consume oxygen and are

---

[12] Mainly leaves, stems, seeds and, sometimes, carbonised trunks that perpendicularly cross coal layers and that still have roots in the ground underlying the bank.

[13] This is in line with the hypothesis made by Alfred Wegener, in his continental drift theory (1915), and confirmed by more recent studies that led to the formulation of the plate tectonics theory.

[14] In northern French–Belgian basin there are up to 400 coal-beds of a thickness of 2,000 m.

[15] The thickest bank known is the one of the Kommentry basin, in central France that in some points exceeds 24 m of thickness.

only available at depths lower than 40 cm. Vegetable materials decompose quickly, and the main chemical reactions include oxidation and hydrolysis; a part of carbon is disposed from the vegetable matter (in the form of carbon dioxide and carbon monoxide), but most of it is kept, and its percentage compared to the other elements increases.

Subsequently, at higher depths, in the absence of oxygen and with pH close to neutral, *anaerobic bacteria* intervene. They survive in the absence of oxygen, and determine a further decomposition of materials—much slower than the previous one—as well as the formation of humic acid or humus[16] through chemical reactions, mainly involving reduction and hydrolysis.

With the gradual sinking of humic deposits, the conditions allowing the lives of fungi and bacteria disappear; in this way, the second phase of transformation begins; it is a merely inorganic phase, involving factors as time, pressure and temperature.

Time only partly influences the entire process; this is highlighted, for instance, by the limited degree of transformation achieved by the coal of a basin identified near Moscow, whose age dates back to the Carboniferous period, that is to say the same period to which date back coals with the highest degree of transformation. In this deposit, a concomitance of other conditions unfavourable to carbonisation occurred. However, in general the most ancient carboniferous formations, given the same other conditions, are more evolved and richer in carbon than recent ones.

Also the pressure due to the load exercised by the overburden is not a particularly relevant cause of the carbonisation process; at the most, this compression may impact on coal porosity. Not even more powerful pressures due to tectonic movements can be considered as directly influencing metamorphosis processes; according to a widespread opinion, the carbonisation of the main basins on the Earth was already completed when, in the Mesozoic Age, the drift of plate tectonics on the earth surface began.

*Temperature* is therefore the key factor in the carbonisation process. In consideration of the thermal gradient of the earth, amounting on average to nearly 3°C every 100 m of depth, and of powerful magmatic intrusions contiguous to the beds of humic deposits, it is believed that the latter—that in the carboniferous period were covered by large layers of sedimentary rocks—might have reached temperatures close to 200°C, sufficient to guarantee the conditions for the carrying out of the metamorphosis process.

#### 2.4.1.3 Physical Characteristics and Technological Merits

The relationship between the chemical composition and the technological features of coal is rather complex; for this reason, it is preferable to directly take into

[16] Complex material, of vegetable origin, mainly made up of carbon.

consideration the parameters characterising fuels according to the different usages they are destined to.

The most frequent evaluations, then, include *the measurement of the calorific value,the immediate analysis*, *fuel behaviour after heating,* and *sulphur analysis.*

The *measurement of the calorific value* is made experimentally, mainly through the use of calorimeters. The calculation is obviously referred to the lower heating value, since all fuels contain a given quantity of water,[17] called *fuel humidity*; when fuels burn, water vapourises at the expense of part of the heat generated.

The value of the moisture degree is one of the determinations of the *immediate analysis.* The other tests of this analysis consist in the evaluation of *ashes*, for which it is important to know the quantity produced during combustion, and the melting point; in this case as well, it is necessary to resort to experimental determinations. The determination of *volatilesubstances* and of *coke*[18] follow, as they are useful to trace back the type of coal examined and the energy use that can be done of it. In this connection, a few distinctions are made among fossil fuels in relation to the result of these tests. For instance, a coal characterised by a high content in volatile substances will be defined as *cannel coal*, in relation to the particular combustion developed; otherwise, the coal in question will be *blind coal.*

The analysis of *burning coal behaviour* is important to make a distinction between coals that burn completely, remaining in an incoherent state, and those that, if warmed, inflate and hinder the passage of air. The latter phenomenon is quantified through a suitable *swelling index.* As regards *sulphur*, finally, it can be found in coal under the form of organic compounds, or of sulphides and sulphates.

#### 2.4.1.4 Composition and Classification

The subdivision introduced in Sect. 2.4.1.1, according to which coals are divided into peat, lignite, bituminous coal and anthracite is not sufficient to rigorously identify the possible energy use.

The classifications adopted for commercial purposes, in fact, were drafted without any reference to univocal and rigorous criteria. This also for the evident purpose by each producing country to use parameters that better highlight and enhance the characteristics of the coals extracted in their own basins. In any case, these classifications are based on the content of volatile matter and, secondarily, on calorific value.

---

[17] Part of water is also formed during combustion.

[18] Coke is the solid residue left after coal warming at 1,000–1,100°C without air, once it lost its volatile substances—either pre-existing or which formed during warming. This material is hard, porous, grayish, with metallic nuances, and is endowed with a high calorific value (up to 7,400 kcal/kg).

In 1949 the United Nations Economic Commission for Europe established an *ad hoc* Committee entrusted with the task of drafting an international classification through which setting the different types of coal, irrespective of their origin. The works were completed in 1956; according to this classification, each coal is characterised by a *three-digit code*, each digit indicating a *class*, a *group* and a *subgroup*, respectively.

In total, 10 classes were identified, which indicate the content in volatile matter; whenever this value exceeds 33%, the higher heating value is introduced as a discriminating value. The groups and subgroups, amounting to four and six, respectively, are determined on the basis of other properties as, i.e., the swelling index and the agglomerating power.[19]

Another important classification is made by ASTM[20] (Specification ASTM D-388) according to which coals are divided into *four classes*:

- *Anthracite*
- *Bituminous*
- *Sub-bituminous*
- *Lignite*

Classes are in turn divided into *groups*, which are characterised by the content in fixed carbon,[21] by the quantity of volatile matter,[22] by the higher heating value and, subordinately, by the agglomerating power.

#### 2.4.1.5 Treatment and Use

In order to improve the collection, transportation and combustion of coal, many researches were carried out. The main hindrance is represented by its *lack of fluidity*—for this reason coal needs expensive transportation infrastructures—linked to the high rate of emissions of polluting substances that need fumes abatement systems, in this case as well entailing outstanding economic commitments.

Various technologies were developed in the attempt of giving coal those peculiar characteristics that allowed oil and natural gas to gain ground in recent times (removing in this way the two "negative" characteristics of solidity and highly polluting combustion):

- *Pulverisation*: presently, this is the most developed and widely used technique in the industrial sector. It allows to burn in steam generators the coal dust injected thanks to the suitable fans. Among its main problems: the easiness of

[19] Coal capacity to agglomerate inert substances.

[20] American Society for Testing and Materials.

[21] Solid residue obtained through distillation and determined with normalized methods without considering ashes.

[22] They represent the *portion transformable into gas through warming* and is determined by heating the fuel inside of a closed oven filled with inert material for a given time and at a predetermined temperature.

self-combustion in deposits and the possible environmental impact in the transportation and storage phase.
- *Gasification and liquefaction*: they entail the addition of hydrogen (through steam) to coal, which in this way becomes a hydrocarbon. The energy necessary for the separation of hydrogen from water shall be provided by coal itself, in order to have a positive economic balance in this operation. Synthetic gas is obtained ($CO + H_2$) which may be further modified by adding steam in order to have a quantity of hydrogen exceeding the amount necessary to convert coal into hydrocarbons.

Some of the processes used and/or under study include:

- Hydrocarbonisation
- Hydrogenation
- Extraction
- Fischer–Tropsch process[23]
- *Coal–oil*: mix of oil and coal dust (all included in the combustion chamber).
- *Coal–water*: creation of an inhomogeneous mix of water and coal particles (not all included in the combustion chamber: need for further links in the technological chain, increasing costs).

#### 2.4.1.6 Reserves and Geographical Distribution

Among fossil fuels, coal is undoubtedly the one that offers more guarantees in terms of future supplies, thanks the huge reserves identified, as shown in Table 2.5.

On the basis of present production and consumption rates, the estimated availability of coal apparently ensures stocks for the next 133 years (Fig. 2.24). This estimate, however, is not entirely exhaustive, first of all because the level of coal exploitation is strictly connected to the other commercial sources, mainly oil and natural gas, whose depletion is expected to occur far earlier. Furthermore, it is necessary to take into consideration the uncertainty in the determination of economically usable reserves, which change according to the variations in market prices.

The localisation of world reserves sees coal sufficiently available in all geopolitical areas; however, a particular concentration is present in North America (29.6%), in ex-URSS (26.7%) and in China (13.5%), as shown in Fig. 2.25.

---

[23] *Fischer–Tropsch process*: method for the synthesis of hydrocarbons and other aliphatic compounds. Synthetic gas, a mixture of hydrogen and carbon monoxide, is reacted in the presence of an iron or cobalt catalyst; much heat is evolved, and such products as methane, synthetic gasoline and waxes, and alcohols are made, with water or carbon dioxide produced as a by-product. An important source of the hydrogen–carbon monoxide gas mixture is the gasification of coal (see *water gas*). The process is named after F. Fischer and H. Tropsch, the German coal researchers who proposed it in 1923.

**Table 2.5** World coal reserves at the end of 2007, in million tons [7]

| | Anthracite and bituminous | Sub-bituminous and lignite | Total | % |
|---|---|---|---|---|
| USA | 112,261 | 130,460 | 242,721 | 28.6 |
| Canada | 3,471 | 3,107 | 6,578 | 0.8 |
| Mexico | 860 | 351 | 1,211 | 0.1 |
| *Total North America* | 116,592 | 133,918 | 250,510 | 29.6 |
| Brazil | – | 7,068 | 7,068 | 0.8 |
| Colombia | 6,578 | 381 | 6,959 | 0.8 |
| Venezuela | 479 | – | 479 | 0.1 |
| Others | 172 | 1,598 | 1,770 | 0.2 |
| *Total Central America* | 7,229 | 9,047 | 16,276 | 1.9 |
| Bulgaria | 5 | 1,991 | 1,996 | 0.2 |
| Czech Republic | 1,673 | 2,828 | 4,501 | 0.5 |
| Germany | 152 | 6,556 | 6,708 | 0.8 |
| Greece | – | 3,900 | 3,900 | 0.5 |
| Hungary | 199 | 3,103 | 3,302 | 0.4 |
| Kazakhstan | 28,170 | 3,130 | 31,300 | 3.7 |
| Poland | 6,012 | 1,490 | 7,502 | 0.9 |
| Romania | 12 | 410 | 422 | 0.0 |
| Russia | 49,088 | 107,922 | 157,010 | 18.5 |
| Spain | 200 | 330 | 530 | 0.1 |
| Turkey | – | 1,814 | 1,814 | 0.2 |
| Ukraine | 15,351 | 18,522 | 33,873 | 4.0 |
| UK | 155 | – | 155 | 0.0 |
| Others | 1,025 | 18,208 | 19,233 | 2.3 |
| *Total Europe and Eurasia* | 102,042 | 170,204 | 272,246 | 32.1 |
| South Africa | 48,000 | – | 48,000 | 5.7 |
| Zimbabwe | 502 | – | 502 | 0.1 |
| Other Africa | 929 | 174 | 1,103 | 0.1 |
| Middle East | 1,386 | – | 1,386 | 0.2 |
| *Total Africa and Middle East* | 50,817 | 174 | 50,991 | 6.0 |
| Australia | 37,100 | 39,500 | 76,600 | 9.0 |
| China | 62,200 | 52,300 | 114,500 | 13.5 |
| India | 52,240 | 4,258 | 56,498 | 6.7 |
| Indonesia | 1,721 | 2,607 | 4,328 | 0.5 |
| Japan | 355 | – | 355 | 0.0 |
| New Zealand | 33 | 538 | 571 | 0.1 |
| North Korea | 300 | 300 | 600 | 0.1 |
| Pakistan | 1 | 1,981 | 1,982 | 0.2 |
| South Korea | – | 135 | 135 | 0.0 |
| Thailand | – | 1,354 | 1,354 | 0.2 |
| Vietnam | 150 | – | 150 | 0.0 |
| Other Asia–Pacific | 155 | 276 | 431 | 0.1 |
| *Total Asia–Pacific* | 154,255 | 103,249 | 257,504 | 30.4 |
| *Total* | 430,935 | 416,592 | 847,527 | 100 |

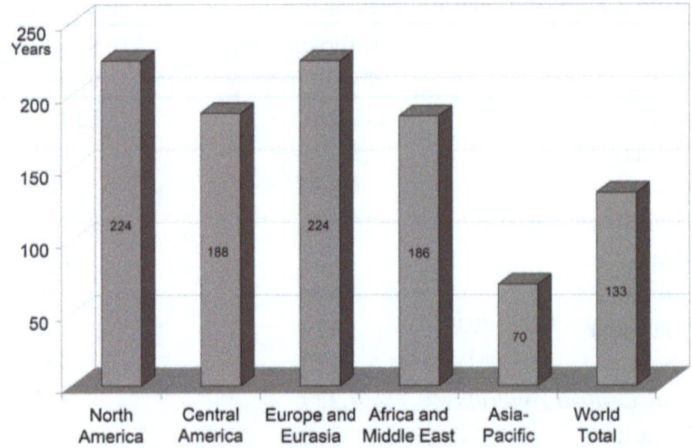


**Fig. 2.24** Coal consumption/production ratio

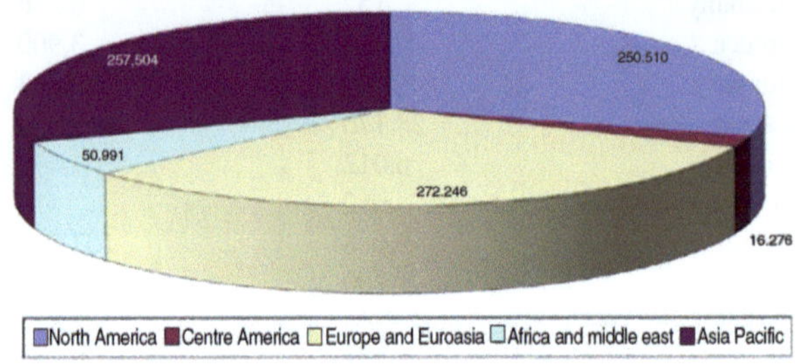


**Fig. 2.25** World distribution of coal reserves

## 2.4.2 *Oil*

### 2.4.2.1 Definition and Structure

The word *petroleum* derives from the two Latin words *petra* and *oleum* that is, to say "stone oil" and, in its general use, it indicates an oily liquid, with a variable density, brown in colour with yellow or fluorescent black shades, insoluble in water. The scientific and technical meaning is wider and includes *all natural mixtures mainly made up of hydrocarbons,*[24] *being either liquid or gaseous or solid*; in this framework, liquid oil (at room conditions) is called *fuel oil* and the gaseous one is called *natural gas*.

Petroleum is the primary energy source presently most used all over the world. The reasons for this spreading are mainly linked to its relatively *high energy density* (on average 10,000 kcal/kg that is around 41,860 kJ/kg), its considerable *chemical–physical stability* and the *limited quantity of solid residues* produced with its combustion. Furthermore, since petroleum is liquid at its normal

[24] *Hydrocarbons* are, by definition, exclusively made up of hydrogen and carbon; strictly speaking, then, the word "hydrocarbons" shall not be used as a synonym of petroleum, since it contains, in lower percentages, other organic compounds made up—besides hydrogen and carbon—by nitrogen, sulphur, oxygen and other elements.

conditions in the environment, it can be *transported relatively inexpensively*; given the same calories transported, it is certainly preferable over both natural gas (aeriform) and coal (solid).

The main disadvantages of this strategic source, which have caused its enduring crisis in the last decades, are linked to the fact that it is characterised by a *geo-politicallyunbalanced distribution*, it is a *non-renewable* source and its combustion produces a *remarkable air pollution.*

In some cases, petroleum may spontaneously reach the earth surface and burst into flames for natural causes; this is how men learned about its existence since the prehistory. Already around 3000 BC, in Mesopotamia, *bitumen* (i.e. thick and oxidized oil) was used as a fuel for lighting, as a binder in the building sector, for the waterproofing of the hulls of boats, or as a preservative for wood and leather, and as a medicine. There are also witnesses indicating the research and extraction of petroleum in the subsoil, through artificial wells, in China as from 250 AD.

In any case, the official date of the beginning of the modern oil industry traditionally dates back to 1859, when the US colonel Drake was entrusted with the task of performing drilling at Oil Creek, in Pennsylvania, with the specific purpose of finding petroleum.

From the morphological viewpoint, its structure is made up of a mix of different hydrocarbons and, to a lower extent, of other organic substances. It is not possible to perform a complete classification of all its components, since they more or less amount to over one thousand; for instance, the American Petroleum Institute (API) has identified 350 different types of hydrocarbons, almost all liquid. The fluctuations of weight percentage values of the main elements include the following:

- C: 83–87%
- H: 11–15%
- O: 0.05–4%
- S: 0.05–7%
- N: 0.05–2%

#### 2.4.2.2 Genesis of Oil

An *oil field*, that is to say the reservoir in which oil is concentrated in such quantities that make its extraction cost-effective, is the final result of a very slow process that determines petroleum formation and, generally, its migration from the place of origin to the place of storage, which are often far from one another. Because of the particular complexity and duration of this process, its phases have not been, until today, sufficiently clarified, although in the last few years satisfactory progresses were obtained, especially in the field of geochemical research.

In the past, the origin of petroleum was ascribed to *the combined action of pressure andtemperature on the remains ofcrustaceans, molluscs and, in any-case,large animals* (both sea and terrestrial animals); present hypothesis, on the contrary, indicate that the initial organic matter originated from sea settlements of

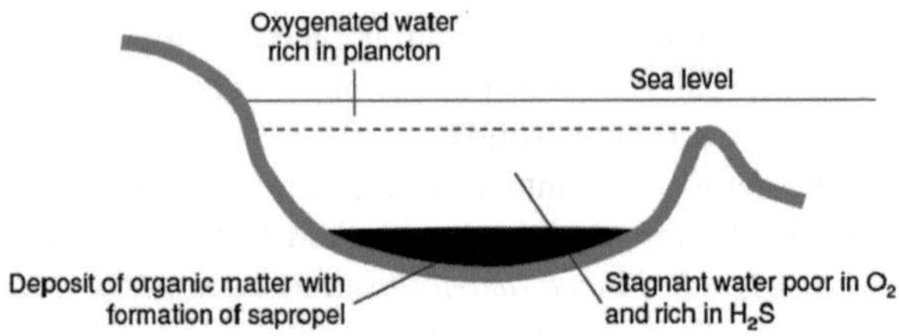


**Fig. 2.26** Area of deposit of oil organic substances

micro-flora and micro-fauna. The *organic origin of petroleum*[25] is in any case ascertained on the basis of some objective elements. For instance, a large quantity of organic matter can be observed in the sedimentary rocks rich in carbon and hydrogen, which are particularly suitable for oil formation; it is also possible to assess that all types of petroleum contain nitrogen and porphyrin pigments of exclusively vegetable and animal origin. Porphyrin, in fact, can derive from both vegetable chlorophyll and from animal haemoglobin. Nitrogen, conversely, varies in sediments directly proportionally to organic carbon.

Animal and vegetable microorganisms, once dead, accumulate at the bottom of aquatic basins, together with small siliceous, calcareous and argillaceous particles, transported as detritus by watercourses and by the wind. The deposit of these sediments takes place in particular in smooth and shallow waters as, for instance, in gulfs, inland seas and lagoons, and it is what still nowadays forms the mud of some basins; this organogenic mud is named *sapropel*. This is the ideal environmental condition to start the petroleum formation process.

Figure 2.26 shows the typical scheme of these ideal environments: it is a lagoon basin, whose limited communication with the open sea leads to the stagnation of the water under the superficial layer with a low concentration of oxygen; hydrogen sulphide ($H_2S$) is developed as a consequence of the limited exchange of aerated water in the water. In this situation, the remains of microorganisms abundantly present in the aerated superficial layer accumulate at the bottom of water.

In normal conditions that is, to say in the presence of oxygen dissolved in water, organic remains undergo the attack by aerobic bacteria,[26] that while oxidizing the organic carbon, bring these remains back to the marine cycle, under the form of various chemical compounds. This does not take place when such organic materials are in an oxygen-free environment, where the action of anaerobic bacteria

---

[25] During the XIX century, many supported the *inorganic origin* of *petroleum*. In fact, it is theoretically possible that the formation of hydrocarbons through chemical processes that occur among inorganic compounds of petroleum, or by polymerisation of primordial methane making up the stellar matter; the presence of methane among volcanic gases is ascertained. It is however incompatible with the mechanisms characterising inorganic chemistry the formation of complex organic substances as those that can be found in petroleum, and is in any case difficult to justify the distribution of deposits, more abundant in the recent layers of the Earth's crust.

[26] *Aerobic bacteria* need free oxygen to develop their action; conversely, anaerobic bacteria only use combined oxygen.

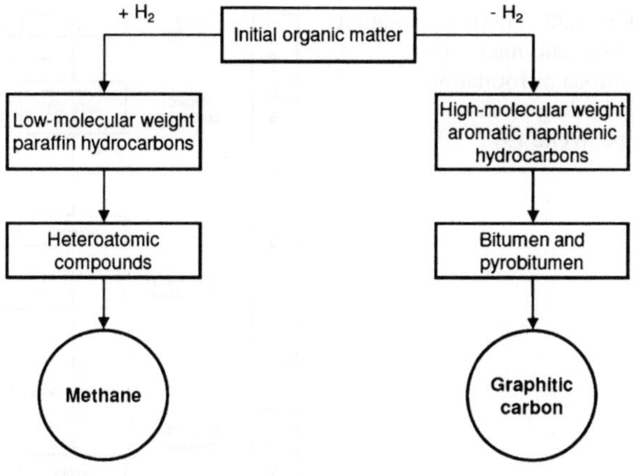


**Fig. 2.27** Simplified diagram of naftogenesis [10]

prevails, and they use the oxygen belonging to organic molecules and, through fermentation processes, mainly produce carbon dioxide and methane.

In environments similar to the one described above, through given biochemical reactions and the influence of some physical factors as temperature and pressure, the so-called *naftogenesis* occurs, which transforms organic raw materials into the compounds that make up petroleum (Fig. 2.27).

Furthermore, in order to allow environmental conditions to remain unchanged, it is necessary that depths undergo a continuous and regular lowering called *subsidence*, able to offset the supply of sediments; in this way, large quantities of organic matter will settle for very long times (several geological eras), without assisting to an interruption of the conditions favourable to deposit.

The slow and progressive subsidence of the Earth's crust, determined by both the attraction action of underlying fluid levels, and by the weight of sediments themselves, allows an accumulation of organic matter included in detritus, which reach several kilometres in thickness, although generally the waters above have limited depths (ranging from a few meters to a few hundreds of meters).

The subsidence of sedimentary sequences is also important since the reaching of higher and higher depths leads to an increase in their temperature, in consideration of the *geothermal gradient*, originated from the endogenous heat of the earth. The action of temperature and its persistence for very long times, represent—as described afterwards—the pre-requirements for most of the chemical transformations of the organic matter.

*Naftogenesis* is a complex process, still not entirely known, from whose interpretation many theories were developed. One of the most exhaustive and accredited theories was presented by Tissot and Welte [9]. According to this theory, the naftogenesis process is divided into four stages:

- Diagenesis
- Catagenesis
- Metagenesis
- Metamorphism

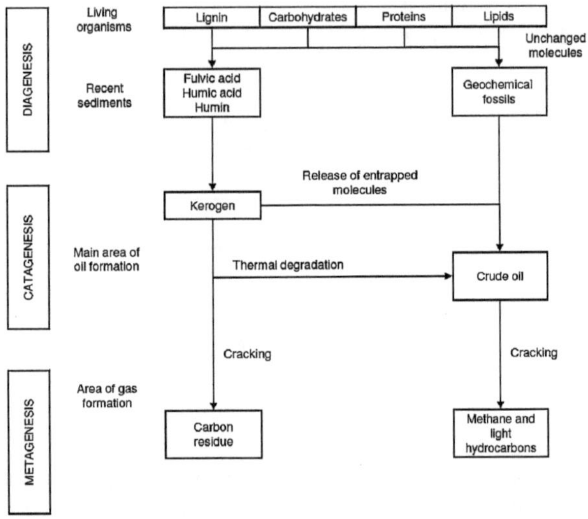


**Fig. 2.28** Possible evolution of organic matter in petroleum formation according to Tissot and Welte [9]

Figure 2.28 shows a simplified evolution of sediments in the first three phases of naftogenesis that is, to say those that properly concern petroleum formation.

*Diagenesis* begins as organic sediments and detritus settle on the bottom of the basin. During this first phase, deposits gradually reach depths of nearly 1,000 m and temperatures of about 50°C. Water, still present in relevant percentages, is partly expelled following *compaction*, a process due to the load exercised by the layers above and that, together with the *cementation* made by the salts dissolved in water, transforms inorganic detritus into sedimentary rock. Furthermore, an intense activity of aerobic and anaerobic bacteria takes place, whose quantity is not sufficient for the decomposition (and thence the disruption) of the organic matter. They operate the separation of *biopolymers*, that is to say the molecules that make up the organic matter, originating new structures made up of smaller and simpler organic molecules. Besides providing nutrition to bacteria, they tend to react among themselves, combining into new large molecular formations called *geopolymers*. In these compounds, the macromolecule with the highest concentration is *kerogen.*[27] This is the substance from which, thanks to the transformations occurring in the subsequent phases, petroleum originates.

Besides kerogen, representing the main derivative of the organic compound, high-molecular weight compounds are formed, defined as *geochemical fossils*, which are synthesised from plants and animals and absorbed in sediments. From a chronological standpoint, they represent the original substance from whose evolution hydrocarbons originate; consequently, they are considered as contributing—

[27] *Kerogen* is an organic matter composed of carbon, hydrogen and oxygen; three main types of kerogen exist, with different evolution and therefore leading to different end product; they are differentiated according to the relative quantities of hydrogen and oxygen compared to carbon.

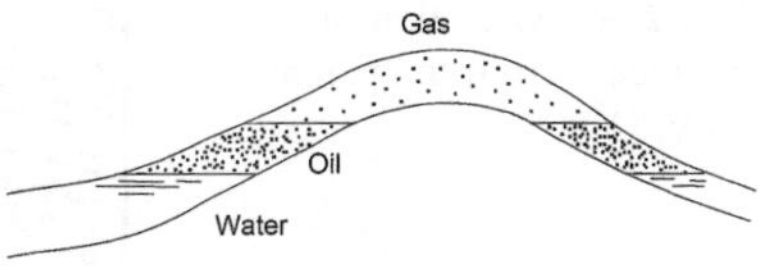


**Fig. 2.29** Schematic diagram of a typical oil field [11]

although marginally—to petroleum formation, since the origin of the latter is mainly ascribed to the transformation that kerogen undergoes with depths.

Sometimes the sediment that settles in basins is particularly rich in vegetable substances, derived from terrestrial plants; this mainly happens in lake basins surrounded by abundant vegetation and in coastal marshes of tropical areas, where a typical thick vegetation grows. In this way, layers of sediments mainly made up of vegetable materials are formed and, at the end of diagenesis, essentially originate peats, lignite and brown coals, in which the structure of the plants of origin is still clearly recognisable.

*Catagenesis* intervenes when, due to subsidence, the thickness of sediments exceeds 2,000 m, and high values of pressure (from 300 up to 1,000–1,500 bars) and temperature (nearly 150°C) are generated. As an effect of pressure, sedimentary rock porosity and permeability decrease, a further quantity of water is expelled, and salinity increases in the remaining portion. In this stage, the major changes in kerogen take place, and lead to the degradation of its molecules.

In this phase, the factors that influence the development of kerogen and the formation of petroleum mainly include *temperature* and *duration*, which characterise the heating and transformation processes of kerogen. These factors are linked proportionally—in the sense that the same result can be achieved at high temperature in short times and, vice versa, in long times at lower temperature. In this phase, methane develops as well, whereas the petroleum produced in the *parent rock* that is to say the rock in which the initial organic compound was contained, once a given quantity is exceeded, is ejected and penetrates the surrounding rocks. This is therefore the beginning of migration that brings crude oil to the so-called *reservoir rocks*, in which real oilfields are localised.

*Metagenesis* represents the last stage of evolution of the organic compound, and takes place at high depths reaching 5–6 km, and at temperatures of 150–200°C. Kerogen molecules lose almost all the oxygen atoms and most of hydrogen atoms, producing limited quantities of methane, whereas—at the end of this phase—they originate a carbon residue, destined to become *graphite*.

Temperature increase also modifies the composition of part of the oil formed during the previous phase of catagenesis; in fact, a *thermal cracking* that is, to say the breaking of those oil molecules with a high-molecular weight, and the formation of smaller and lighter molecules that are destined to make up *light oil* and *gas* (mainly *methane*).

As a consequence of the high temperatures involved, several mineral sediments undergo solution and recrystallisation processes. Furthermore, they exercise a catalytic action in chemical reactions that, as far as hydrocarbons are concerned, lead to the formation of methane alone.

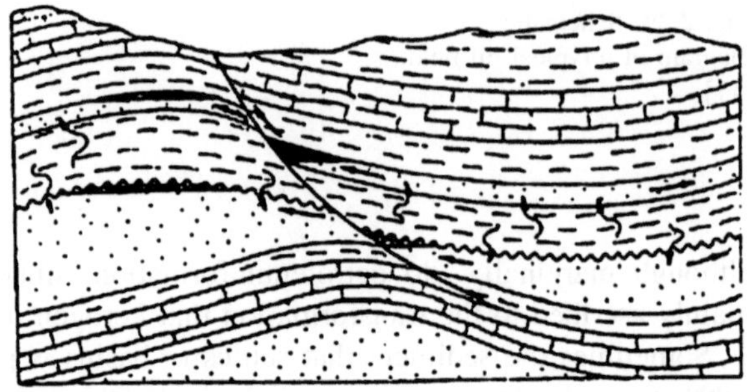

**Fig. 2.30** Migration of petroleum from the parent rocks [12]

*Metamorphism*, finally, determines the total loss of the original features of organic sediments. Hydrocarbons are not involved in this stage, whereas what remains of the original constituents of kerogen is transformed into graphite.

As already mentioned, *temperature* is the guiding factor of naftogenesis, since some of the chemical-physical reactions that transform organic compounds into petroleum only occur at temperatures close to 250–300°C. *Pressure*, on the contrary—which increases with the depth of sedimentary layers—mainly intervenes in determining the removal of fluids from the parent rock and their migration. In the final analysis, a typical petroleum formation is made up of a set of rocks impregnated of liquid and gaseous hydrocarbons, mixed with salty water (Fig. 2.29).

These formations are generally divided into *primary* and *secondary*, respectively, when petroleum is kept in the rocks of origin (*parent rocks*) or when because of the effect of compaction, it migrates to other rocks called *reservoir rocks*, different from those in which it was formed. In Fig. 2.30, which summarises the migration movement of petroleum that abandons parent rocks, arrows represent the direction of petroleum migration from parent rocks to the adjacent permeable rocks, allowing it to reach the reservoir rocks that represent the areas of accumulation (oilfields), and, in some cases, up to the surface. Parent rocks can have different natures: argillaceous, calcareous, siliceous, and carbonaceous. Impregnated rocks are generally sedimentary, sandstones or sands.

The geological eras which the formation of petroleum dates back to include, in order of importance: Cenozoic, Paleozoic and Mesozoic[28]; only a few formations date back to more recent times, and more precisely to the Pleistocene period.

[28] This is the historical sequence of these eras:
Palaeozoic era (from 570 to 225 million years ago);
Mesozoic Era (from 225 to 65 million years ago);
Cenozoic Era (from 65 to 2 million years ago).

**Table 2.6** Conversion values of API degrees

| API degrees | Relative density |
|---|---|
| 0 | 1.076 |
| 10 | 1.000 |
| 15 | 0.966 |
| 20 | 0.934 |
| 25 | 0.904 |
| 30 | 0.876 |
| 35 | 0.850 |
| 40 | 0.825 |
| 45 | 0.802 |
| 50 | 0.780 |

#### 2.4.2.3 Physical Properties and Technological Merits of Petroleum

In nature, it is not possible to find two oilfields providing petroleum with the same composition. They are different to such an extent that sometimes there are changes in the petroleum of the same oil well at the moment of extraction. It is therefore necessary to carefully assess the characteristics of each crude oil, in order to identify the appropriate refining processes.

From the energy viewpoint, the main property is the *lower calorific value*. In case of petroleum, different from what one might expect because of the variability of its composition, the lower heating value has a little variable value (9,600–10,800 kcal/kg)[29]; the *higher heating value* is generally higher by nearly 10%. The high energy density of petroleum is the main reason for the very large spread of this substance as a primary energy source.

Another important characteristic is *density* that is, to say the weight of the substance by unit of volume, which obviously varies according to temperature. From density it is possible to derive the first information on the refining processes to be followed; high-density crude oils, for instance, contain petrol in lower quantities and, consequently, are less valuable. Furthermore, as density increases, the calorific value decreases.

Usually, reference is made to *relative density*, that is to say the ratio between the oil mass under consideration and the mass of the same volume of water, both at a temperature of 60°F (15.6°C).[30]

In density determination, API degrees are very much used, and they vary according to the relative density, based on the formula:

$$1^\circ\text{API} = \frac{141.5}{\text{relative density}} - 131.5$$

[29] Usually, a medium calorific value of 10,000 kcal/kg is considered, which is also the reference value for ton of oil equivalent. A t.o.e. has then a thermal content of $10^7$ kcal, amounting to nearly $41.9 \times 10^9$ J = 41.9 GJ.

[30] Relative density differs from *specific gravity* since, although being a ratio between equal values of matter and water, it is calculated at a temperature of 4°C.

**Table 2.7** General classification of hydrocarbons

| | Structure | Series | Parent | Homologous |
|---|---|---|---|---|
| Saturated | Acyclic | Paraffin ($C_nH_{2n+2}$) | Methane ($CH_4$) | Ethane<br>Propane<br>Butane<br>Pentane |
| | Cyclic | Naphthenic ($C_nH_{2n}$) | Cyclopropane (trimethylene) ($C_3H_6$) | Cyclobutane<br>Cyclopentane |
| Unsaturated | Cyclic and polycyclic | Aromatic or Benzenic ($C_nH_{2n-6}$) | Benzole (Benzene) ($C_6H_6$) | Toluene |
| | Acyclic | Olefinic Acetylenic ($C_nH_{2n}$) | Ethylene ($C_2H_4$) | |

Table 2.6 shows some values of API degrees with the corresponding relative density.

In general, petroleum has a relative density ranging from 0.950 to 0.750 kg/dm$^3$ (17–57 API), with the most frequent values included between 0.950–0.850 (17–35 API).

The other properties of petroleum include *viscosity*, whose determination is necessary for the extraction procedures.

For safety purposes, during transportation and storage, important items include the *flash point* (that is to say oil temperature at which the vapours released from it, in the presence of a flame, catch fire without creating combustion to petroleum itself) and the *burning point*, which corresponds to the temperature of oil at which vapours develop at such quantities that, if triggered, are also able to fire the liquid petroleum. As to the transportation of crude oil through the pipelines crossing cold regions, it is also important to take into account the *freezing point (pour point)*, that is to say the temperature below which petroleum is completely coagulated and does not flow any longer. As regards petroleum containing paraffin solutions, instead of the pour point, the temperature at which paraffin starts to solidify (*cloud point*) is taken into consideration, producing a "cloud of wax". Finally, the *boiling point* of the different components of crude oil, sets the reference temperatures for *fractional distillation*.

#### 2.4.2.4 Composition and Classification of Petroleum

As already said, petroleum is made up of mixtures and solutions of many hydrocarbons (liquid, gaseous and solid), with relatively scarce quantities of other organic, oxygenated, sulphured and nitrogenous compounds. Because of this complex composition, there is no any systematic classification system. In any case, in the determination of the various types of crude oil, the hydrocarbons composing them are mainly taken into consideration.

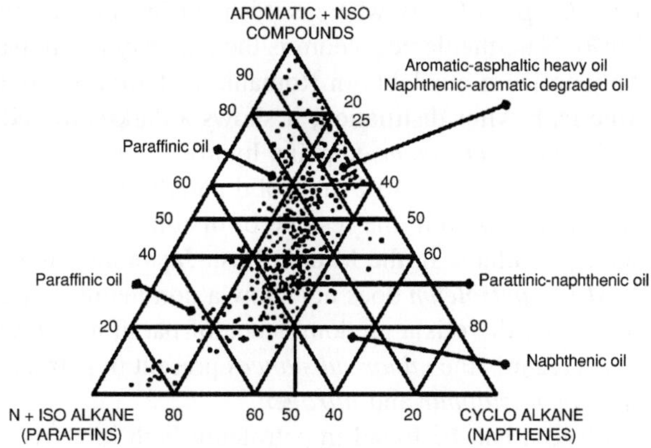


**Fig. 2.31** Ternary diagram of petroleum composition [9]

A basic criterion consists in the subdivision of hydrocarbons into *saturated* (carbon atoms saturated by simple bonds) and *unsaturated* (double or triple bonds between carbon atoms).

A second classification, based on the molecular structure, divides the hydrocarbons contained in crude oil into:

- *Acyclic* (or *aliphatic*) with carbon atoms bonded as an open chain;
- *Cyclic* with carbon atoms bonded as a closed chain;
- *Polycyclic* with the chain containing several links.

As regards these two classifications, the main hydrocarbons can be divided into three series characterised as follows:

- *Paraffin series*, saturated hydrocarbons, acyclic.
- *Naphthenic series*, saturated hydrocarbons, cyclic.
- *Aromatic series* (or *benzene*), cyclic and polycyclic unsaturated hydrocarbons.

There are then two other categories less important in the oil sector: *olefins* and *acetylene series*; both of them are characterised by acyclic unsaturated hydrocarbons.

Table 2.7 shows a schematic subdivision of hydrocarbons according to the criteria mentioned above.

Usually, petroleum contains at the same time the three main series mentioned above; Fig. 2.31 shows the ternary diagram of petroleum composition.

According to the prevailing series of hydrocarbons, petroleum can be divided into *paraffinic*, *naphthenic*, *aromatic* and *mixed*.

*Paraffinic petroleum* contains abundant dissolved quantities (up to 10%) of solid paraffin (called *petroleum wax*). Many oil fields with paraffinic petroleum are located in the USA (Pennsylvania, Ohio, Oklahoma and Texas) and in Europe (Galicia and Romania). The residues of distillation are characterised by a light colour.

*Naphthenic petroleum* contains high percentages of cyclopentane and dicycloesane (naphthenic series) and large quantities of asphalt (for this reason called *asphaltic petroleum*), whereas it is poor in solid paraffin (maximum accounting for 1.5%). Naphthenic petroleum is the majority of oil available in Russia, and part of the petroleum available in Romania and America (California, Mexico and South America). After distillation, it shows a dark-coloured residue.

*Aromaticpetroleum* contains hydrocarbons of the aromatic series for a maximum percentage of 40% and solid paraffin is present in quantities lower than 5%. In this case as well, the residue of distillation is dark-coloured. This petroleum is mainly available in the Middle East, Japan and in South-East Asia.

*Mixed petroleum* does not show a dominating series of hydrocarbons, although very often there is a predominance of paraffin and naphthene.

As regards the *minor substances* present in petroleum, they are mainly made up of *oxygen*, *sulphur* and *nitrogen*.

*Oxygen* can be found in petroleum both in a free state and as a component of several organic compounds; furthermore, due to air exposure, several oxygenated substances are formed. In general, oxygen content grows with the specific gravity of petroleum; in asphalts, for instance, it may also reach 5%.

*Sulphur* is present in all crude oils, both in a free state and as a component of $H_2S$, and of many sulphurated organic compounds. Usually, in crude oils with a low specific gravity there is a low sulphur content, in the same way as the lightest fractions of whatever type of petroleum are, in general, poor in sulphur. Crude oils with a sulphur content higher than 0.5% are called *sulphorous* (*high-sulphur crudes*). In asphalts, finally, contents are very high, even exceeding 10%.

*Nitrogen* is contained in all types of petroleum, generally in quantities lower than 0.5%, with maximum peaks of 2%. Also in this case, it is available in the free state, or in organic compounds. Among nitrogenous substances, particular importance is ascribed to *porphyrines*, not for their features, but rather because they provide indirect confirmation of the theory on the organic origin of petroleum; these substances are, in fact, exclusively derived from haemoglobin and chlorophyll.

The microscopic study of petroleum also showed that it may contain different materials of organic origin, like, for instance, flakes of siliceous shells, fragments of insects, spores, algae and unicellular organisms (also living ones).

Finally, petroleum has an inorganic content that is shown in combustion-originated ash. Ashes, mainly made up of salts, generally represent 0.01–0.05% of crude oil weight, and their level increases as the weight density of oils increases; in asphaltic ones, it is possible to reach 1%.

#### 2.4.2.5 Treatment and Use

The petroleum extracted from oil fields is generally subsequently refined (some Arab countries also use it directly), to obtain products with a higher energy content and a better combustion capability.

From crude oil it is possible to derive:

- Gas
- Gasoline
- Kerosene
- Diesel oil
- Lubricant bases
- Fuel oil
- Bitumen
- Other products.

#### 2.4.2.6 Oil reserves and Geographical Distribution

The definition "oil reserves" generally indicates the quantity of petroleum that can be cost-effectively drilled from oil fields either through flowing wells or by means of suitable interventions (repressurisation, fluidization, pumping, etc.).

The recoverable quantity of the petroleum originally present in the oil field depends on many factors; in particular, it varies according to the hydrocarbons composing it and to the type of rock containing it. This percentage may range between 10 and 50% of the real quantity of oil originally present in the field, called *original oil in place* (OOIP).

The determination of the quantity of petroleum still present in a production basin or in a newly discovered oil field implies a careful geological and chemical analysis. It is however necessary to underline that very often the official assessment is heavily influenced by political and economic interests.

However, thanks to the improvement of surveying and production methodologies, the crude oil reserves which can be really exploited are destined to increase. Furthermore, despite the objective difficulties in the determination of the real capacity of an oil field, large areas of the Earth—first of all seabed—are still nowadays almost entirely unexplored, and this justifies a certain optimism.

Table 2.8 shows the data on the distribution of ascertained petroleum reserves. It shows that these reserves are almost exclusively concentrated in some geographical areas. In fact, over 60% of oil fields are located in the Middle East. In particular, Saudi Arabia—the richest country in petroleum—hosts over one-fifth of world reserves, followed by Iran, Iraq, Kuwait, United Arab Emirates, Venezuela (the first non-Middle East country), and the Russian Federation (Fig. 2.32).

Based on the present production and consumption rates, the estimated availability of petroleum apparently guarantees stocks for the next 42 years (Fig. 2.33).

**Table 2.8** Ascertained oil reserves at the end of 2007 [8]

| | $10^9$ tonnes | $10^9$ barrels | % |
|---|---|---|---|
| USA | 3.6 | 29.4 | 2.4 |
| Canada | 4.2 | 27.7 | 2.2 |
| Mexico | 1.7 | 12.2 | 1.0 |
| *Total North America* | 9.5 | 69.3 | 5.6 |
| Argentina | 0.4 | 2.6 | 0.2 |
| Brazil | 1.7 | 12.6 | 1.0 |
| Colombia | 0.2 | 1.5 | 0.1 |
| Ecuador | 0.6 | 4.3 | 0.3 |
| Peru | 0.1 | 1.1 | 0.1 |
| Trinidad and Tobago | 0.1 | 0.8 | 0.1 |
| Venezuela | 12.5 | 87.0 | 7.0 |
| Other Central America | 0.2 | 1.3 | 0.1 |
| *Total South and Central America* | 15.8 | 111.2 | 9.0 |
| Azerbaijan | 1.0 | 7.0 | 0.6 |
| Denmark | 0.1 | 1.1 | 0.1 |
| Italy | 0.1 | 0.8 | 0.1 |
| Kazakhstan | 5.3 | 39.8 | 3.2 |
| Norway | 1.0 | 8.2 | 0.7 |
| Romania | 0.1 | 0.5 | 0.0 |
| Russian Federation | 10.9 | 79.4 | 6.4 |
| Turkmenistan | 0.1 | 0.6 | 0.0 |
| Great Britain | 0.5 | 3.6 | 0.3 |
| Uzbekistan | 0.1 | 0.6 | 0.0 |
| Others | 0.3 | 2.1 | 0.2 |
| *Total Europe and Eurasia* | 19.5 | 143.7 | 11.6 |
| Iran | 19.0 | 138.4 | 11.2 |
| Iraq | 15.5 | 115.0 | 9.3 |
| Kuwait | 14.0 | 101.5 | 8.2 |
| Oman | 0.8 | 5.6 | 0.5 |
| Qatar | 3.6 | 27.4 | 2.2 |
| Saudi Arabia | 36.3 | 264.2 | 21.3 |
| Syria | 0.3 | 2.5 | 0.2 |
| United Arab Emirates | 13.0 | 97.8 | 7.9 |
| Yemen | 0.4 | 2.8 | 0.2 |
| Others | – | 0.1 | 0.0 |
| *Total Middle East* | 102.9 | 755.3 | 61.0 |
| Algeria | 1.5 | 12.3 | 1.0 |
| Angola | 1.2 | 9.0 | 0.7 |
| Chad | 0.1 | 0.9 | 0.1 |
| Republic of Congo | 0.3 | 1.9 | 0.2 |
| Egypt | 0.5 | 4.1 | 0.3 |
| Equatorial Guinea | 0.2 | 1.8 | 0.1 |
| Gabon | 0.3 | 2.0 | 0.2 |
| Libya | 5.4 | 41.5 | 3.4 |
| Nigeria | 4.9 | 36.2 | 2.9 |

(continued)

**Table 2.8** (continued)

| | $10^9$ tonnes | $10^9$ barrels | % |
|---|---|---|---|
| Sudan | 0.9 | 6.6 | 0.5 |
| Tunisia | 0.1 | 0.6 | 0.0 |
| Others | 0.1 | 0.6 | 0.0 |
| *Total Africa* | 15.5 | 117.5 | 9.5 |
| Australia | 0.4 | 4.2 | 0.3 |
| Brunei | 0.2 | 1.2 | 0.1 |
| China | 2.1 | 15.5 | 1.3 |
| India | 0.7 | 5.5 | 0.4 |
| Indonesia | 0.6 | 4.4 | 0.4 |
| Malaysia | 0.7 | 5.4 | 0.4 |
| Thailand | 0.1 | 0.5 | 0.0 |
| Vietnam | 0.5 | 3.4 | 0.3 |
| Others | 0.1 | 0.9 | 0.1 |
| *Total Pacific Area* | 5.4 | 41.0 | 3.3 |
| *Total world* | 168.6 | 1,238.0 | 100.0 |

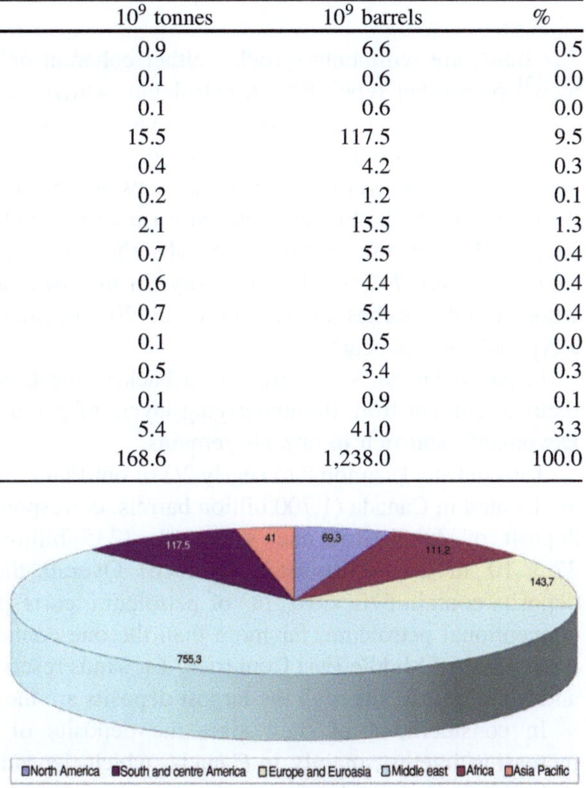


**Fig. 2.32** Distribution of world oil reserves [8]

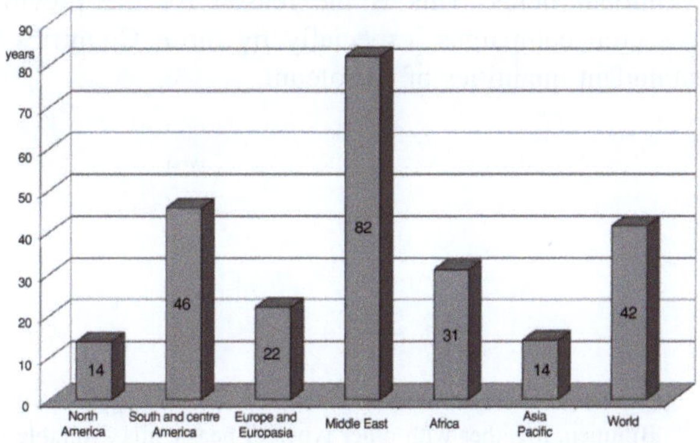


**Fig. 2.33** Oil consumption/production ratio [8]

### 2.4.3 Tar Sands

*Tar sands* are sedimentary rocks (either coherent or incoherent) containing bitumen[31] or another type of thick petroleum, which cannot be drilled with the conventional methods used for petroleum with an API[32] density below 12°.

These sands, usually quartzous, originated from lake and river sediments, and deposited on the layer of parent rocks between the end of the Upper Devonian and the beginning of the Lower Cretaceous (in a time leg that lasted nearly 225 million years). Through this contact, they absorbed the underlying oil and gas, hence saturating over 90% of their porosity. In this way, an oily content was created, whose weight reaches a maximum of 18–20% of sand mass, and that therefore has a typically *wet* appearance.

In particular, tar sands are traced back to the Lower Cretaceous[33] and derive their tar content from the underlying layers of parent rocks dating back to Upper Devonian[34] and rich in organic remains.

Tar sand stocks amount to nearly 2/3 of our Planet's oil stocks; the main deposits are located in Canada (1,700 billion barrels, corresponding to $270 \times 10^9$ m$^3$ in the deposit of Athabasca) and Venezuela (235 billion barrels, corresponding to $37 \times 10^9$ m$^3$ in the deposit of Orinoco). Overall, the Canadian and Venezuelan deposits contain $570 \times 10^9$ m$^3$ of petroleum, corresponding to $280 \times 10^9$ m$^3$ of conventional petroleum, far more than the one contained in the subsoil of Saudi Arabia and of Middle East Countries. Tar sands reserves are also present in Russia and Middle East, although the largest deposits are the two ones mentioned above.

In consideration of their size, the deposits of this energy source arouse increasing interest, mainly in Canada, which depends upon a concrete technical and economic difficulty for the exploitation in percentages comparable to conventional fields. This is the reason for the previous reluctance in promoting research campaigns, especially by those Countries that have at one's disposal abundant quantities of petroleum.

[31] Bitumen, together with other types of heavy oils available in these sands, belongs to solid or semisolid hydrocarbons. They are composed of high-molecular weight complex blends, whose chemical knowledge is imperfect because of analysis difficulties. This category also includes the asphalts used for road paving.

[32] Cf. Sect. 2.4.2.3.

[33] Geological period belonging to the Mesozoic Era, dating back to nearly 135 million years ago.

[34] Geological period belonging to the Palaeozoic Era, dating back to nearly 360 million years ago.

**Table 2.9** Typical composition of natural gas

| Gas | | Content (%) |
|---|---|---|
| Methane | $CH_4$ | 70–90 |
| Ethane | $C_2H_6$ | 0–20 |
| Propane | $C_3H_8$ | |
| Butane | $C_4H_{10}$ | |
| Carbon dioxide | $CO_2$ | 0–8 |
| Oxygen | $O_2$ | 0–0.2 |
| Nitrogen | $N_2$ | 0–5 |
| Hydrogen sulphide | $H_2S$ | 0–5 |
| Rare gases | A, He, Ne, Xe | Traces |

*Source* http://www.naturalgas.org

## *2.4.4 Natural Gas*

### 2.4.4.1 General Description and Genesis

Mostly made up of gaseous hydrocarbons and by other gases in variable quantities, natural gas can be found both in its own deposits, and combined to petroleum. Therefore, the following different configurations can be identified, which often coexist;

- Gas is *free* and occupies the upper part of the deposit, whereas below there is petroleum or, in the absence of petroleum, it overhangs the areas of parent rock soaked with generally salty water;
- Gas is *dissolved in oil* and the quantity in solution varies according to pressure and temperature conditions in the deposit, and with the physical characteristics of oil and of the gas mixture;
- Gas is *dissolved in water*, whenever it is in the gas–water interface of free gas deposits, or completely dissolved in gas-waters deposits, as those in the Po Delta. In consideration of the high pressure and temperature levels detected in these deposits, the water solubility of natural gas reaches remarkable values, also amounting to some hundreds of percentage points in volume;
- Gas is *liquefied* when, at depths exceeding 2,000 m and in conditions of high pressure and temperature, natural gas and petroleum form a single liquid phase, physically indistinguishable (*condensate*).

There are then still undetermined quantities—or determined with a large approximation—of natural gas "entrapped" in complex crystalline structures, consisting a water molecules linked through hydrogen, methane and/or other gas bonds, including, in particular, gaseous hydrocarbons. In order to remain stable, these crystals must be in particular temperature and pressure conditions.

According to many researchers—also on the basis of studies carried out as from the '60s on this particular energy source made up of the so-called *gas hydrates* (or *methane hydrates* or *clathrates*)—this might represent one of the primary energy sources of the future. In fact, in free and in suitable conditions the gas entrapped in the cavities of crystal lattice and originated from water molecules, it is possible to derive natural fuel gas.

The *genesis* of gaseous hydrocarbons representing the energy source called *naturalgas* is strictly related to the modes and conditions of formation of liquid hydrocarbons (see Sect. 2.4.2.2), since both derive from a common matrix.

The chemical analysis of natural gas is by far less complex than petroleum, because of the lower number of its compounds and their simplicity; it is performed through distillation by means of mass spectrometers and gas chromatographs.

As to its composition, *methane* is the hydrocarbon present in the highest percentage; for this reason, people tend to use the word *methane* as a synonym of natural gas. The other compounds generally belong to the paraffin series and to olefins (ethylene in particular). If the deposit also contains petroleum, then the quantity of hydrocarbons other than methane (ethane, propane, butane and higher) may reach 50%.

Natural gases are classified into *dry*, *lean and wet*, according to the content in liquid hydrocarbon vapours. Vapour content represents a point of strength, since they can be separated,[35] condensed and used for the production of *light diesel oils*.

Another characteristic that enhances natural gas is the presence, more or less marked, of *propane* and *butane* in quantities that make its separation economically convenient. These hydrocarbons, whose liquefaction is simple and occurs at a relatively low pressure, are very much used as fuels and marketed in easily transportable cylinders, at a pressure of 4–6 bars, with a composition of 60–70% butane and 30–40% propane; their product name is *liquefied petroleum gas* (LPG).

The other main gases that can be identified in natural gas include nitrogen ($N_2$), carbon dioxide ($CO_2$), hydrogen sulphide ($H_2S$) and, rarely, hydrogen ($H_2$), carbon oxide (CO) and helium (He) (Table 2.9).

As regards the lower heating value, the average value of 12,000 kcal/kg can be considered, that is to say 8.200 kcal/$m^3$.

Among the main values linked to natural gas, besides *density (*both *absolute* and *relative)* and specific gravity,[36] particular importance is also ascribed to *pressure*, *temperature*, *volume* and *flow rate*. The latter, very much used especially in transportation phases, is generally expressed in $m^3$/h and, when referred to normal conditions, is measured in $Nm^3$/h.

Over the last decades, natural gas gained a growing importance in energy markets.

Although the areas rich in crude oil are generally also rich in natural gas, not always the importance of the latter in the scale of production of primary sources has been considered as the former. Its main *drawbacks* include the *need for major financial investments* and *technological hindrances*, mostly linked to the transportation problems due to its gaseous physical state. The transportation of natural

[35] Through the so-called "*degasolination*" process.

[36] As regards absolute density and specific gravity in natural gas compared to liquid fuel, the dependency from pressure and temperature is higher. Methane relative density amounts to 0.57.

gas, in fact, demands large infrastructures, gas pipelines, sometimes long thousands of kilometres, with pumping and decompression plants and widespread distribution networks.

The political stability of producing Countries as well as of the Countries of transit of gas pipelines is key for supplies. Furthermore, natural gas demands particular caution during distribution, due to the risk of gas leaks and explosions. Finally, another of its drawbacks is the fact that combustion—that is triggered at a high flame temperature—entails a high development of nitric oxides. In the past, this aptitude determined some marketing difficulties.

Another drawback of natural gas is represented by the fact that *it is not possible to have a constant flow rate of gas for the entire life of the deposit*; the production profile, in fact, rapidly peaks during the first years of gas extraction, and then slowly decreases for the rest of the life of the deposit.[37]

However, natural gas has many of the *advantages of petroleum* and, in addition, *further positive aspects* as, for instance, a limited global production of polluting substances and of $CO_2$ following its combustion. With respect to the already mentioned development of nitric oxides, methane combustion determines:

- A *reduction of* $CO_2$ *emissions* accounting for 50% compared to the $CO_2$ produced by coal and 30% compared to petroleum[38];
- *Total or nearly total absence of sulphur oxides, particulate,metals and non-combustible residues.*

Other advantages of natural gas include the higher efficiency of combustion and its high calorific value (12,000 kcal/kg).

Particularly noticeable is also the *availability* ensured by the diversification of stocks, possible thanks to a geopolitical distribution far less critical than petroleum.

The setting up of gas storage reservoirs, possibly in areas close to highly consuming regions, can be carried out through the utilisation of:

- Gas deposits close to depletion (during the period of low consumption, generally in summer, the gas coming from distant deposits or from import is pumped in the storage reservoir in order to be extracted during the peaks of demand).
- Reservoirs in the liquefied natural gas regasification plants, which allow a rapid conversion and availability of natural gas.
- The transportation routes of the network of gas pipelines, in which the operating pressure is higher than the distribution pressure, can also be considered as storage reservoirs to be used in the periods of daily peaks.

[37] Sometimes this behaviour can be seen—although more limitedly—also in crude oil fields.

[38] Given the same energy produced through combustion.

**Table 2.10** Natural gas reserves ascertained at the end of 2007 [8]

| | $10^{12}$ m$^3$ | % |
|---|---|---|
| USA | 5.98 | 3.4 |
| Canada | 1.63 | 0.9 |
| Mexico | 0.37 | 0.2 |
| *Total North America* | 7.98 | 4.5 |
| Argentina | 0.44 | 0.2 |
| Bolivia | 0.74 | 0.4 |
| Brazil | 0.36 | 0.2 |
| Peru | 0.13 | 0.1 |
| Colombia | 0.36 | 0.2 |
| Trinidad & Tobago | 0.48 | 0.3 |
| Venezuela | 5.15 | 2.9 |
| Others | 0.07 | 0.0 |
| *Total South and Central America* | 7.73 | 4.4 |
| Azerbaijan | 1.28 | 0.7 |
| Denmark | 0.12 | 0.1 |
| Germany | 0.14 | 0.1 |
| Italy | 0.09 | 0.1 |
| Kazakhstan | 1.90 | 1.1 |
| The Netherlands | 1.25 | 0.7 |
| Norway | 2.96 | 1.7 |
| Poland | 0.11 | 0.1 |
| Romania | 0.63 | 0.4 |
| Russian Federation | 44.65 | 25.2 |
| Turkmenistan | 2.67 | 1.5 |
| Ukraine | 1.03 | 0.6 |
| Great Britain | 0.41 | 0.2 |
| Uzbekistan | 1.74 | 1.0 |
| Others | 0.43 | 0.2 |
| *Total Europe and Eurasia* | 59.41 | 33.5 |
| Bahrain | 0.09 | 0.1 |
| Iran | 27.80 | 15.7 |
| Iraq | 3.17 | 1.8 |
| Kuwait | 1.78 | 1.0 |
| Oman | 0.69 | 0.4 |
| Qatar | 25.60 | 14.4 |
| Saudi Arabia | 7.17 | 4.0 |
| Syria | 0.29 | 0.2 |
| United Arab Emirates | 6.09 | 3.4 |
| Yemen | 0.49 | 0.3 |
| Others | 0.05 | 0.0 |
| *Total Middle East* | 73.22 | 41.3 |
| Algeria | 4.51 | 2.5 |
| Egypt | 2.06 | 1.2 |
| Libya | 1.50 | 0.8 |
| Nigeria | 5.30 | 3.0 |
| Others | 1.21 | 0.7 |

(continued)

**Table 2.10** (continued)

| | $10^{12}$ $m^3$ | % |
|---|---|---|
| *Total Africa* | 14.58 | 8.2 |
| Australia | 2.51 | 1.4 |
| Bangladesh | 0.39 | 0.2 |
| Brunei | 0.34 | 0.2 |
| China | 1.88 | 1.1 |
| India | 1.06 | 0.6 |
| Indonesia | 3.00 | 1.7 |
| Malaysia | 2.47 | 1.4 |
| Myanmar | 0.60 | 0.3 |
| Pakistan | 0.85 | 0.5 |
| Papua New Guinea | 0.44 | 0.2 |
| Thailand | 0.33 | 0.2 |
| Vietnam | 0.22 | 0.1 |
| Others | 0.37 | 0.2 |
| *Total Pacific Area* | 14.46 | 8.2 |
| *Total world* | 177.38 | 100.0 |

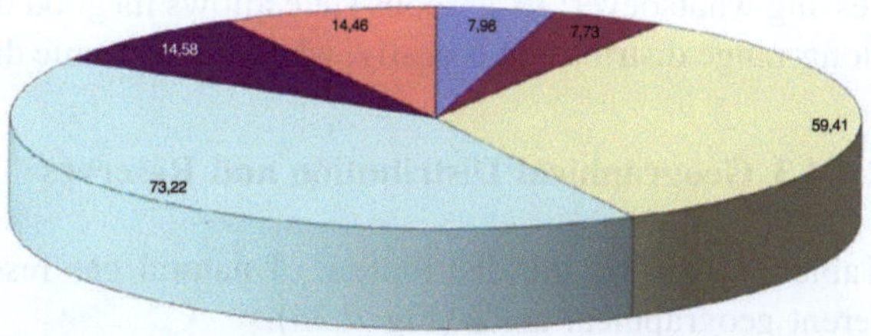


**Fig. 2.34** Distribution of natural gas reserves all over the world

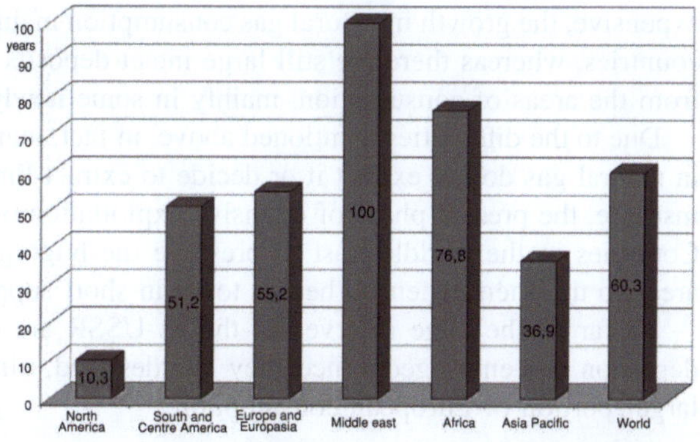


**Fig. 2.35** Natural gas production/consumption ratio

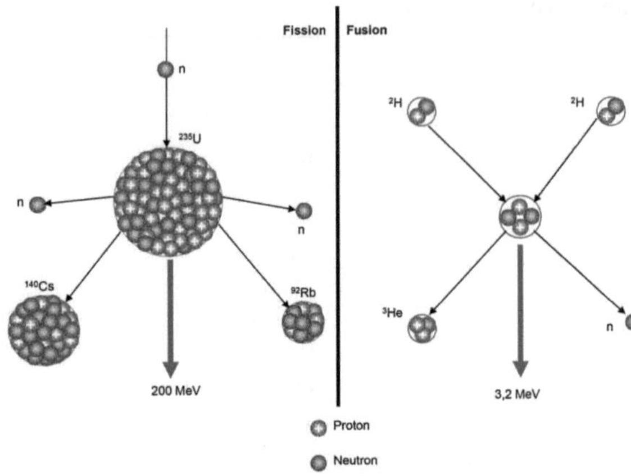


**Fig. 2.36** Nuclear fission and fusion

#### 2.4.4.2 Processing and Utilisation

Very often, methane is used in the form in which it is extracted, without any processing whatsoever; its gaseous state allows its good distribution in the short range (long-range distribution is costly, while transoceanic distribution is highly difficult).

#### 2.4.4.3 Geographical Distribution and Reserves

Table 2.10 shows the distribution of natural gas reserves ascertained in the different geographical areas (Fig. 2.34).

On the basis of present production rate, stock would cover nearly 60 years (Fig. 2.35).

Because of the technological problems that make its transportation more expensive, the growth in natural gas consumption mainly occurred in industrialised countries, whereas there are still large intact deposits in regions located far away from the areas of consumption, mainly in some newly industrialising countries.

Due to the difficulties mentioned above, in fact, many Countries potentially rich in natural gas do not extract it or decide to extract limited quantities of it. So, for instance, the present phase of intensive exploitation of petroleum allows the Arab Countries in the Middle East to preserve the huge gas reserves located in their areas to use them when oil begins to be in short supply.

As far as the huge reserves of the ex-USSR are concerned, shorter times of depletion are envisaged, since they are destined, already nowadays, to cover a larger portion of European consumption.

### *2.4.5 Nuclear Power*

Nuclear power is released, in huge quantities, by reactions that determine transmutations of given elements. The quantity of energy derived from these reactions can be calculated through the well-known Einstein relation that links mass and energy.[39] The calculation is based on the *massdefect* that the products of the reaction show compared to the elements involved in the reaction itself.

The remarkable difference between the quantities of energy that develop through these reactions and those linked to combustion is explained with the fact that in the latter the forces keeping electrons are exploited, hence involving the outer levels of the atom, whereas in nuclear reactions the nucleus of atom is *decomposed or in any case it undergoes a transformation*, and during these actions higher forces come into play. For instance, the energy generated through fission by 1 g of uranium is equivalent to nearly $8.4 \times 10^{10}$ J, whereas the one released through combustion by 1 g of gasoline is equal to nearly $4.84 \times 10^{4}$ J; given the same masses, then the ratio between the two energy capacities amounts to nearly $24 \times 10^{6}$.

The chemical elements involving spontaneous nuclear reactions are called *radioisotopes* and can be either *natural* or *artificial*.

The processes that make this form of energy available include the *fission* of nuclei with a high atomic weight and the *fusion*[40] of light nuclei.

#### 2.4.5.1 Energy Use of Nuclear Power

Nuclear power can be exploited in two ways:

1. Through the *fission* of the *uranium* atom
2. Through the *fusion* of two *hydrogen* atoms.

If, on the one hand, the fission technology has been available for a few decades, fusion, on the other hand, is still undergoing a pre-commercial development phase.

The uranium used for fission purposes is contained in its minerals in quantities ranging between 0.1 and 0.5%, and is 0.71% made up of isotope $^{235}U$, 99.28% of isotope $^{238}U$ and the remaining 0.01% by isotope $^{234}U$; the "fissile" one that is, to say useful for nuclear fission, is $^{235}U$, present in a much lower percentage. For this reason, uranium generally undergoes "enrichment" to obtain higher percentages of the fissile isotope.

---

[39] This is the famous equation $E = mc^2$, where *e* is, by definition, the value of nuclear power owned by a mass, *m* is the so-called *mass defect* and c is the speed of light ($c \cong 3\times 10^8$ m/s). The mass value of a nucleus, theoretically calculated by adding up the masses of neutrons and protons that compose it, is higher than the one experimentally calculated, and the difference between the two quantities is precisely the mass defect.

[40] This is a reaction in which, through the fusion of nuclei of light elements, a mass nucleus lower than the sum of the initial masses is formed, and energy is generated in a quantity proportional to this mass defect.

From fission, through one neuron, of a $^{235}U$ atom (Fig. 2.36) it is possible to obtain one cesium atom, one of rubidium, two neurons[41] and a quantity of energy amounting to 200 MeV.[42]

From the fusion of two hydrogen atoms, it is possible to obtain one helium atom and a quantity of energy amounting to 3.2 MeV.

In consideration of molecular weights and by comparing the energy released given the same mass, it is possible to derive that from the fission of 1 kg of $^{235}U$ 18.6 GWh are obtained, whereas from the fusion of 1 kg of hydrogen, 5,262 GWh are obtained.

## Nuclear fission

In theory, all the elements with a sufficiently high atomic weight can undergo the fission process, that is to say the splitting of the nucleus through a neutron bombardment with a suitable energy level; in practice, however, only a few elements are *fissile* that is, to say able to really generate this reaction.

In nature, the only fissile element known is *uranium* (U), which is, therefore, the only primary source presently available for the fission process. *Natural uranium* is made up of a *mixture of three isotopes*, with an atomic weight of 234, 235 and 238, present in the percentages of 0.01, 0.71 and 99.28%, respectively. Isotope $^{235}U$ is the only fissile out of the three of them; the other fissile isotope of uranium, $^{233}U$, is artificial and can be derived, through neutron bombardment, from isotope 232 of thorium ($^{232}Th$), that for this reason, is called *fertile* isotope, and can be numbered among the primary sources of nuclear power. Also uranium isotope 238 is fertile since, when undergoing suitable processing, originates plutonium 239 isotope ($^{239}P$), fissile as well. The next Chapter provides a description of the methods used to artificially increase the percentage of fissile isotopes.

Minerals containing uranium[43] and thorium are rather widespread throughout the surface of the Earth, but the concentrations of these minerals hardly make deposits economically interesting.

In any case uranium, out of these two elements, arouses the main interest. In the main deposits, minerals have an average uranium content of 1%; they are located in Australia, Nigeria, Brazil, Canada and United States.

---

[41] Neuron emissions allow the fission reaction to be self-sustained (chain reaction); in fact, the neurons produced will affect other uranium atoms by causing other fissions that in turn produce other neurons, etc.

[42] MeV: MegaelectronVolt. 1 MeV = $1{,}60217646 \times 10^{-13}$ J.

[43] Those with the highest uranium content are *pitchblende* and *uranites*, made up of blends of uranium oxides in variable percentage. They are also called "primary minerals", as opposed to "secondary minerals", with a minor uranium content.

Nuclear Fusion

As far as *nuclear fusion* is concerned, there are no fine-tuned technological processes allowing the peaceful use of the energy produced by the relevant reactions, and consisting in the development of energy following the fusion of atoms or molecules of light elements. The main aim, in particular, is to reproduce in a much lower scale the nuclear reactions occurring in the sun. The most interesting chemical species for these reactions include some low-atomic weight elements, such as *deuterium* (D), *tritium* (T) and *lithium* (Li).

*Deuterium* and *tritium* are two hydrogen isotopes. They are present in nature in very low percentages compared to hydrogen.[44] The large spread of the latter element on earth makes in any case deuterium relatively abundant; in considering that only 1% of this isotope is concretely available, then the total world resources are considered as amounting to nearly $50 \times 10^9$ t.

Lithium is available in nature mainly in the form of oxide ($Li_2O$), but it can also be extracted from sea water when its concentration reaches 0.1 ppm, or it can be derived from those rocks where a quantity amounting to 20 ppm is present. The most suitable isotope for nuclear fusion is $^6Li$, whose world availability amounts to nearly $67 \times 10^4$ t.

Finally, it is suitable to recall *natural radioisotopes*. They are grouped into four families and are characterised—besides by the number of neutrons present in the nucleus—also by their *half-life*.[45] Each of the four series starts from a *parent* that, with subsequent decays, generates the other radioactive isotopes of the family, until reaching a stable element (lead or bismuth) that closes the series.

### 2.4.5.2 Reserves and Geographical Distribution[46]

Recent figures are those collected by the Nuclear Energy Agency (NEA) and are referred to 2003 estimates. The quantity of Uranium available in deposits, which can be extracted at the presently competitive cost of 80 \$/kg, amounts to around 2.5 Mt. Conversely, if a highest extraction price amounting to 130 \$/kg is considered as acceptable, then the reserves of Uranium increase to the value of 3.2 Mt.[47] Thus, at the today rate of consumption:

Enriched U derived from the one at 80 \$/kg R/P = 36 years.

Enriched U derived from the one at 130 \$/kg R/P = 47 years.

---

[44] The deuterium–hydrogen report is nearly 1–6,500; the percentage of tritium is, on the contrary, negligible (nearly $10^{-15}$%).

[45] Is the time needed so that 50% of the atoms considered turn into stable elements through the natural decay process.

[46] *Source* http://www.aspoitalia.it/documenti/coiante/nucleareidrogeno.html

[47] Every 6 tons of natural U generate 1 ton of enriched U.

It is then necessary to stress that the strategic importance of nuclear power does not consist in the possibility of "controlling" uranium basins (that are also rather uniformly distributed from the geopolitical viewpoint), but rather in owning the updated "know-how" on the most economically-effective and efficient processes for the energy exploitation of the primary source.

## References

1. Energy Policies of IEA Countries (Compendium) (1998) Review, ISBN: 92-64-16146-5, p 319
2. Nakícenovíc N, Grubler A, MacDonald A (1998) Global energy perspectives. Cambridge University Press, Cambridge p 299 ISBN 0-521-64569-7
3. Dunn PD (1986) Renewable energies: sources, conversion and application. Peter Peregrinus Ltd, London
4. World Energy Outlook (2010) International Energy Agency, Paris, International Energy Agency ISBN 978-92-64-086241
5. World energy assessment (2000) United Nations Development Programme. Washington, DC. Available at http://www.undp.org/energy/activities/wea/drafts-frame.html, New York, ISBN: 92-1-126126-0
6. McKendry P (2002) Energy production from biomass (part 1): overview of biomass. Bioresour Technol 83(1):37–46
7. World Energy Assessment (2004) UNDP. New York, ISBN: 92-1-126167-8
8. BP Statistical Review of World Energy (2008) BP p.l.c., London
9. Tissot B, Welte DH (1991) Petroleum formation and occurrence. A new approach to oil and gas exploration, 2nd edn. Springer, Berlin
10. Ceron P (N. 4, Febbraio 1976) Energia Geotermica—Stato attuale e prospettive, con particolare riguardo alla produzione di energia elettrica" in "L'energia geotermica, Studi e Ricerche ENEL; also: World Energy Council 2010, Survey of Energy Resources, London ISBN: 978-0-946121-021
11. Cavinato C (1964) Giacimenti minerari. UTET, Torino; Also: Taylor TN, Taylor EL, Krings M (2009) Paleobotany. The biology and evolution of fossil plants, ISBN 9780-123739-728
12. Barnaba FP (1990) Geologia degli Idrocarburi. AGIP. Milano; also: Boggs S Jr (2006) Principles of sedimentology and Stratigraphy, Pearson Hall, 4th edn, ISBN 0-13-154728-3
13. Raabe J (1985) The design, use, and function of hydromechanical, hydraulic, and electrical equipment. VDI-Verlag, Dusseldorf, ISBN 3184006166

# Chapter 3
# Energy Vectors

## 3.1 The Need for Energy Transportation and Storage

While considering the human use of energy, "where" and "when" are often more important than "how" and "how much". Human beings do not simply need energy: they need it wherever and whenever they decide. An energy system is therefore satisfactory when it is able to guarantee the right quantity and form of energy in the right moment (for all required time) and at the place of need.

From the energy market viewpoint, to succeed in meeting the demand in the place and at the moment requested is often considered more important than the way in which energy is supplied and also than the quantity of primary energy needed to make the quantity and form requested available in the place and at the moment of the demand.

The process of energy production and supply is dominated by the economic conditions of demand, largely more than by its energy balance.

A common example, familiar to all, can easily clarify the issue: we all know that the sunlight is available during the daytime; however, if from a social and economic viewpoint it is considered more convenient to play a sports match in the evening—although from an energy viewpoint this choice is absolutely unjustified—the match is played in the evening with artificial lighting and subsequent energy spending. Although lightning would be cheaper from an energy perspective in a place rather than another, the match would take place in any case in the place privileged by the socio-economic reality, irrespective of the necessary energy spending.

The modern concept of *energy system*, therefore, has to be able to meet—even better to "foresee"—these needs. The concepts of time and space therefore forcefully fall within energy analysis and in the capacity to develop, design and implement energy systems. Energy transportation and storage become in this way the main attributes to be considered while standardising systems and usage technologies; in a word, the way itself of reasoning in terms of energy.

F. Orecchini and V. Naso, *Energy Systems in the Era of Energy Vectors*,
Green Energy and Technology, DOI: 10.1007/978-0-85729-244-5_3,

A primary energy resource is generally available in a place different from the energy demand. This implies the need for energy *transportation*; and the need for continuously and instantly meeting the energy demand makes the *storage* necessary.

The concepts of transportation in space and storage in time of a given quantity of energy are nowadays predominant to correctly meet the needs relating to global development and access to technologies of an increasing number of human beings. To express at best the concepts of energy transportation and storage and to effectively integrate them in the analysis of energy systems, it is necessary to introduce a new concept: the *energy vector* (or career).

On the basis of the characteristics highlighted in Fig. 1.6 and of the analysis of energy systems, the definition of energy vector is the following:

*An energy vector allows transfer, in space and time, a given quantity of energy, hence making it available for use distantly in time and space from the point of availability of the natural source.*

The interpretation of the entire world energy system radically changes if considered from the perspective of energy vectors. Already today, the success and the failure of different modes of exploiting primary energy resources and making their use available for end users, is regulated by the growing need of the energy market to avail itself of standardised energy vectors always endowed with specific characteristics, and not only of energy sources able to meet their demand from the quantity viewpoint. This characteristic of the energy market and of energy systems is destined to become increasingly important in the decades to come.

Energy vectors must allow to transfer energy, to store relevant quantities of it for a deferred use over time, in ways that are not always foreseeable in advance. These vectors have to guarantee higher and higher performance levels in the conversion and transformation processes, in transportation and storage, with a limitation in the environmental impact of the entire cycle and particular attention must be paid to the utilisation phase.

*Many primary sources are also vectors*, in the sense that they are able to allow energy storage and transportation. Conversely, for other sources this ability has to be carried out through the production of specific vectors obtained by converting or changing the natural source.

The main energy vectors presently known and in many cases already used are the following:

- Fossil fuels
  - Oil and its derivatives
  - Natural gas
  - Coal
- Electricity
- Heat-transfer fluids
- Hydrogen
- Mechanical transmission systems
- Radiative transfer.

The main energy characteristic of a vector is the form of energy that allows it to transport and thence to make it available at the "delivery", in a place and time other than those available at the source. Knowing the form of energy transported and stored allows the correct inclusion of each energy vector in the framework of each complex energy system.

The energy vectors presently most known and used transport the following forms of energy:

- Fossil fuels → chemical energy
- Electricity → electric power
- Heat-transfer fluid → thermal energy
- Hydrogen → chemical energy
- Mechanical transmissions → mechanic energy
- Radioactive transfer → radiant energy

As it can be easily inferred, some of these vectors are also primary sources (this is the case of fossil fuels), others (mechanical transmissions and thermal vectors) carry forms of energy demanded and directly used in the form transported for end use. In particular, vectors as electricity and hydrogen must be the object of specific production processes; they require technologies within the system that might be useful for their final conversion, allowing to provide users with the useful effect needed (thermal, electric, mechanic and luminous).

## 3.2 Attitude to Energy Transportation in Space and Time and Range of Application

It is very important to asses in detail the characteristics of each energy vector as concerns its capacity to transfer, in space and time and the quantity of energy transported.

Fossil fuels owe their extraordinary energy success to the large availability in nature, to limited or zero extraction costs, to the lack of interest shown until nowadays by the human society vis-à-vis the environmental impact entailed by their use and the practical impossibility of natural recovery of the resource taken away from the ecosystem; but most of all to their extraordinary characteristics as energy vectors. The zero extraction cost, a concept not much tackled in the literature, is strictly related to each natural resource and particularly evident in mineral resources. The removal of a resource from nature, if not linked to particular economic mechanisms linked to its formation (for instance the cultivation of fields for grain collection), is mainly free of charge. Nature deals with the formation and conservation resources, men only deal with its "collection" from the place of availability. The technical cost to access fossil fuels is therefore simply linked to the cost of research, the cost to reach the deposit and the extraction cost.

Their availability has no cost whatsoever, since it was previously arranged and "paid" by nature.

The lack of interest for the new formation of the natural resource freely taken is commonly accepted by the human society, particularly as concerns mineral resources, whose extreme difficulty (in almost all cases, impossibility) of re-formation through human work has not discouraged its extensive use, without considering its inevitable depletion. In different periods, the lack of resource replenishment capacity has been, only partly mitigated by techniques destined to reuse products and their components, and in recent times by the end-of-life recycling of part of raw materials.

The extraordinary capacity of keeping the quantity of energy stored—the real access key to the world spreading and the commercial success of fossil fuels—is inborn in their structure. Fossil resources, in fact, are available to us because it was possible, for millions of years, to thoroughly preserve the quantity of energy that had been fixed and stored in them in past geological eras.

Once extracted from the subsoil, coal and oil in particular maintain, at atmospheric conditions, also the quantity of energy they contain in an extremely stable way, hence favouring the storage in deposits rather easy to create and the transportation through simple and inexpensive means.

The superiority of oil over coal from the viewpoint of easiness of storage and transportation in terms of energy density, and easiness of handling (far better for a liquid substance rather than for a fragmented solid one) is the main reason for the success that allowed the former to replace the latter as a leading fuel in the world energy scenario of the 20th century.

The characteristics of fossil fuels as energy vectors are, therefore, at the basis of their global success and are the ones with which each new vector has to be compared in order to show its competitiveness.

Putting (for the time being) any other considerations aside—as for instance the obtainability from different primary sources, the renewability of the source, the level of emission of undesired and polluting substances at the point of use, which are certainly penalising for fossil fuels—and focusing on their capacity to allow the movement of a quantity of energy in time and space—that is to say their best qualities—it is easily possible to understand what are the characteristics requested to each energy vector that intends to play a major role in world's energy system.

## 3.3 Duration and Range of Transfers

In order to better understand the characteristics of the main energy vectors, it is suitable to divide the aptitude to transfer energy into three levels for space and three levels for time.

*Transfer in short range* indicating the capacity of a vector to transfer energy for distances of tens and hundreds of metres, until extreme cases of some kilometres.

*Transfer in medium range* indicating the capacity of a vector to transport energy for distances of tens and hundreds of kilometres, up to extreme cases of some thousands of kilometres.

*Transfer in long range* that is capacity of a vector to transport energy for distances of thousands of kilometres.

*Transfer in short term* that is the capacity of a vector to store energy for a few days up to extreme cases of a few weeks or months.

*Transfer in medium term* that is the capacity of a vector to store energy for a few months, up to the extreme case of a few years.

*Transfer in long term* that is the capacity of a vector to store energy for years, up to the potential capacity of storing it for thousands or millions of years.

*Fossil fuels* are able to transfer energy in the short, medium and long range, and also for a short, medium or long period with transportation solutions absolutely simple, sufficiently safe and inexpensive (or reasonably costly). Each time, for each single characteristic and for each technical solution of storage and transportation, the other energy vectors are inevitably compared to the equivalent costs (energetic, technological, economic) of transportation or storage of the same quantity of energy for the same order of magnitude in terms of distance or time, as the main fossil fuels.

*Electricity* allows to transfer energy efficiently, often more convenient than fossil fuels, in a short and medium range in terms of kilometres. Although the transportation infrastructure and the related infrastructures involve major investments, the low energy and economic costs linked to transportation and the high level of reliability of solutions turn electricity into a competitive energy vector, sometimes outperforming fossil fuels, although considering the transportation capacity only, and not the versatility of production, storage, utilisation and the limited environmental impact achievable.

As far as long-range transfers are concerned, for the time being the transportation losses, the size of investments, the technical difficulty of solutions in case of crossing of oceans, chains of mountains and places with extreme weather conditions, make electricity less competitive.

The lack of competitiveness in long-range transportation is also linked to the characteristic of the electricity vector of *not allowing storage over time* of the energy transported (for this reason, electricity can be defined an "incomplete vector"). This makes it necessary a connection in controls and in the management systems of interconnected and chained transmission networks, which becomes increasingly difficult with the geographical distance and the decision-making independence between the network operators and the political framework of the countries involved.

The indirect methods for the storage of electric power over time are increasingly competing with fossil fuels, especially for mobile applications, in portable electronics and in transportation vehicles. The complexity and the cost to implement the different systems, besides the cost for management, maintenance and duration over time, make increasingly difficult the competition between indirect electricity storage and storage of equal quantities of energy with fossil fuels. However, in this case, the growing request for zero-emission solutions at the point of use (presently generally adopted for electronics and telephony or portable electronics, although gradually implemented also for mobility, especially in urban areas) can turn out to be able to drive the market.

Indirect storage methods allow to achieve the two most important conditions for the economic enhancement of an energy vector that is, to say the short and medium term storage capacity. And in many cases long-term storage as well is not beyond the technological and economic range.

*Hydrogen*, while allowing energy storage and transportation in the chemical form, and since it can be employed in all energy conversion technologies using fossil fuels, besides in other new generation technologies (as fuel cells), is very often immediately associated to fossil fuels, and also considered—for these similarities—as a possible "new universal fuel".

In theory, hydrogen allows a short, medium and long-term storage when in gaseous form and in a form absorbed or chemically bonded with other substances. As concerns, on the contrary, the storage in a cryogenic liquid form, it is possible to state that it is only possible in the short term and gradually less and less satisfactorily as the timeframe extends. Yet, in any case, storage is never possible with a zero-energy spending and always needs committing technological conditions as concerns the solutions for thermal insulation or the presence of ancillary components that envisage a continuous heat subtraction.

*Synthetic fuels* basically have the same features as fossil fuels, with which they share applications; therefore, they easily allow storage and transportation in the short, medium and long term/range.

*Heat transfer fluids* allow both transportation and storage, but transportation is limited to the short range and storage to the short term.

*Mechanical transmissions*, as well as oleodynamic and pressure-dynamic transmissions, like the heat-transfer fluids with which they share several applications for the simultaneous transfer and thermal and mechanic energy, allow both transportation and storage, but only in the short range and short term.

*Radiation* allows, on the contrary, transportation and not storage. Transportation can take place at a very long range (interstellar and interplanetary distances), in the vacuum of space, and in a very short range, in the atmosphere.

Table 3.1 sums up the aptitudes of the different vectors mentioned above.

## 3.4 Characteristics of Energy Vectors

For each energy system, it is possible to identify the vector that allows energy transportation and storage, the way in which storage and transportation–distribution are possible, and the means that allows its storage, transportation and distribution.

### *3.4.1 Ways and Means of Storage, Transportation and Distribution of Vectors*

This paragraph contains an analysis of the definition of vector and of its main properties and functions.

**Table 3.1** Energy transportation and storage capacity of the main energy vectors in the short, medium and long range and in the short, medium and long term

| Energy vector | Transportation | Storage |
|---|---|---|
| Fossil fuels | Short, medium, long range | Yes, short, medium, long term. |
| Electricity | Short and medium range | No (indirect methods) |
| Hydrogen | Short, medium, long range | Yes, short, medium, long term. (cryogenic exception) |
| Synthetic fuels | Short, medium, long range | Yes, short, medium, long term. |
| Heat exchanging fluids | Short range | Yes, short term |
| Mechanical, oleodynamic and pressure-dynamic transmissions | Short range | Yes |
| Radiation | Very long range in space Short range in the atmosphere | No |

#### 3.4.1.1 Vector

*The energy vector is the minimum unit of the system: it is the substance, the chemical compound or element, or the physical phenomenon that allow the transportation of a given quantity of energy*. More precisely, a vector (substance or phenomenon) allows the *transportation of an energy potential*, since what is transferred in time and space is the "potential" capacity of a given substance or of a given phenomenon to allow the achievement of a useful energy conversion.

By convention, therefore, reference is made to *energy vector* rather than energy *potential vector*, although assuming that the correct and complete definition would impose to refer to the potential.

An energy vector is therefore the *fossil fuel* (chemical compound), *electricity* (more correctly: the electronic flow, physical phenomenon), *hydrogen* (chemical element), *synthetic fuel* (chemical compound), *heat-transfer fluid* (substance with given physical characteristics), *the shear stress* or *tensile-compressive stress*, *oleodynamic pressure*, *aerodynamic pressure* (physical phenomenon), *solar radiation* (physical phenomenon).

#### 3.4.1.2 Transportation Procedures

The modalities for the transportation and distribution of vectors can vary, in different applications, also for the same vector. The possibility of different ways of storage, transportation and distribution is an indirect index of the good quality of a vector. Having a higher number of possible choices makes, in fact, the vector more easily usable in different frameworks and for different solutions of energy systems.

Transportation procedures necessarily envisage the use of a network of infrastructures that can be:

- *Dedicated network*
  - Oil pipelines
  - Gas pipelines
  - Long-distance power lines
  - Mechanical transmission
  - Oleodynamic transmission
  - Pressure-dynamic transmission
  - Heat-transfer fluid distribution systems
- *Non-dedicated network* (also existing for other purposes of transportation of goods and persons):
  - Water (sea, river)
  - Road
  - Railway
  - Air
  - Remote transmission

#### 3.4.1.3 Means of Transport

A means of energy transport is, conversely, the means aimed at containing—during transportation operations—the energy unit which is the object of transfer of availability in time and space.

Each transportation system envisages one or more possible means, whose characteristics relate to the quantity of energy to be transferred, the distances to be covered, and the energy system in which they have to be included; these means are:

- *Dedicated network*
  - Oil pipelines → pipe.
  - Gas pipelines → pipe.
  - Long-distance power lines → electric cables.
  - Mechanical transmissions → shaft, connecting rod, crank.
  - Oleodynamic transmission → pipes.
  - Pressure-dynamic transmission → pipes.
  - Heat-transfer fluid distribution system → pipes.
- *Non-dedicated network*
  - Water (sea, river) → ship.
  - Road → road vehicle.
  - Railway → train.
  - Air → aircraft, helicopter.

In the same way, for the use of suitably developed technological structures or solutions the storage modality provides:

- Tanks and pressure tanks
- Electrochemical applications
- Springs, flywheels
- Thermal-insulating tanks.

## 3.5 Analysis of the Main Energy Vectors

This chapter deals with the characteristics of the main energy vectors and primary and secondary energy sources in their function as energy vectors (that is to say useful for energy transportation and storage) as well as the relevant processes and technologies.

### 3.5.1 *Oil as Energy Vector*

Among fossil fuels, oil is the vector that shows the best transportability characteristics, thanks to its physical features (it is liquid at ambient temperature and pressure).

In general, the transportation and distribution of oil and its derivatives takes place through the modes described in Fig. 3.1.

Crude oil is transported for long/very long distances through oil pipelines or by sea in oil tankers (in this case, it reaches the port through an oil pipeline where it is loaded on oil tankers and taken to the port of destination; from there, through an oil pipeline, it is transported to refineries). Once arrived to the user Country, it is refined and oil products (petrol, gasoil, LPG, fuel oil and others) are transported, for short/medium ranges to end users or to the distribution facilities, normally by means of oil road tankers.

When calculating the energy spending due to a given energy quantity transportation through vectors starting from the data relating to the transportation of goods in general, it is necessary to consider that the mass transported is the sum of the mass of the vector plus the mass of the container. This means that per each kg of energy vector, the total mass transported is higher.[1] Therefore, indicating with $m_v$ (kg) the mass of the energy vector transported and with LHV (kJ/kg) its lower heating value, the energy transported is:

$$E_v = m_v \cdot \mathrm{LHV} \tag{3.1}$$

[1] For solid and liquid substances (at room temperature) this aspect is little relevant: conversely, it is very much significant for low temperature liquefied and pressurised gases, since in these cases the weight of the container is often of the same order of magnitude as the gas therein contained.

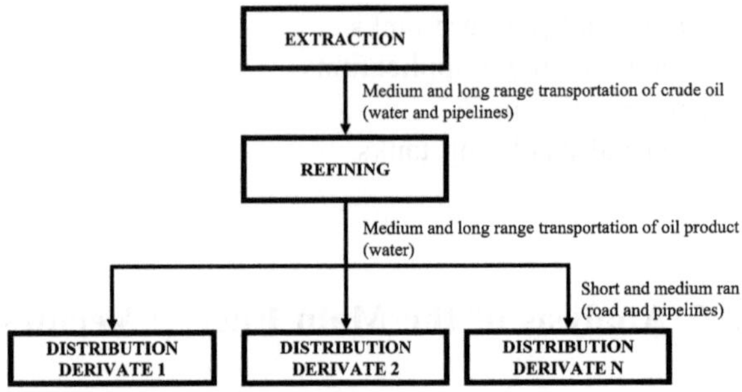


**Fig. 3.1** Chart describing the transportation and distribution of oil and its derivate

Indicating with $R_p$ the ratio between the mass of the vector and the overall mass $m_{\text{tot}}$ (vector + container), it follows that:

$$m_v = R_p \cdot m_{\text{tot}} \tag{3.2}$$

Therefore, the energy transported $E_v$ is

$$E_v = R_p \cdot m_{\text{tot}} \cdot \text{LHV} \tag{3.3}$$

Now, indicating with $E_t$ the energy spent for the transportation of the mass unit ($m_{\text{tot}} = 1$) (kJ/kg km) shown in Table 3.2, the energy spending for the transportation of a fuel along n km, expressed as a percentage of the energy transported ($E_{t\%,n}$) can be calculated as follows:

$$E_{t\%,n} = \frac{E_t \cdot n}{E_v} = \frac{E_t \cdot n}{R_p \cdot \text{LHV}} \tag{3.4}$$

Based on (3.4), since oil lower heating value amounts to 41,868 kJ/kg, assuming $R_p = 0.9$ and according to the data shown in Table 3.2 it is possible to calculate the energy costs for the different types of transport.

Table 3.3 shows the energy spending, expressed as a percentage of the energy transported, along a distance of 1,000 km.

As already said, oil can be transported through oil pipelines. In this case, energy spending can be calculated as follows.

If we consider the generalised form of the first law of thermodynamics[2] for open systems and for adiabatic transformations (without heat exchanges, that is to say $Q = 0$), indicating with $\Delta H$ the enthalpy variation (per mass unit), $\Delta E_p$ the variation of potential energy (per mass unit) and $\Delta E_c$ the variation of kinetic energy (per mass unit) and with $\xi$ the energy losses due to the load loss (per unit of length of the pipeline), the work $L$ necessary to transport an oil mass $m$ for a distance $D$, is the following:

[2] Kinetic energy and potential energy are also taken into consideration.

**Table 3.3** Energy spending for oil transportation through non-dedicated networks

| Transportation system | Energy spending (%/1,000 km) |
|---|---|
| Road | 7.0 |
| Railway | 3.4 |
| Water (sea, river) | 2.0 |

**Table 3.2** Energy spending for freight transportation [1]

| Transportation system | Energy spending (kJ/kg km) |
|---|---|
| Road | 2.638 |
| Railway | 1.298 |
| Water (sea, river) | 0.754 |

$$L = m \cdot \Delta H + m \cdot \Delta E_p + m \cdot \Delta E_c + \xi \cdot D \tag{3.5}$$

Energy losses $\xi$ can be calculated according to the load losses $J$ (expressed in metres of load losses per metre of length of the pipeline[3]) using the Darcy–Weisbach equation, applied to incompressible fluids in permanent motion

$$J = \frac{\lambda \cdot v^2}{2 \cdot g \cdot d} \tag{3.6}$$

where $\lambda$ is an adimensional coefficient of resistance function, in general, of the relative roughness factor of the pipeline $\varepsilon$ and of the Reynolds number, $d$ is the diameter of the pipeline, and $g$ is the gravity acceleration. The coefficient $\lambda$ can be calculated by using the Colebrook–White equation:

$$\frac{1}{\sqrt{\lambda}} = -2 \cdot \log\left(\frac{2.51}{Re \cdot \sqrt{\lambda}} + \frac{\varepsilon}{d \cdot 3.71}\right) \tag{3.7}$$

where $\varepsilon$ is the relative roughness factor and $Re$ is the Reynolds number as expressed by the ratio:

$$Re = \frac{\rho \cdot v \cdot d}{\mu} \tag{3.8}$$

where $\rho$ is density and $\mu$ the dynamic viscosity.

Once $J$ is known, the energy loss is ($J/m$):

$$\xi = J \cdot m \cdot g \tag{3.9}$$

Thence, as concerns the mass unit, the following unitary work $L_u$ is obtained:

$$L_u = \Delta H + \Delta E_p + \Delta E_c + J \cdot g \cdot D \tag{3.10}$$

Indicating with $\rho$ fluid density, with $p$ pressure, with $h$ height and with $v$ velocity, the following is obtained:

[3] Physically, then, $J$ is dimensional.

$$\Delta H = \frac{\Delta p}{\rho} \tag{3.11}$$

$$\Delta E_p = g \cdot \Delta h \tag{3.12}$$

$$\Delta E_c = \frac{1}{2} \Delta v^2 \tag{3.13}$$

Pressure remains constant ($\Delta p = 0$, thence $\Delta H = 0$); furthermore, the height variations can be neglected[4] ($\Delta h \cong 0$, thence $\Delta E_p \cong 0$). The initial velocity is null.

The unitary work, therefore, is the following:

$$L_u = \frac{1}{2} \cdot v^2 + J \cdot g \cdot D \tag{3.14}$$

In general, oil pipelines are made of steel pipes with a diameter varying between 0.3 and 1.4 m; fluid velocity ranges between 1 and 6 m/s.

For example, considering a section of a steel oil pipeline (for which a relative roughness $\varepsilon = 0.01$ mm can be assumed) with a diameter of $d = 1$ m and a velocity of crude oil of $v = 4$ m/s, the following is obtained[5]: Re = 16.000, $\lambda = 0.02735$ and J = 0.02231.

Therefore

$$L_u = \frac{1}{2} \cdot v^2 + J \cdot g \cdot D = 4 + 0.21883 \cdot D[J] \tag{3.15}$$

For a distance of 1,000 km ($D$ = 1.000.000 m), the unitary work $L_u$ (J) is[6]:

$$L_u = \frac{1}{2} \cdot v^2 + J \cdot g \cdot D = 4 + 218.827 = 218,831 \tag{3.16}$$

In loss calculation, only distributed losses were taken into consideration; as a matter of fact, it is also necessary to consider the concentrated losses due to bends, joints, valves, gates, etc. These losses are calculated by considering for each element an equivalent length. Therefore, for the calculation of total losses, a fictitious length of the pipeline is considered, given by the sum of the real pipeline and the equivalent one due to the elements generating concentrated losses. It is readily evident that to calculate the equivalent length it is necessary to know in detail the pipeline in question: in a first approximation, it is possible to consider a fictitious length amounting to 1.2 times the real one. Therefore putting aside the

[4] It is possible to assume that positive and negative height variations along the pathway are equivalent, that is to say the height of the starting point is the same as the end point.

[5] Considering for crude oil a dynamic viscosity of 0.215 Pa s and a density of 0.86 kg/l.

[6] In general, for long distances (typical of oil pipelines) the term relating to kinetic energy becomes negligible compared to load losses:
$J \cdot g \cdot D > > \frac{1}{2} \cdot v^2 \Rightarrow L \cong J \cdot g \cdot D$.

**Fig. 3.2** Oil pipelines

level relating to kinetic energy variation, the unitary work for the transportation along a section of length $D$ is:

$$L_{u,D} = J \cdot g \cdot 1.2 \cdot D \tag{3.17}$$

Then, to calculate the real energy spending for transportation, it is necessary to consider the efficiency of pumping stations ($\eta_{SP}$). Considering a 40% efficiency, the following is obtained:

$$L_{\mathrm{eff},D} = 1.2\frac{J \cdot g}{\eta_{SP}}D \tag{3.18}$$

For a section of 1,000 km: $L_{u,\mathrm{eff}} = 656.48$ kJ/kg.

Considering oil lower heating value (41,868 kJ/kg), energy spending for transportation as a percentage of the energy transported is nearly 1.6% for a distance of 1,000 km.

In general, oil pipelines (Fig. 3.2) as a means for oil transportation, as well as gas pipelines, show the following characteristics

- Disadvantages
  - High construction and laying costs
  - Cost for pumping and distribution stations
  - Difficult conversion in case of oilfield or deposit exhaustion or depletion
- Advantages
  - Low energy transportation costs
  - Good level of safety[7]

[7] Safety and damage risks vary according to the Countries and the regional areas crossed.

In general, oil pipelines are made up of steel pipes with a diameter ranging between 30 and 140 cm, possibly laid aboveground in order to make inspections easier. The liquid is channelled through pumps that give the flow a velocity ranging between 1 and 6 m/s.

The most famous oil pipelines nowadays include:

- *The Friendship Pipeline*. It is the longest pipeline in the world and transports oil along nearly 4,000 km from Russia, crossing Ukraine and Hungary, and reaching Germany. Today it is the main route for the transportation of Russian and Kazak oil to Europe. Its range amounts to 1.2–1.4 million barrels per day. The diameter of pipelines ranges between 420 mm and 1,020 mm. 20 pumping stations are active.
- *Baku-Tbilisi-Ceyhan (BTC) pipeline*. It is used to transport the oil extracted at the oilfield of the Caspian Sea to the Mediterranean Sea shores. It is endowed with eight pumping stations and its total length is 1.776 km. It stretches to Azerbaijan for 440 km, then to Georgia for 260 km and finally to Turkey for 1,076 km, and it is able to transport one million barrels per day.
- *Trans-Alaska Pipeline System*. It connects Alaska with the heart of America through a pipeline of nearly four thousand kilometres.

As far as storage is concerned, crude oil is stored in large quantities and for long times at a national level. These storages represent the *strategic stocks* of importing Countries.

#### 3.5.1.1 Derivatives of Petroleum

As already said, petroleum is not used as crude oil, but rather it is processed to obtain different types of fuels. After some pre-processing stages (as water elimination), it undergoes continuous distillation processes in tray columns whose pressure is close to atmospheric pressure. As shown in Fig. 3.3, the most volatile products are released from the top of the column, whereas less volatile products are collected at decreasing heights.

The quantity of refining products obtained hugely varies with the quantity of oil processed. The quantities of products obtained are rather predefined, and, in any case, the refining process always produces—from a given type of petroleum—all the different products. On average, nearly 20% of crude oil provides petrol, nearly 55% diesel oil and fuel oil, 7% covers internal refining consumption, 7% oil, 3% LPG, 3% bitumen, 1% lubricant bases, and the remaining 4% includes other products as petroleum coke, paraffin, petrolatum, etc. Petrol, diesel oil and fuel oil totally account for a large portion of refining products (75%). In order to change the petrol and fuel oil percentages that, as already said, are on average predefined, according to the quality of crude oil processed, and adjust them according to their demand, further processing is necessary to obtain light hydrocarbons (petrol, made up of hydrocarbons containing 5–10 carbon atoms) starting from heavy hydrocarbons (diesel oil and fuel oil, made up of hydrocarbons containing 12–30 carbon atoms).

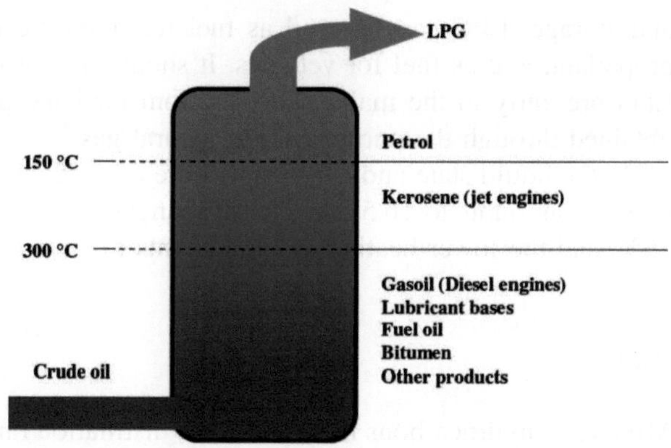


**Fig. 3.3** Diagram of a distillation tower

This processing consists in the breakage of large molecules into small ones: this processing is called "cracking", and can be either thermal or catalytic.

The main refining products include:

- Gas (in general LPG, liquefied)
- Petrol
- Kerosene
- Diesel oil
- Lubricant bases
- Fuel oil
- Bitumen
- Other products

It is important to recall that it is not possible to obtain one single distillate from refining, since it produces all the different fuels at the same time.

The quantity of the different distillates can vary through subsequent processing of:

- Cracking
- Alkylation
- Reforming.

From the energy viewpoint, the *refining process efficiency* can be defined as the ratio between the energy content of fuel products alone (LPG, petrol, kerosene, diesel oil and fuel oil, without considering those products destined to non-energy use) and the energy content of crude oil. In this way, the efficiency of the distillation process accounts for nearly 87%.

LPG (Liquified Petroleum Gas)

LPG, one of the most widespread oil gases, is mainly made up of propane, butane and their blends; it can be liquefied, under pressure, to facilitate its transportation and storage. LPG can be used as fuel for domestic use and heating, as aerosol propellant, and as fuel for vehicles. It should be noted that a relevant part of the LPG presently in the market derives from methane production processes, and is obtained through the purification of natural gas.[8]

At the liquid state and at a temperature of 40°C, the absolute maximum vapour pressure amounts to 16.5 bar. Density ranges between 0.508 and 0.585 kg/dm$^3$ at 15°C and the lower heating value amounts to nearly 46,100 kJ/kg.

## Petrol

Mixture of hydrocarbons included in the distillation range between 20° and 215°C, used to fuel spark-ignition internal combustion engines for the propulsion of vehicles.

*Leaded petrol.* Gasoline complying with UNI 20156 regulation, with a minimum octane-level of 97 RON, whose maximum level allowed of lead compounds shall not exceed 0.15 g Pb/1.

*Lead-free petrol.* Gasoline complying with European Regulation 228, with a minimum octane-level of 95 RON, whose contamination with lead compounds shall not exceed 0.013 g Pb/1

On average, density amounts to 0.74 kg/dm$^3$ and LHV amounts to nearly 43,350 kJ/kg.

## Kerosene

Blend of hydrocarbons included between the distillation range of 180 and 230°C. It is used as a propellant for jet engines and as heating and lighting fuel. In addition to the aviation turbine kerosene (ATK), Deodorised kerosene provides, other products such as:

- *Domestic kerosene* Product that, while burning, releases a limited amount of smoke, has no odour emission, and is therefore suitable for domestic use and heating purposes.
- *Power kerosene* Distillate with sufficient antiknock properties, used in special internal combustion engines;
- *Tractor kerosene* Fuel for agricultural vehicles. In countries where tax benefits are available for this class of products, it undergoes a specific colouring.

Density amounts to 0.8 kg/dm$^3$ and the average lower heating value amounts to nearly 43,200 kJ/kg.

---

[8] Propane and butane are contained in natural gas (see Sect. 2.5.4.2).

### Diesel Oil

Blend of paraffin and aromatic hydrocarbons that distil between 150 and 400°C, is obtained from the atmospheric distillation of crude oil and from processes including cracking and hydrocracking. Its viscosity at 40°C ranges between 2.0 and 4.5 $mm^2/s$, an average density of about 0.83 $kg/dm^3$ and a LCV of nearly 42,000 kJ/kg.

*Engine diesel oil* is used to fuel spontaneous ignition internal combustion engines. It is characterised by a high number of cetane, a low-sulphur content (0.005%, 50 ppm decreasing to 10 ppm and lower in many markets) and by a distillation range which also provides for a limit at 95%. The limits relating to cold performance must be met.

*Heating diesel oil* is used for heat and energy production in residential buildings. It must comply with less strict constraints in terms of viscosity, distillation, cold performance, ignition features and sulphur content (0.2% in weight).

### Lubricant Bases

Blends of predominantly paraffinic hydrocarbons ($C_{12}$–$C_{50}$), at different degrees of viscosity are obtained through solvent refining of distillates and vacuum residues. They are used for the production of engine oils, industrial lubricants and lubricating greases. The latter are obtained by adding suitable thickening agents to lubricant bases.

### Fuel Oil

*Fuel Oil* (*HS*) This term includes all those products, including blends, whose viscosity exceeds 53 $mm^2/s$ at 50°C, and with a maximum sulphur content of 4% in weight.

*Fuel oil* (*LS*)–(*SF*) This term includes all those products and blends, whose viscosity exceeds 53 $mm^2/s$ at 50°C and with a maximum sulphur content of 1% in weight.

On average, its density amounts to 0.83 $kg/dm^3$ and its LHV to nearly 42,000 kJ/kg.

### Bitumen

Blends of hydrocarbons (either saturated or aromatic > $C_{25}$), solid or highly viscous at room temperature. They have a colloidal structure, thermoplastic characteristics, adhesive properties and are obtained as residues from vacuum distillation. Some of them undergo further processes.

Fig. 3.4 Coal transportation and distribution chart

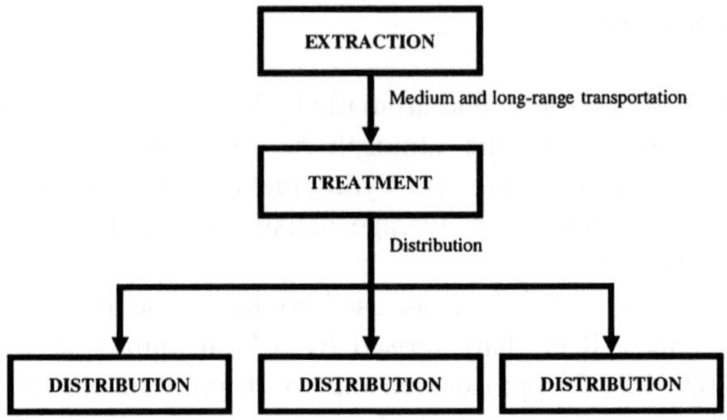


Other Products

*Petroleum coke*. Granular or needle-shaped substance mainly made up of coal and obtained from the pyrolysis of heavy fraction. It may contain high-molecular-weight hydrocarbons.

*Paraffin and petrolatum*. Solid or semi-solid substances at room temperature obtained from the extraction through solvent of vacuum distillates (paraffin) and vacuum residues (petrolatum).

### *3.5.2 Coal as Energy Vector*

Among fossil fuels, coal is the one that shows the less favourable characteristics as energy vector. In fact, since it is a solid fuel, its transportation is more burdensome from the energy viewpoint, and, given the same energy stored, its storage involves larger volumes than oil (coal: 68.24 $dm^3$/GJ; oil: 28.10 $dm^3$/GJ).

The long-range transportation of solid coal mainly takes place via water. In many cases, in order to give coal the best possible characteristics for its transportation, it undergoes several processes, the same already described in Sect. 2.4.1.5 (Fig. 3.4), that is:

- *Pulverisation*
- *Gasification and liquefaction*
- *Hydrocarbonisation*
- *Hydrogenation*
- *Fischer–Tropsch process*
- Coal–oil
- Coal–water

The calculation of energy spending can be made similar to the case of oil, considering that the average lower heating value of coal amounts to 29,308 kJ/kg and assuming $R_p = 0.9$. These costs are shown in Table 3.4. They are expressed as a percentage of energy transported along a distance of 1,000 km.

Table 3.4 Energy spending for coal transportation

| Transportation system | Energy spending (%/1,000 km) |
|---|---|
| Road | 10.0 |
| Railway | 4.9 |
| Sea | 2.9 |

**Fig. 3.5** Siberian gas pipeline

## 3.5.3 Natural Gas as Energy Vector

As far as transportation and storage are concerned, natural gas shows intermediate characteristics between oil (liquid) and coal (solid). It can be suitably transported in gas pipelines (Fig. 3.5), but, although it is in a gaseous state, transportation is more expensive from the energy view point. The strategic importance of this vector encouraged the resort to the diversification of supply sources and technologies. In particular, the transportation via sea of liquefied gas (LNG) at cryogenic temperatures and its regasification at the delivery ports is a more and more widespread practice. This solution is adopted for transportation via sea through the so-called "methane tankers" or LNG tankers, ships endowed with cryogenic tanks: liquefaction hugely reduces volume[9] and this allows to transport natural gas in an energy-efficient as well as safer way (natural gas is not inflammable at the liquid state). Normally transportation takes place at atmospheric pressure at a temperature of nearly—163°C (at atmospheric pressure, the boiling point is—161.4°C), and in some cases pressure tanks are used, therefore with higher temperatures.

Once arrived to the port, the gas, still liquefied, is transferred from the ship to a storage tank inside the regasifiers. It is subsequently sent to a vaporiser, that acting on temperature performs, gasification through gas expansion, with gas that goes back to its natural state. Temperature variation generally takes place through thermal exchange in tube bundles between liquid gas and seawater, which releases

[9] The density of gaseous methane at 75 bar and room temperature amounts to 55.9 kg/m$^3$, whereas at the liquid state it amounts to 424.7 kg/m$^3$.

its heat to the gas; pressure, on the contrary, is reduced through gas expansion in suitable tankers.

Instead of unnecessarily dispersing the "cold" in the sea, regasifiers can be associated to plants that envisage the use of low temperatures (for instance, food-freezing facilities, typically present in ports), "recovering" in this way the freezing energy with remarkable saving.

Natural gas is stored in a gaseous state at high pressure in cylinders or at low pressure in gasometers. In many Countries, the use of natural gas as energy vector is very much diffused also thanks to a widespread distribution network that reaches an extremely high number of users, including residential users.

#### 3.5.3.1 Transportation of Natural Gas

For the calculation of the energy spending related to transportation in gas pipelines, the same procedure as oil transportation in oil pipelines applies. Then:

$$L_u = \Delta H + \Delta E_p + \Delta E_c + J \cdot g \cdot D \tag{3.19}$$

In case of gas, the variation of potential energy can be neglected; therefore the enthalpy variation relates to the adiabatic compression of gas:

$$\Delta H = c_p \cdot \Delta T \tag{3.20}$$

where $c_p$ is the constant pressure specific heat and $\Delta T$ is temperature variation due to compression. In order to have an adiabatic compression, the following applies:

$$\frac{T_2}{T_1} = \left(\frac{p_2}{p_1}\right)^{k-1/k} = \beta^{k-1/k} \Rightarrow T_2 = T_1 \cdot \beta^{k-1/k} \tag{3.21}$$

where $k$ is the ratio between constant pressure specific heat ($c_p$) and constant volume specific heat ($c_v$):

$$k = \frac{c_p}{c_v} \tag{3.22}$$

In case of high compression ratios, as in the case of compression of natural gas for transportation purposes, compression takes place in several inter-refrigerating stages.[10] Indicating with $n$ the number of stages, the compression ratio of the single stage[11] $\beta_s$ is:

$$\beta_s = \sqrt{[n]}\beta \tag{3.23}$$

[10] Between the exit of one stage and the access to the next, the liquid is cooled at constant temperature. In this way the needed compression work decreases.

[11] Considering the compression ratio as equal in the different stages.

**Table 3.5** Polynomial coefficients for the calculation of constant pressure specific heat (kJ/kg)

| Polynomial coefficients | | | | |
|---|---|---|---|---|
| Gas | A | B | C | D |
| Natural gas | $1.57 \times 10^{-8}$ | $2.22 \times 10^{-5}$ | $6.74 \times 10^{-3}$ | 2.68 |
| Hydrogen | $6.60 \times 10^{-8}$ | $8.44 \times 10^{-5}$ | $3.61 \times 10^{-2}$ | 9.31 |

As an example, it is possible to consider a gas pipeline with a 1 m diameter transporting gas at the pressure of 75 bar (assuming a compression in three interrefrigerated stages) with a velocity of 10 m/s.

The constant pressure ($c_p$) specific heat and the constant volume ($c_v$) specific heat and, therefore, $k$, depend on temperature; the calculation can be made, with a good approximation, considering constant $c_p$, $c_v$ and $k$ with values calculated at a mean temperature between the initial and the final one. In any case, also with this simplification, it is necessary to know the final compression temperature. The constant pressure specific heat can be expressed with a good approximation according to temperature through a third-degree polynomial:

$$c_p = A \cdot T^3 + B \cdot T^2 + C \cdot T + D \tag{3.24}$$

Table 3.5 shows the polynomial coefficients to calculate the constant pressure specific heat (kJ/kg) for some gases.

The end-of-compression temperature[12] is:

$$T_2 = T_1 \cdot \beta_s^{k-1/k} = 298 \cdot 4,22^{0,15} = 370\,\mathrm{K} \tag{3.25}$$

The ideal enthalpy variation (isoentropic compression) of the single stage is

$$\Delta H_{s,i} = c_p \cdot \Delta T = 3,480 \cdot 72 = 247,900\,\mathrm{J/kg} \tag{3.26}$$

The total enthalpy variation is obtained by multiplying for the number of stages:

$$\Delta H_i = \Delta H_{s,i} \cdot n = 743,700\,\mathrm{J/kg} \tag{3.27}$$

It is also possible to take into account irreversibility (compression is not isoentropic) considering a thermodynamic compression efficiency $\eta_c$ defined as follows:

$$\eta_c = \frac{\Delta H_i}{\Delta H} \tag{3.28}$$

[12] The $k$ value used concerns the average temperature between the initial and the final one. The calculation can be made through the following procedure: a value of the end temperature is hypothesised and the average temperature is calculated, followed by the calculation of the value of $k$ and the value of end temperature that is compared to the hypothesised one. If a marked difference is obtained, the value of $T_2$ is assumed as the one calculated, and the same procedure is followed until the value hypothesised and the value calculated of end temperature coincide (or, better, the difference is not limited to a given value considered as acceptable).

Assuming a thermodynamic compression amounting to 87%, the following applies:

$$\Delta H = \frac{\Delta H_{s,i}}{\eta_c} = \frac{781,717}{0.87} = 854,840\,\text{J/kg} \tag{3.29}$$

The variation of kinetic energy is[13]:

$$\Delta E_c = \frac{1}{2} \cdot \Delta v^2 = \frac{10^2}{2} = 50\,\text{J/kg} \tag{3.30}$$

As in the case of oil, the variation of potential energy is neglected. For loss loads, Re = 50,827,273 is obtained, from which $\lambda = 0.00829$ and J = 0,04228.

Considering a section of 1,000 km, the following is obtained: $L_u =$ 743,700 + 50 + 414,778 = 1,158,528 J/kg.

In this case, the compression activity and the load losses for 1,000 km are of the same order of magnitude. This compression is in any case convenient from the energy viewpoint. It should be noted that, in transporting the gas at atmospheric pressure, the load losses (always for a distance of 1,000 km) per kg of gas transported would increase by over three times.

With considerations similar to oil in relation to concentrated losses and the efficiency of pumping stations, the following applies:

$$L_{u,\text{eff},D} = \frac{\Delta H}{\eta_{\text{SP}}} + \frac{J \cdot g \cdot 1,2}{\eta_{\text{SP}}} D \tag{3.31}$$

In a section of 1,000 km, $L_{u,\text{eff},D} =$ 3,490,803 J/kg is obtained. This value includes over 7% of the energy content of gas transported. To offset the load losses, compressor stations are planned every 100–150 km. These stations are fuelled by the gas itself, therefore this consumption corresponds to gas consumption. In practice, for a transportation distance of 1,000 km, nearly 90% of the gas initially sent reaches its destination; the rest is consumed to supply compression stations.

Compression stations are mainly made up as follows:

- *Compressor unit*, normally centrifugal compressors activated by gas turbines, equipped with own control systems.
- *Pipes and the relevant mechanical equipment*, for complete gas processing (filters, valves, air-coolers etc.).
- *Plant control systems*, able to perform the entire management of turbo-compressors and of the plant processing systems.
- *Electric system*, able to guarantee the necessary power supply according to the various needs and priorities.
- *Civil infrastructures*, for plant-engineering use (switch room, control room, etc.), personnel offices, workshop and storehouse.

---

[13] It is evident how the variation of kinetic energy is negligible compared to the compression work.

**Table 3.6** Energy spending for the transportation of natural gas

| Transportation system | Energy spending (%/1,000 km) | |
|---|---|---|
| | Compressed (200 bar) | Liquefied (−163°C) |
| Road | 11.6 | 7.9 |
| Railway | 5.7 | 3.9 |
| Water | 3.3 | 2.3 |

The most famous gas pipelines presently used include:

- *Bluestream* It is one of the main international gas pipelines transporting natural gas from Russia to Turkey, through the Black Sea, with a total length of 1,213 km (373 of which on the Russian territory, 396 offshore and 444 on the Turkish territory). The maximum transportation capacity is nearly $16 \times 10^9$ $m^3$ of gas per year
- *Greenstream* It is a gas pipeline connecting Libya to Italy. The quantity of gas transported is $24 \times 10^9$ $m^3$ per year. Its length amounts to 520 km and reaches the compressor stations of Mellitah and the gas terminal in Gela after passing through sea depths of more then 1,000 m (maximum depth: 1,127 m)
- *Gasdotto Enrico Mattei* It connects Italy to Algeria via Tunisia. The gas pipeline originates in Hassi R'Mel, in the Algerian desert, and after crossing Tunisia, it plunges in the Mediterranean Sea reaching the Sicilian coasts of Mazara del Vallo. From here it sails up Italy and reaches the destination point of Minerbio, in the Po Valley, where gas is stored in one of the largest storage centres in Italy. Overall, its length exceeds 2,200 km, 370 km of which in Tunisia and 380 km offshore in the Sicily Channel

In general, the gas transportation and distribution system is made up of pipes with a diameter ranging between 25 and 1,200 mm, transporting pressure gas (relative pressure[14]) between 0.5 and 75 bar besides the pumping stations and the storage systems. In addition to the gas pipelines for international transportation, there are also local networks (typically regional) and the networks for gas distribution and delivery to end users.

Similarly to the descriptions in the previous chapters, it is possible to calculate the energy spending for the transportation of natural gas as well through LNG carriers in non-dedicated networks.

The lower heating value of natural gas amounts to 48,180 kJ/kg; considering $R_p$ as amounting to 0.47 for compressed gas at 200 bar[15] and 0.69 for liquefied gas at −163°C[16] the values shown in Table 3.6[17] are obtained.

[14] Referred to atmospheric pressure ($p_r = p_{ass}$–1 bar).

[15] Gas density at 200 bar amounts to 159.5 kg/$m^3$.

[16] The density of liquefied methane gas at −163°C amounts to 427.4 kg/$m^3$.

[17] The values shown relate to transportation only and do not consider the energy spending for compression, liquefaction and regasification.

**Table 3.7** Storage capacity of a gas cylinder compressed at 75 bar

| Gas cylinder compressed at 75 bar | | |
|---|---|---|
| Storage capacity (% in weight) | 4.7 | % |
| Energy density in volume | 2,062 | kJ/litre |
| Energy density in mass | 2,369 | kJ/kg |

The energy spending for the transportation of liquefied gas via sea accounts for 2.3% of the energy content of the gas transported; therefore, also considering the energy spending for liquefaction and regasification (which, on the contrary, can be recovered), this procedure is energy efficient.

#### 3.5.3.2 Pressure Storage of Natural Gas

As described in the previous chapter, the energy spending for a 75 bar[18] compression amounts to 855 kJ/kg. Considering the efficiency of electric engines moving compressors amounting to 93%, it is possible to obtain a compression (electric) work of 919 kJ/kg, corresponding to 1.9% of the energy content of the gas stored.

Besides the necessary energy spending, other relevant characteristics for the storage systems include:

- *Storage capacity* It is the ratio (expressed as a percentage) between the mass of the stored vector and the total mass of the storage system (including the vector).
- *Energy density, in volume* It is the energy stored per unit of volume of the storage system;
- *Energy density, in mass* It is the energy stored per mass unit of the storage system (including the vector)

A cylinder of a 40 l capacity contains 2.31 kg of gas[19] (with an energy content of 111,352 kJ), has a size of nearly 54 l and weighs 47 kg. On the basis of the descriptions made, Table 3.7 shows the storage characteristics for natural gas.

### 3.5.4 *Electricity as Energy Vector*

The role of electric power and of its applications in the energy field is to be considered as absolutely important from the Twentieth century onawards, in consideration of the huge changes occurred in the social and productive activities of the economic systems that availed themselves of the use of electricity. The use of electric power spread in developed countries first, and subsequently in all the others, permeating any form of individual and social activity.

---

[18] The energy density (kJ/m$^3$) of natural gas at 75 bar is similar to the one that hydrogen owns (described in detail afterwards) at a pressure of 350 bar.

[19] Gas density at 75 bar amounts to 57.9 kg/m$^3$.

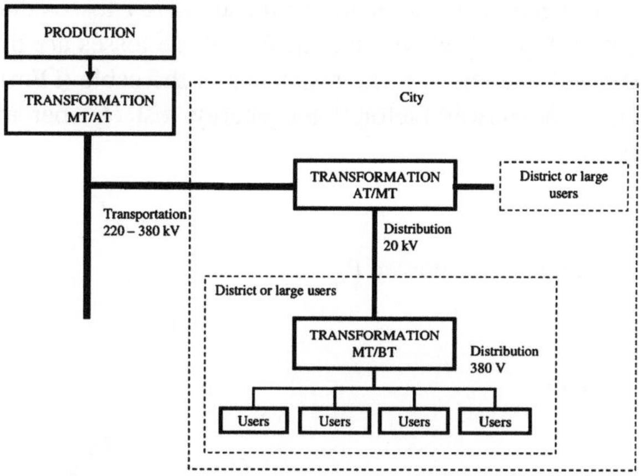


**Fig. 3.6** Electric power transportation and distribution diagram

Thanks to the success of electric power, there was a progressive improvement of living conditions (both from the sanitation viewpoint and from the perspective of social relations, with the easy access to an increasing number of information, culture and leisure sources), an increase in technological development and in the related productive activities, as well as a reduction in manual activities.

Electricity has become a primary good and its strategic weight is destined to grow, with electric services continuously increasing in terms of quantity and quality as well as purposes. As concerns the electric grids of industrialised countries, it is important to consider the huge importance of transmission voltage that, in order to allow the service to be technically and economically feasible, shall increase as the distance of users increases.

Electric energy shows some extraordinary characteristics as energy vector, mostly thanks to its features in terms of "attitude to transport": in fact, it is possible to transport electric power also for medium-long distances with a limited energy spending. As far as storage is concerned, on the contrary, electricity cannot be directly stored, but there are different types of indirect storage, some of which as sufficiently efficient for applications of major interest and large market distribution.

#### 3.5.4.1 Energy Transportation Through the Electricity Vector

The transportation[20] of electric power, from production plants to users, takes place at different voltages in the different sections of the pathway (Fig. 3.6).

[20] Often, in relation to electric power, reference is made to transportation and distribution: transportation means high or very high voltage transportation for long distances from the production plants to the AT/MT transformation plants, whereas distribution means the medium–low voltage transportation for limited distances from the MT/BT transformation plants to end users.

In general, in order to minimise energy losses during transportation, it is necessary to use high voltage: in fact, these losses are due to the Joule effect. Having defined $\rho$ as the specific resistance of the cable (Ω/m), $V$ the voltage, $I$ the current, $\cos\varphi$ the power factor,[21] the energy lost ($Q$) per metre of electric cable is the following

$$Q = \rho \cdot I^2 \tag{3.32}$$

The electric power $P_E$ is:

$$P_E = V \cdot I \cdot \cos\varphi \tag{3.33}$$

therefore

$$Q = \rho \cdot \frac{P_E^2}{V^2 \cos^2\varphi} \tag{3.34}$$

Considering the power factor ($\cos\varphi$) as constant, being $\rho$ a characteristic of the cable, the following applies

$$Q = K \cdot \frac{P_E^2}{V^2} \tag{3.35}$$

The distribution and utilisation voltage is relatively low (100–120, 220, 380 V according to the different cases and of geographical areas) for safety purposes as well as for technical–economic reasons (low unit power requested by household appliances), and makes it necessary a differentiation in the voltage levels, in the production, transportation and utilisation phases, respectively.

It would be suitable to further increase the operating voltage of the electricity transmission grid, but unfortunately there are practical limits due to the high cable insulation costs, to the problems of mechanical instability of cables, as well as to the presence of dissipative phenomena, typical of high voltage (corona discharge). Furthermore, beyond a certain voltage threshold, the costs of foundations, transformers, and of the other elements of the transmission line rapidly increase. It is also necessary to recall that, since electric power cannot be stored in its form, the excess electricity introduced in the grid and not immediately used is lost, with a subsequent financial loss.

An important characteristic of electricity transmission grids is the creation of a mesh scheme, hence allowing the continuity of power supply also in case of failure in some lines.

As already said, the voltage of the main electricity transmission grids varies according to production, transportation and utilisation needs. Generally, high-voltage lines for large distances are run at 220, 400 or at the so called "Extra high voltage" of 765 kV (Fig. 3.7). The nodes of meshes include *step-up transformers* in case of energy input from a power plant, and *step-down transformers* if energy is

[21] Represents the phase shift between voltage and current.

supplied through regional or provincial dispatch grids. The latter, at a voltage of 60–150 kV, supplies medium-voltage primary distribution networks (10–20 kV), whose range does not exceed tens of km. Representing a maximum limit in rural areas. The last transformation finally takes place in switching stations, in order to supply to users electric energy at the voltage—in general—of 100-12, 220 and 380 V.

Due to its strategic role in the overall energy framework, the electric power distribution grid shall comply with strict requirements in terms of continuity of supply, reduction within close limits of frequency variations in the grid, as well as minimisation of delays in the nominal voltage at the delivery point.

As far as frequency variations are concerned, they can produce oscillations in the rotation speed of synchronous or asynchronous motors, whereas voltage differences may cause voltage drops in various elements of the plant (lines, transformers, etc.).

These requirements are suitably met by the mesh networks, endowed with great flexibility in the management of faults to the lines and in the lack of uniformity between electric production and consumption.

For instance, to prevent problem in case of a fault to the line or to a generator, it is sufficient to oversize plants by 10%, in order to face the total load of the network. Furthermore, at least two lines always converge to each node, hence making it possible the overload of one line to face the demand in case of fault of the other. There are then systems to automatically regulate the frequency and the voltage, through which it is possible to obtain excellent supply standards. For instance, the frequency oscillation of the large network interconnected all over Western Europe is of 0.1 Hz only, for a nominal frequency of 50 Hz (0.2%).

The already mentioned technical problems have to be carefully analysed and solved in the most cost-effective way, so as to reduce the environmental impact of transmission lines. In fact, it is necessary to carefully analyse the clearance between conductor and support, mainly linked to the likelihood of line discharge in switching transients, and the clearance between conductor and ground, in order to avoid discharges towards objects placed under the line. Conductors, furthermore, will have to be chosen not only on the basis of losses due to the Joule effect, but also according to the losses due to the corona discharge and radio interferences. The technical–economic solution of these problems might entail huge advantages. With a line of 1,050 kV it is possible to supply a power of 5,600 MW with a corridor of 70 m rather than 280 m, necessary if a 380 V line was used (Fig. 3.7). This extraordinary result can be obtained thanks to the higher power density that can be reached through a voltage of 1,050 V (80 MW/m instead of 20 MW/m).

The characteristics of the distribution network vary according to the electric standards in force in different Countries: in particular, the distribution system and the voltage and frequency of electricity supplied to users vary, besides the type of electrical outlet for civil/household use.

The distribution system can be:

- M: single-phase
- S: three-phase star with neutral

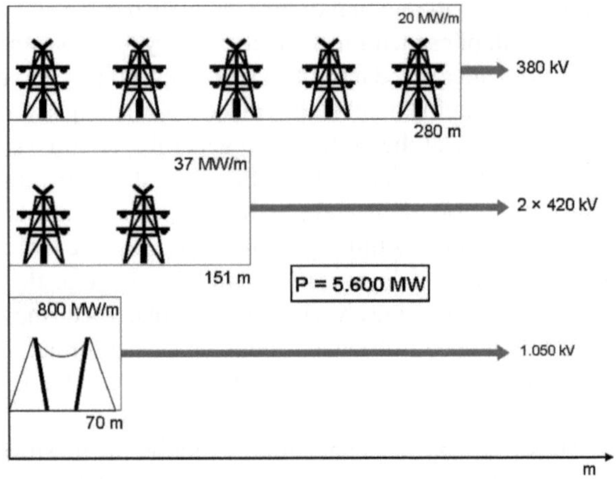


**Fig. 3.7** Transmissible power and size of transmission lines

**Table 3.8** Distribution characteristics for some countries

| Country | Distribution | Household voltage (V) | Frequency (Hz) |
|---|---|---|---|
| Italy | S 230/400 | 230 | 50 |
| United Kingdom | S 240/415 | 240 | 50 |
| Japan (East)[22] | D 100/200 | 100 | 50 |
| Japan (West)[23] | D 100/200 | 100 | 60 |
| China | S 220/380 | 220 | 50 |
| United States | S 120/208 | 120 | 60 |
| | D 120/240 | | |
| | T 460 | | |

- D: delta with a fourth wire in the middle of a winding
- T: three-phase three-wire with eventual two-phase distribution.

Table 3.8 shows the distribution characteristics for some Countries.

Figure 3.8 shows the world map with electricity voltage and frequency in the distribution plants for household use.

Therefore, given the same electric power transported, losses are inversely proportional to squared voltage.

For these reasons, as already said, the transportation, until possible, takes place at very high voltage, 220 and 380 kV or more; in case of distribution in urban centres, for safety reasons the transportation voltage is lower (20 kV), whereas the supply, exception made for large users, takes place at the utilisation voltage. In any case, there is the attempt of minimising the distances covered with low voltage.

The percentage power lost ($P_{d\%}$) during transportation is proportional to the power transported and to the distance ($D$) of transportation:

[22] Tokyo, Kawasaki, Sapporo, Yokohoma and Sendai.

[23] Osaka, Kyōto, Nagoya, Hiroshima.

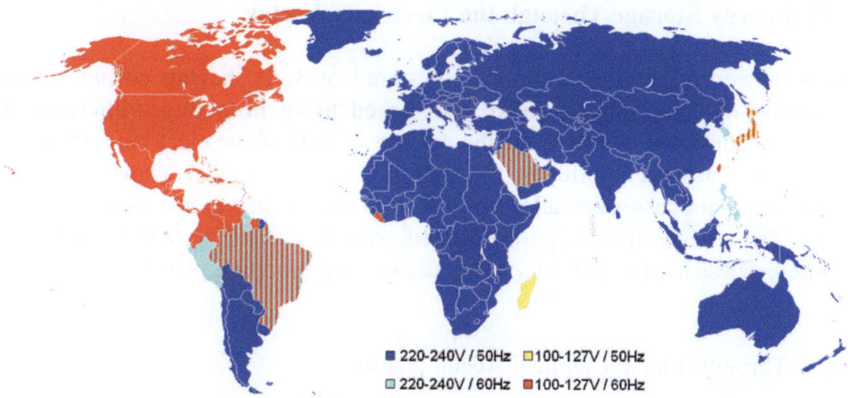


**Fig. 3.8** Electricity voltage and frequency map in distribution plants for domestic use (*sources* Standard elettrici nel mondo, http://it.wikipedia.org/wiki/Standard_elettrici_nel_mondo)

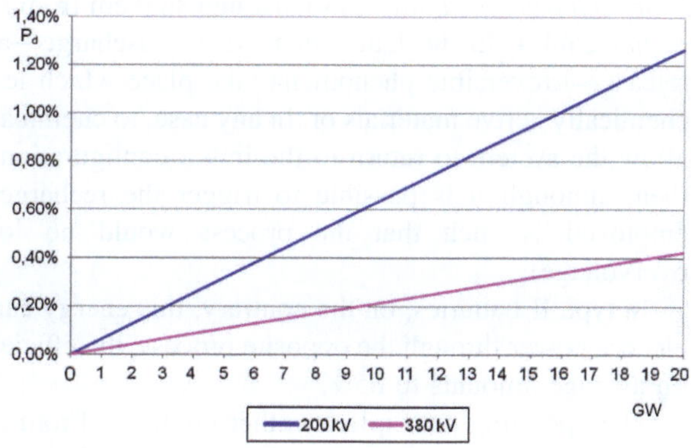


**Fig. 3.9** Power loss on the HV electricity transmission grid along 1,000 km according to the power transported

$$P_{d\%} = \frac{Q}{P_E} = \frac{\rho}{V^2 \cdot \cos^2 \varphi} P_E \cdot D \tag{3.36}$$

In case of copper cables, resistivity is $\rho = 1.72 \times 10^{-8}$ Ω/m

Considering an average power factor of 0.75, the power losses in the network along 1,000 km are shown in Fig. 3.9.

Given all the same parameters (including the power transported) the ratio between power losses amounts to the ratio of squared voltages; this explains the different voltage transformations although they obviously entail power losses due to the efficiency of transformers (Fig. 3.6). For instance, it is possible to consider that, while transporting an electric power of 10 GW along a distance of 500 km, the losses on the line at 380 kV (HV) amount to 0.1% of the power transported, whereas at 20,000 V(MV) these losses amount to 38.2% of the power transported.

### 3.5.4.2 Energy Storage Through the Electricity Vector

Electric power cannot be economically stored as it is, having considered its intrinsically dynamic characteristic, also linked to an instability—difference in voltage—primarily responsible for the existence itself of the electronic flow.

There are, however, different methods of *indirect storage*, namely through transformation processes that allow to shift from electric power to a form of energy suitable for storage that, when requested, restore electric power through the opposite process as, for instance, accumulators and pumping plants.

#### Storage Through Electrochemical Accumulators

An extensively used technology for the indirect storage of electric power is represented by *electric accumulators*: these devices transform electric power into chemical energy through an electrochemical reaction, and are also defined as *type II or rechargeable batteries* to distinguish them from *type I batteries* (that are non-rechargeable). In the latter, in fact, the discharge—and therefore during energy release—irreversible phenomena take place which lead to a structural change of chemically active materials or, in any case, to chemical transformations that do not allow the system to return to the initial configuration any longer. In some cases, then, although it is possible to trigger the recharge process, the energy to be employed is such that the process would no longer be convenient (high overvoltage).

In type II batteries, on the contrary, this energy can be newly transformed into electric power through the opposite process; the efficiency of this storage process[24] on average amounts to 85%.

The operating principle is rather complex. From the structural viewpoint, the "classical" accumulator can be described as follows: two electrodes are plunged in an electrolytic fluid and act as poles, one positive and one negative, which can be connected to the circuit to be powered or to a circuit they are powered from (Fig. 3.10).

When the accumulator acts as a generator and provides direct current, it goes out of the positive pole and goes back to the negative pole.[25]

It is generally said that accumulators are electrochemical generators of electric power because when they release current, during the discharge phase, the chemical energy available is transformed into electric power through electrochemical reactions. Obviously, the opposite happens in the charging phase; the electric

---

[24] Namely the ratio between the electric power available from the accumulator for use and the electric power spent to charge the accumulator.

[25] It is widely known, however, that *positive charges* are so called by convention; as a matter of fact, it is negative charges that enter, electrons, and therefore the terms should be used the other way around!

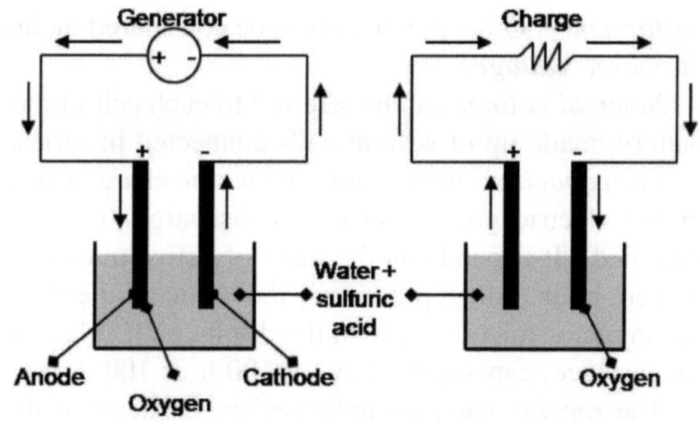


**Fig. 3.10** Double operation of the accumulator: charge and discharge phases

power received from the grid is stored through the opposite electrochemical reactions, under the form of chemical energy.

Therefore an accumulator is a voltaic cell whose structure is similar to that of a battery and, as all the other electrochemical generators, it is able to supply part of the chemical energy of its active matter in the form of direct current circulating in the user circuit. The recharge of the accumulator is obtained through its connection to a suitable generator of direct current.

As already said, an accumulator, in its simplest form, mainly includes two electrodes, one positive and one negative, and an electrolyte. Besides the two electrodes—containing the active matter—accumulators include other components: "current collectors", carrying electric power from and to electrodes, and separators (soaked with electrolyte), which prevent the two electrodes, the positive and negative one, from coming into contact inside of the voltaic cell, causing a short circuit.

An accumulator may consist of a single element, although usually it is formed by several elements connected in series and, in this second case, it is called *secondary battery* (commonly speaking, the word "secondary" is generally omitted). The voltage between the battery clips is therefore determined, besides by the electrochemical system used, also by the number of cells connected in series; for instance, a common 12 V lead battery for automobiles is formed by six lead cells, connected in series, each one of 2 V. More generally, the appliances to be powered operate at different voltages and demand different power levels; the sizing of an accumulator in terms of voltage and power supplied, is determined by selecting both the electrochemical system to be adopted and the number of cells. In principle, the capacity of an accumulator to supply electric power should not be influenced by the number of charge and discharge cycles made. In practice, on the contrary, in each charge and discharge cycle there are changes in the structure and in the volume of electrodes that gradually reduce the performance of the accumulator. For this reason, while designing each accumulator, it is necessary to take this effect into consideration, and to offset it as much as possible.

In order to evaluate the performance of an accumulator, various parameters must be considered: *nominal voltage*, *capacity*, *energy*, *power*, *lifetime* and *performance*. Other parameters to be considered include: *working voltage* and *final discharge voltage.*

*Nominal voltage* can be referred to each cell of the accumulator or to the entire battery, made up of several cells connected in series.

The *capacity* is the quantity of electric charge that the accumulator can circulate in the electric circuit during its discharge (in conditions that shall always be specified). It depends on the mass of active matters and is generally expressed in ampere-hour (Ah) representing the value of theoretical current intensity that the accumulator might supply in the discharge if it lasted 1 h (a 100 Ah accumulator, for instance, can supply 1 A for 100 h or 100 A for 1 h).

The *specific capacity* indicates the capacity of the accumulator per mass unit (for instance Ah/kg) or per volume unit (for instance Ah/dm$^3$).

The *energy*, that the accumulator is able to supply, is given by the capacity multiplied by the average discharge voltage and is generally expressed in watt–hour (Wh).

The *specific energy* indicates the energy that can be supplied by the accumulator per mass unit (Wh/kg)

*Energy density* indicates the energy that can be supplied by the accumulator per volume unit (Wh/dm$^3$).

The *power* that can be supplied by the accumulator is given by the average discharge voltage multiplied by the current and is expressed in Watt (W).

The *specific power* is expressed in Watt/kilogram (W/kg)

*Power density* is expressed in Watt/cubic decimetres (W/dm$^3$).

As for efficiency, the following is considered:

- *Amperometric efficiency* (ratio between the number of ampere-hour through the accumulator during the discharge and during the previous charge).
- *Energy efficiency* (ratio between the energy supplied during the discharge and the one absorbed during the previous charge).

The *lifetime* mainly depends on the conditions of work of the accumulator, and therefore can be indicated only for given charge and discharge steady states that must be specified. In particular, when the accumulator is subject to consecutive charge and discharge cycles, the *cycle lifetime* is considered, which indicates the aptitude of an accumulator to resist to consecutive charge and discharge cycles; in this case, the lifetime is generally indicated with the number of charge and discharge cycles that originate a lowering in the capacity at a pre-determined percentage (generally 80%) of the original value.

The *working voltage* is the average voltage present between the two plugs, positive and negative, when the accumulator supplies current.

The *end voltage of the discharge* is the working voltage at which, for either technical and/or economic reasons, it is worth stopping the discharge.

Accumulators are classified according to the electrochemical system used and it is therefore possible to identify different types of accumulators. The most diffused industrialised systems today include:

- Lead batteries
- Alkaline batteries (cadmium–nickel or nickel–metal hydride)
- Lithium batteries

As regards applications, accumulators can be divided into *fixed systems, traction, starting and portable accumulators.*

*Accumulators for Fixed Systems (Stationary Accumulators)*: are the accumulators (lead or nickel–cadmium) for fixed or even floating systems, that are installed for emergency power supply (lighting or other services) at electric power stations, telecommunications stations, hospitals, etc.;

*Traction Accumulators* are destined to supply energy for the traction of electric-driven vehicles (city centres factories, golf courses, railway stations and airports, etc.) or for the propulsion of boats. Often, in these accumulators, the active matter is contained in electrolyte-permeable pipes, grouped in a chassis. Their main requirements include:

- Long discharge duration.
- Good resistance to mechanical stress.

Plate data shall include: nominal voltage and discharge capacity in 5 h.

*Starting Accumulators (APU-Auxiliary Power Units);* are used in vehicles for a set of functions: starting and sparking of internal combustion engines, lighting and auxiliary services.

*Portable Accumulators*; are characterised by a high compactness and a good resistance to crashes and vibrations. Another fundamental requirement is tightness, to prevent electrolyte leakage and the corrosion of the components of the equipment in which the accumulator is included.

*Lead–Acid Accumulators*

Lead–acid batteries were the first secondary batteries that were commercialised, and their production gradually increased as the demand for portable energy grew, as well as its possible destination to different applications: from the sparking of boats and aircrafts, to the power supply of the electrical system and the injection system of these vehicles; from energy storage in electric power plants to uninterruptible power supply for emergency energy in general; from the use in communications devices to the power supply to the memory circuits of computers.

Lead–acid accumulators, both for the simplicity of reactions, and for the easiness of realisation, have been developed in different types and for several destinations of use.

Their main characteristics include:

- Advantages
  - Cheapness and easiness of production
  - Various sizes, ranging from small one (1 Ah) to large size (>100 Ah)
  - Good performance for high discharge velocity (starting)

- Good performance both at high and low temperature
- Good electric efficiency (~60%)
- High cell potential: 2.2 V (it is the highest among the various batteries using aqueous electrolyte)
- Easy indication of the state of charge
- Good capacity retention for applications with intermittent charge.

• Disadvantages

- Moderate number of cycles
- Limited energy density (30–40 Wh/kg)
- Low charge retention (sulphation)
- The storage of batteries in discharged conditions may lead to irreversible electrode bias
- Evolution of hydrogen and oxygen (explosive mixture).

The reactions in a Lead–acid accumulator are the following:

$$\text{Negative electrode} \quad Pb + HSO_4^- \underset{\text{charge}}{\overset{\text{discharge}}{\rightleftarrows}} PbSO_4 + H^+ + 2e^-$$

$$\text{Positive electrode} \quad PbO_2 + HSO_4^- + 3H^+ + 2e^- \underset{\text{charge}}{\overset{\text{discharge}}{\rightleftarrows}} PbSO_4 + 2H_2O$$

$$\text{Global reactions} \quad Pb + PbO_2 + 2H_2SO_4 \underset{\text{charge}}{\overset{\text{discharge}}{\rightleftarrows}} 2PbSO_4 + 2H_2O$$

As it can be seen, at the basis of electrode processes, both on the negative and on the positive side, a mechanism is in place relating to the formation and disappearance of the lead sulphate in which the sulphuric acid present in the electrolytic solution is involved, and is consumed during the discharge through water production.

The excessive formation of $PbSO_4$ (sulphation) can produce rather large crusts that are removed from electrodes, leading to the loss of active material in the next charge.

The electrolyte ($H_2SO_4$) can be considered as a full active material as Pb and $PbO_2$ to such an extent that in some applications it can act as a limiting agent.

When the cell is close to be completely charged, $PbSO_4$ is totally transformed into Pb and $PbO_2$ and the cell potential reaches high values (2.39 V), and water electrolysis processes begin, that is to say:

$$\text{Negative electrode} \quad 2H^+ + 2e^- \rightarrow H_2$$

$$\text{Positive electrode} \quad H_2O \rightarrow \tfrac{1}{2}\, O_2 + 2H^+ + 2e^-$$

$$\text{Global reaction} \quad H_2O \rightarrow \tfrac{1}{2}\, O_2 + H_2$$

In sealed batteries, this reaction is controlled so that the hydrogen and oxygen that are formed react as follows:

$$Pb + HSO_4^- + H^+ + 1/2\,O_2 \rightarrow PbSO_4 + H_2O$$
$$PbO_2 + HSO_4^- + H^+ + H_2 \rightarrow PbSO_4 + 2\,H_2O$$

As already said, this reaction only takes place during the charging phase and only if the cell is sealed. As a matter of fact, these types of cells have a safety valve that opens when the internal pressure reaches 4 bar.

The nominal potential of the cell is 2 V in open-circuit, although this depends on the concentration of the electrolytic solution (for instance $V = 2.125$ V for a cell whose solution has a specific gravity of 1.28 g/ml and $V = 2.05$ V when the specific gravity is 1.21 g/ml). At the end of the discharge, if carried out at low-current, it amounts to 1.75 V although it can reach a value of 1.0 V per cell in case the discharge takes place at a high velocity and low temperature.

The cell potential depends on the concentration of the solution, and the choice of the specific gravity is made according to the applications that the batteries are destined to. In principle we can say that the concentration of the solution must be such as to have a very good ion conductivity and to be neither too aggressive vis-à-vis separators, nor corrosive for some components of the cell, as the metallic grids supporting electrode materials.

The Open Circuit Voltage (OCV) for a lead–acid battery is a function of temperature and of the concentration of the electrode solution, as expressed in the Nernst equation:

$$E = 2.047 + \frac{RT}{F}\ln\frac{a_{H_2SO_4}}{a_{H_2O}} \tag{3.37}$$

where $E$ is the open circuit voltage (V) of the cell, $R$ is the gas constant (8,314 J/K mole), $T$ is temperature (K), F is the Faraday constant (96.490 C) and $a_i$ is the activity (concentration) of the mixture (mole/litre).

Since the electrolyte concentration varies during the discharge, the relative activity of $H_2SO_4$ in Nernst equation changes, and consequently the OCV of the battery changes. Since the change in the concentration of the solution leads to a proportional variation of its specific gravity, it might be possible to trace back the OCV value when the density of the electrolyte is known. However, this is not generally true for the different types of batteries.

The OCV is also more or less markedly influenced by temperature, and it can even shift from a direct to an inverse proportionality according to the concentration of the electrolyte. For instance, the OCV temperature coefficient $dE/dT$ is positive for concentrations of the solution exceeding 0.5 M, whereas below 0.5 M, the temperature coefficient is negative.

Normally, lead–acid batteries operate with concentrations exceeding 2.0 M (corresponding to a density of 1,120 g/ml) and with a temperature coefficient of nearly +0.25 mV/°C.

The self-discharge rate (loss of capacity, or state of charge, when an external charge is not applied) of the lead–acid cell is rather quick, but it can be significantly reduced through manufacturing devices. The self-discharge rate depends on different factors. Pb and $PbO_2$ are thermodynamically unstable in the sulphuric acid solution and, in an open circuit (accumulator not connected to the charge as for instance accumulators stored in a warehouse), they react with the electrolyte chemically and not electro-dynamically as it happens in the operation with the charge. Oxygen is developed at the positive electrode, whereas hydrogen is developed at the negative electrode, with a velocity that depends on the overvoltage relating to the formation of the two gases as well as on concentration (the quantity of gas that develops increases as concentration increases):

$$\text{Positive pole} \quad PbO_2 + H_2SO_4 \rightarrow PbSO_4 + H_2O + 1/2\ O_2$$

$$\text{Negative pole} \quad Pb + H_2SO_4 \rightarrow PbSO_4 + H_2$$

In the first of the two reactions, the formation of $PbSO_4$ through self-discharge is slow, normally below 0.5% per day at 25°C. As concerns the self-discharge taking place on the negative electrode (the second reaction), it is generally quicker, especially when the cell is contaminated by various metallic ions (for instance, the antimony released by the positive grid due to corrosion). In fact, it can spread towards the negative pole, where it settles, causing a local oxidation–reduction in which cation Sb reduces and Pb oxidises, converting into $PbSO_4$. Since $H_2SO_4$ reacts with electrodes forming (in both ones) $PbSO_4$, on the one side the quantity of active electrolytic material decreases (electrodes are "consumed") and on the other side the concentration of electrolyte and its density decrease. In some batteries, containing grids made with an alloy Pb–Sb, they lose nearly 1% of charge per day at 25°C, and the capacity loss increases when shifting to a factor 2 or 5 as the battery gets older. Batteries with low Sb content lose less than 0.5% of charge per day and do not get worse as they get old.

"*Sealed*" batteries (or without maintenance), and *charge retention* batteries, in which the self-discharge was minimised, use grids of low antimony or no Sb grids, as for instance the grids made with a Ca–Pb alloy. In any case, since there are other positive effects deriving from the use of alloys containing antimony (for instance mechanical stability), this element was not banned from grids, but a useful compromise was found by using low Sb alloys.

Self-discharge also depends on temperature, in particular it increases as T increases, but below –10°C it remains rather constant with temperature. Self-discharge can be reduced by keeping batteries in a range of temperatures between 5 and 15°C.

### *NiMH–NiCd Batteries*

NiMH batteries are an evolution of nickel–cadmium batteries (NiCd) and compared to the latter, they have the advantage of a higher energy density by 30–40% (Wh/kg or $Wh/dm^3$), and they do not have the disadvantages due to the use of heavy metals.

For NiCd, semi-reactions are:

Negative pole $Cd + 2OH^- \Leftrightarrow Cd(OH)_2 + 2e$

Positive pole $2NiOOH + 2H_2O + 2e^- \Leftrightarrow 2Ni(OH)_2 + 2OH$

Global reaction $Cd + 2NiOOH + 2H_2O \Leftrightarrow Cd(OH)_2 + 2Ni(OH)_2$

For a NiMH accumulator, the charge electrochemical process, at the negative pole, made up of a metal alloy of nickel and rare earth (for instance M = Nickel/Lanthanum alloy—$LaNi_5$), is conversely represented by the following reaction:

Negative pole $MH + OH^- \Leftrightarrow M + H_2O + e$

Positive pole $NiOOH + H_2O + e^- \Leftrightarrow Ni(OH)_2 + OH^-$

Global reaction $MH + NiOOH \Leftrightarrow M + Ni(OH)_2$

In particular, the metal alloys used are able to store and subsequently release a quantity of hydrogen one thousand times higher than their volume. The electrolyte used is a diluted solution of potassium hydroxide (KOH) to which other chemical compounds—although in lower quantities—are added to improve the battery performance.

The separator between anode and cathode, which prevents any electric contact between electrodes, although allowing a sufficient ion exchange, is made of a thin nylon film. The construction of an NiMH (nickel–metal hydride) element is very similar to NiCd elements (more precisely: elements with an alkaline electrolyte nickel–cadmium system). The main difference lays in the structure of the negative electrode. The metallic cadmium (in charge conditions) is replaced by a metallic alloy able to absorb big quantities of hydrogen, without increasing pressure. While the chemical processes at the positive electrode are the same for both types of batteries, with the negative electrode things change.

In a NiMH battery, hydrogen atoms are stored in a metallic grid where, during the charge, metal hydride is produced. Atoms move again outside of the grid during the discharge, leaving the basic metallic alloy alone. In NiCd elements, cadmium is conversely transformed into cadmium hydroxide during the discharge. In both types of batteries, an oversized negative electrode (compared to the positive electrode) prevents the damages to the element in case of total discharge or overcharge.

Figures 3.11 and 3.12 show how the performance of batteries changes as temperature and current change. In particular, as current intensity—expressed as C-rate[26]—increases, the value of the discharge potential sharply decreases, and even more markedly when temperature decreases. As regards capacity, temperature increase is generally positive, but this value shall not exceed the threshold beyond which processes as decomposition, evaporation, degradation, etc. that

[26] When current has the value of C (Ah), that is to say the value of capacity, the battery is discharged after 1 h. If the value of current is 5C, the battery is discharged in 1/5 of an hour (20 min), whereas finally if current amounts to C/10 the battery will be discharged in 10 ore.

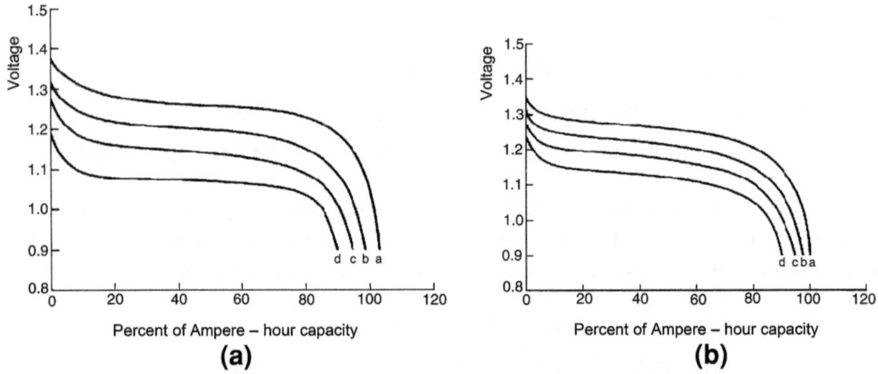


**Fig. 3.11** Discharge performance in Ni-MH cells at 20°C (**a**) and 45°C (**b**)

**Fig. 3.12** Charge and discharge characteristics of Ni–Cd cells

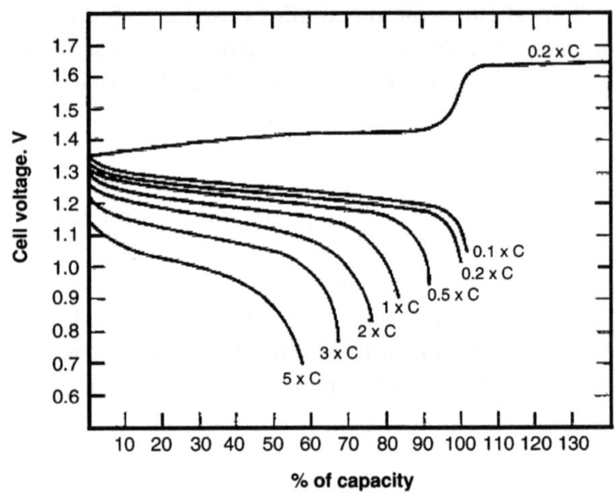


influence the electrochemical process itself are triggered. Remarks on the capacity are more evident in Fig. 3.14 concerning lithium–ion batteries.

*Lithium–Ion Batteries*

Lithium–ion batteries, put in the market in the 90s, have higher energy density, number of charge–discharge cycles and discharge performance than the other solutions described above, but also higher costs. The charge electrochemical process at the positive pole, made up of graphite, is represented by the following reaction (Fig. 3.13):

$$C + yLi^{+} + ye^{-} \Rightarrow CLi_{y}$$

during which lithium is inserted in the graphite layers, whereas at the negative pole, made up of oxide mixed with lithium and cobalt, the following electrochemical reaction takes place:

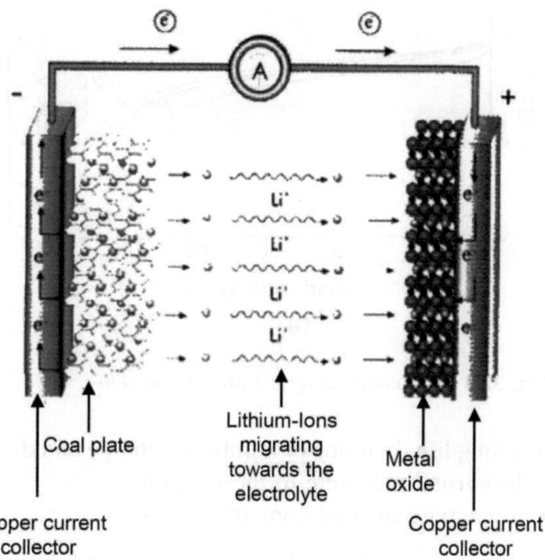


**Fig. 3.13** Scheme of an electrochemical process in a Lithium–Ion accumulator

$$LiCoO_2 \Rightarrow Li_{1-y}CoO_2 + yLi^+ + e^-$$

The cell global reaction therefore consists in the migration of lithium–ions between the two electrodes; there is no metallic lithium in the system. The discharge process obviously takes place the other way around, and the polarities of electrodes are inverted.

When manufactured, these batteries are completely discharged; it is therefore necessary to impose from the outside a suitable potential to transfer lithium from one compartment (Lithium Cobaltite—$LiCoO_2$) to the other (graphite) before using it. These devices can easily lose their characteristics when both the discharge potential and the charge potential are not suitably controlled.

This type of batteries can use two different families of electrolytes. The first one made up of non-aqueous solutions with a high dielectric constant (propylene carbonate, ethylene carbonate, dimethyl sulphoxide etc.) in which lithium salts are dissolved ($LiPF_6$, $LiBF_4$, $LiClO_4$ and $LiAsF_6$) and subsequently other organic compounds are added (tetrahydrofuran, diethyl carbonate etc.) in order to have low viscosity and increase the ion conductivity of solutions. The second is made up of lithium conductor polymers, obtained by immobilising lithium salt solutions in suitable gelled polymeric matrixes.

The lithium–ion accumulator shows different interesting characteristics for many applications:

- high voltage (3.6 V), amounting to nearly as much as the double of the voltage present in the clips of alkaline batteries (Fig. 3.14);
- high mass and volume specific energy (particularly important in mobile phones where the weight of the accumulator represents an important part of the total weight);

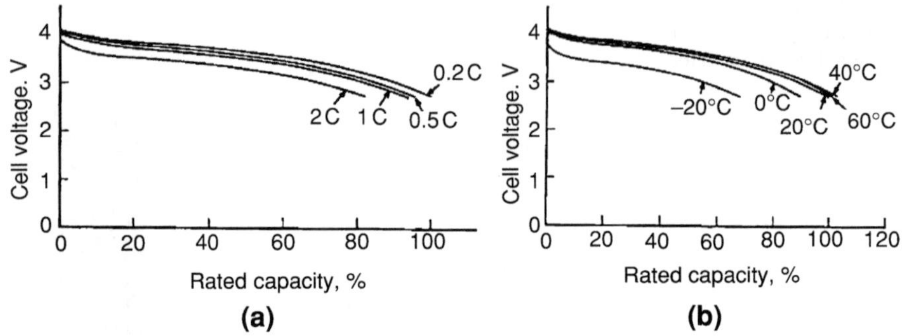


**Fig. 3.14** Discharge curve of Lithium–Ion cells

- a long life, in terms of number of charge and discharge cycles, with low capacity loss from one cycle to the next one;
- low environmental contamination.

Having considered that the capacity of lithium-ion accumulators is lower than the one obtained through negative electrodes in metallic lithium, researches were launched on *lithium accumulators* using solid polymeric electrolytes.

In these accumulators, the anode is made of a metallic lithium film that is applied on a thin layer of polymeric solid electrolyte, in turn with a metallic contact with the cathode (formed by transition metallic oxides, as magnesium, nickel or cobalt). The layer of polymer, with a very low ion conductivity, also acts as a separator between anode and cathode.

In these batteries, electrodes are made up of a special lithium mixture. The Li-Ion battery is activated when ions move between electrodes following the charge–discharge process. Thanks to the special lithium compound, the battery maintenance is easy.

The Li-Ion battery is characterised by a high performance, which makes it useful in a wide range of applications. These batteries can be charged over 1,000 times and are free from the so-called "memory effect".

Energy density is very high as well: almost two times as much as higher than the NiCd battery. Li-Ion batteries support considerably well overcharges.

The almost continuous charge that is necessary, for instance, for the battery of a cordless telephone shows not problems for a Li-Ion pack, whereas many NiCd packs do not survive more than a few months.

The loss due to self-discharge of these new batteries is nearly 50% lower than NiCd and NiMH batteries, and automatically entails a longer storage time. Li-Ion batteries, however, have a high voltage per cell (3.6 V), therefore they are not compatible in voltage with dry batteries or NiCd batteries. Their main application therefore takes place in battery packs, where each Li-Ion cell replaces three NiCd.

*Memory Effect and Lowering of the Discharge Potential*

If a battery is repeatedly charged before it is completely exhausted, it "forgets" it has further energy capacity in addition to the one provided until that moment. In other words, if starting from a completely charged battery, only 70% of its energy capacity is used and subsequently a recharge is made, the electrochemical device becomes unaware of 30% of energy the potential that becomes at that point unusable.

This phenomenon is generally detected in NiCd batteries only in some aerospace applications; it is possible to exclude that it takes place in whatever other land application, exception made for some extremely unusual circumstances.

This "real" memory effect is often confused with a more common and similar one called *discharging potential decrease*, which can easily and usually occur in NiCd and NiMH batteries.

In NiCd batteries, the discharging potential decrease is due to the increase in the size of cadmium crystals. The material forming electrolytes is made up of small-sized crystals. Until these crystals remain of a small size, electrochemical cells work appropriately; when the size increases, a sharp reduction in the superficial area of electrode materials occurs, with subsequent decrease in voltage and therefore in the performance of the electrochemical device.

In case crystals increase excessively, it is also possible that their edges penetrate through the separator and short-circuit the two electrodes. In these conditions, it is even possible to see a self-discharge of the battery. The effect of the growth in the size of crystals is more marked if the battery is charged for several days, or if it is repeatedly discharged in a non-complete way. In order to avoid this effect, it is necessary to cycle (charge and discharge) completely the battery at least once every 2 or 3 weeks.

The same phenomenon of growth (self-growth) of the size of crystals takes place although the battery is not used for a long period of time.

In this case, to recover the initial characteristics it is necessary to have a very low and deep discharge process able to completely remove the energy remained in the electrochemical cell.

In NiMH batteries, the lowering of the discharging potential originates following an overcharge process that modifies the crystalline structure of the nickel hydroxide from beta gamma shape; the latter has an electrode potential nearly 50 mV lower than beta. Also in this type of batteries the phenomenon can be removed through a process of complete charge and discharge of the electrochemical device.

The most recent technology seems to have virtually eliminated this problem; some manufacturers presently state the absence of any type of "memory" effect in NiMH batteries. Lithium batteries do not have the inconvenient of the lowering of discharging potential since during the charge and discharge process (intercalation and deintercalation of lithium ions) there is no change neither in the size of grains nor in the crystalline structure of electrode materials.

**Table 3.9** Characteristics of the different types of accumulators

| Characteristics | Lead | NiCd | NiMH | Li-Ion |
|---|---|---|---|---|
| Energy density | Medium | Medium | Very high | Very high |
| Cyclic behaviour | Fair | Excellent | Good | Excellent |
| Self-discharge | Low | Low | Low | Very low |
| Quick charge | Fair | Excellent | Good | Fair |
| High-current charge | Good | Excellent | Good | Fair |
| Reliability | Excellent | Good | Good | Fair |
| Costs | Very low | Low | High | High |
| Compatibility in voltage | Fair | Excellent | Excellent | Scarce |
| Environmental impact | High | Very high | Very low | Basso |
| Discharge voltage stability | Fair | Excellent | Excellent | Fair |

### Comparison Between Possible Solutions For Electrochemical Accumulators

On the basis of the above, it is now possible to make a comparison between the performance expected for the main types of accumulators (Table 3.9)

The rating of a battery includes:

- Mechanical characteristics
  - Size
  - Material of the container
  - Type of electrodes
  - Metal alloy
  - Vents
  - Gaskets
  - Handles
  - Cable terminations
  - Electrolyte
- Electric features
  - Nominal voltage $U_n$ (V)
  - Capacity: $C_n$ (Ah)
  - Nominal current: $I_n$ [1 A (100 h of current)]
  - Charge voltage: $V_{max}$ (V a 25°C)
  - Final charge voltage: $V_{min}$ (V a 25°C)
  - Limit temperature (range of operation in °C)
  - Recommended temperature (Range of operation in °C)
  - Self-discharge (in % per month)
  - Minimum duration: (number of years—depends on temperature, on the discharge cycles, and their intensity)
  - Discharge allowed (in % of $C_n$)
  - Maintenance (number of times per year—i.e.: filling-up)

**Table 3.10** Characteristics of the different types of accumulators in numbers

| Characteristics in numbers | Lead | NiCd | Ni-MH | Li-ion |
|---|---|---|---|---|
| Discharge potential (V) | 2.1 | 1.2–1.0 | 1.2–1.0 | 3.7–3.5 |
| Specific energy (Wh/kg) | 30–40 | 60–85 | 80–100 | 90–120 |
| Energy density (Wh/l) | 75–100 | 140–180 | 240–300 | 300–380 |
| Cost per cycle ($/cycle) | 0.25 | 0.06 | 0.1 | 0.08–0.05 |
| Cycles | 500 | 1,000 | 450 | 600–1000 |
| Self-discharge (%/month) | 21 | 15 | 20 | 12 |
| Power density (W/l) | 210 | 1,000 | 800 | 500 |
| Temperature interval | −15, +40 | −40, +45 | −40, +45 | −20, +50 |

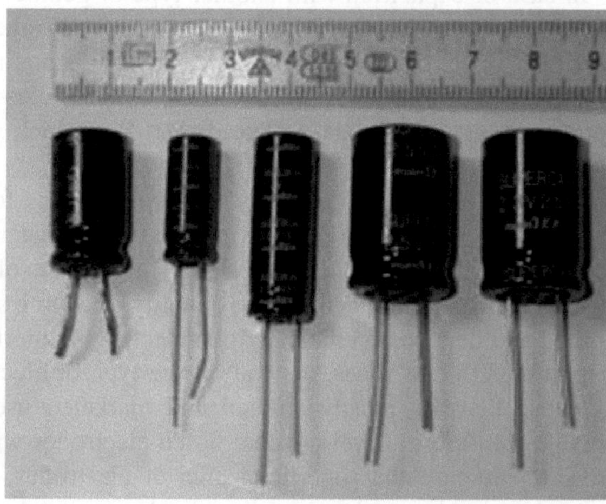


**Fig. 3.15** Supercapacitors

Tables 3.10 and 3.12 show how lithium batteries overall have the best performance. As energy accumulators, they have the least self-discharge, nearly 12% per month although, in absolute terms, this value always remains rather high. As energy generators, they show the best performance, having a value of working potential tripled compared to Ni–Cd and NiMH batteries, and nearly doubled compared to lead accumulators. Their specific energy and energy density are the highest ones, therefore as energy vector they own by far the best performance. The number of cycles for lithium batteries is high, thence although they are more expensive compared to the other batteries, the cost per cycle is low and this is what matters from the economic viewpoint. From a general perspective it is therefore possible to understand the market success of this type of batteries.

## Storage with Supercapacitors

Supercapacitors (Fig. 3.15) store electric power in two supercapacitors in series with an—eletrochemical double layer (EDL), without the chemical oxido-reduction processes that occur in accumulators (rechargeable batteries).

Supercapacitors are therefore "*electrophysical*" accumulators of electricity. The simplest supercapacitor is made of two polarisable electrodes,[27] a separator and one electrolyte; the electric field is stored at the interface between electrodes and the electrolyte.

The capacity $C$ of a capacitor (F) with flat and parallel plates depends on the surface of plates S ($m^2$), their distance d [m] and on the dielectric constant of the dielectric ε (F/m) according to the following relation:

$$C = \varepsilon \cdot \frac{S}{d} \tag{3.38}$$

In case of capacitors with another type of geometry, the capacity of $S$, $d$ and ε has another form, but the functional dependency is the same. Then, in general, it is possible to state that to increase the capacity it is possible to act both on the surface of electrodes—increasing it as much as possible—and on their distance—by minimising it—and finally on the dielectric—which needs to have a high relative *dielectric constant* $\varepsilon_r$.[28]

This type of energy storage system is interesting for its high power density and its durability; furthermore, energy storage is easier and more reversible when compared to conventional batteries. The reverse of the medal of the physical process resides in the fact that the quantity of charge storable in a supercapacitor is limited and depends on the electrode/electrolyte interface surface. Supercapacitors can be of different types, depend on the type of electrode or electrolyte.

Those that are mostly studied and marketed use electrolytes in aqueous or organic solutions, as well as coal-based electrodes with a high superficial area; in order to increase the superficial area of electrodes, materials containing carbon nanotubes are being developed, and other researches target to obtain electrodes made up of nanostructured carbon films. The presence of vacuums and furrows in the grains of this film suggests the high porosity and the low density of the material.

The high porosity of nanostructured carbon films stored in this way allows the active surface available (1.400 $m^2$/g) to reach the following values:

- specific capacity 75 F/g:
- maximum energy density 76 Wh/kg;

---

[27] An *ideally polarisable electrode* is an electrode with an infinite polarisation resistance and, at the same time, the curve in diagram i/V is a horizontal straight line. Conversely, for an electrode ideally not polarisable, the polarisation resistance is null, and the line in diagram i/V is vertical. In the former case the potential can be varied without having any current leaking in the circuit; in the latter case, the potential cannot be shifted from the balance value for the very high current values (infinite) that would be produced. In practice, polarisation resistance takes finished values that can range from fractions of Ω to several MΩ.

[28] The relevant dielectric constant indicates the number of times by which the electric field is reduced between the plates of the condenser compared to the case of vacuum. For instance, water is used ($\varepsilon_r = 80$) as dielectric, between the plates there will be an electric field 80 times lower than in the case of vacuum.

- maximum power density 506 kW/kg.

Compared to electrochemical batteries, double layer capacitors are not subject to wear: they support over 500,000 charge/discharge cycles with a minimum lifetime of 10 years, without any change in the capacity due to time.

Particularly important is their capacity to be charged and discharged at very high current. For this reason, they represent a particularly attractive storage solution for the so-called "regenerative braking" in hybrid-electric and battery electric vehicles and for all those situations in which the charge and discharge at high quantity of energy have to take place very quickly. In case of *energy recovery braking*, or *regenerative braking*, the kinetic energy of vehicles is converted into electric power, which is quickly stored and released when needed, in order to be used again during the subsequent acceleration.

In urban traffic in particular, whose driving cycles are characterised by continuous accelerations and breakings, it is possible to recover in this way up to 25% of the energy involved.

Similar saving potentials relate to railway local traffic, through which it is possible to store the energy during braking in order to release it at the subsequent restart. In considering the several stops and starts of these means of transport, the energy that can be recovered is easily understandable. Appropriate systems are already available in the market for these applications.

The most interesting solution among those used in the sector of mobility concerns the devices made up of solid components, using—instead of the traditional liquid electrolytes—a polymeric solid electrolyte. In the medium term, this device is expected to offer improved performance and higher safety and duration than conventional systems.

Some applications for supercapacitors include:

- *backup power supply*, household appliances (pocket calculators, cameras, videocameras) to protect clocks, programme memories or functions while electrical appliances are switched off. For instance: change of batteries, power output etc;
- *main power*, power button of household appliances, charging systems for household appliances;
- *alternative power*, for combinations with solar cells, they are charged during the periods of insulation and used as power pack during the night or in non-sunny days.

### *Storage with Hydroelectric Pumping Plants*

An indirect system that is very effective for the indirect storage of electric power is represented by pumping plants.

In this case, several conversions apply; the electric power is converted into mechanical energy through an electric motor and, in turn, into gravitational potential energy through a pump that brings water from the after-bay reservoir located at a certain height to a water basin at a higher height, as shown in

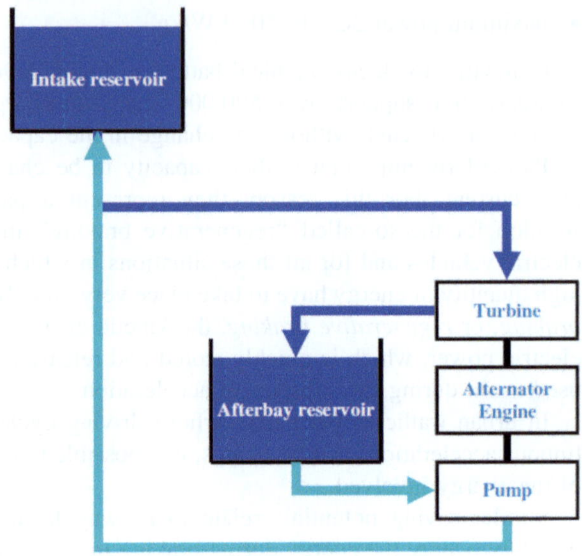


**Fig. 3.16** Scheme of a pumping plant

Fig. 3.16. In the opposite process, the potential gravitational power of water is converted, within pipelines, into kinetic energy and then into mechanical energy by a hydraulic turbine; this energy, in turn, generates electric power thanks to an electric generator.

The overall efficiency[29] of the process amounts on average to 75%.

### Forms of Thermal and Magnetic Storage

Very encouraging results are recorded in this field. As regards the former, successful tests were made with high-density polyethylene (HDPE) in the range of temperatures between 100 and 150°C, whereas as to the latter, the threshold of superconductivity operation was increased, hence allowing to hope in a future electric power storage in an electromagnetic field.

### Forms of Kinetic Storage

Energy storage in this form is, on the contrary, a relatively ancient technology, which is also widely used in combustion engines. In order to level out the diagram of consumption, the installation of flywheels that can be mechanically or inductively charged and discharged, is being increasingly considered when designing energy systems.

---

[29] Namely the ratio between the electric power produced by the potential gravitational energy of a given quantity of water and the electric power spent for its storage.

The technical lifecycle expected is about 20–30 years, with a low level of maintenance, and the duration and density of electricity storage do not substantially depend on charge and discharge trends. This makes the flywheel particularly interesting for those applications in which there is a slow charge and rapid discharge cycle, or viceversa.

In consideration of these positive elements, several problems emerged, linked to the dynamic balancing of rotating blades, to the lubrication of supporting bearing and to the minimisation of power losses. A possible solution is represented by magnetic-rotor suspensions, although their management is still very complex. Different materials were tested, ranging from kevlar to nylon and to steel, but the present trend goes towards composite materials due to their higher potential energy density.

### 3.5.5 Hydrogen as Energy Vector

Hydrogen can be transported and stored in a gaseous, liquid or solid state, and in particular it can be absorbed on special materials. Each form shows advantages and disadvantages, and all of them, although mostly already used, demand significant efforts in terms of research and development in order to be entirely reliable and competitive from the economic and commercial viewpoint.

In case of transportation of gaseous hydrogen, it is possible to develop a system similar to methane pipelines; some problems relating to the materials used have to be solved, as well as the need for a higher compression energy (the lower heating value in volume of hydrogen compared to methane[30] demands a higher range for the same quantity of energy, given the same operating pressure). The transportation in the liquid form shows more complex problems and, perspectively, it seems to be convenient only for large quantities and high mileage.

Hydrogen distribution to users, in case of a more extensive use (i.e., in the transportation sector), entails the problem of a suitable distribution network infrastructure as well as the need for huge investments necessary for its implementation. Storage technologies depend on the applications considered and are decisive especially for the use onboard of vehicles, where a high energy density is necessary. The present solutions (cylinders, hydrides, liquid hydrogen) might not be able to ensure a large-scale distribution in the medium-long term. Their further development, or completely new solutions, are therefore necessary, and are the object of several research and development programs all over the world.

---

[30] Hydrogen: 10,692 kJ/Nm$^3$; Methane: 34,535 kJ/Nm$^3$.

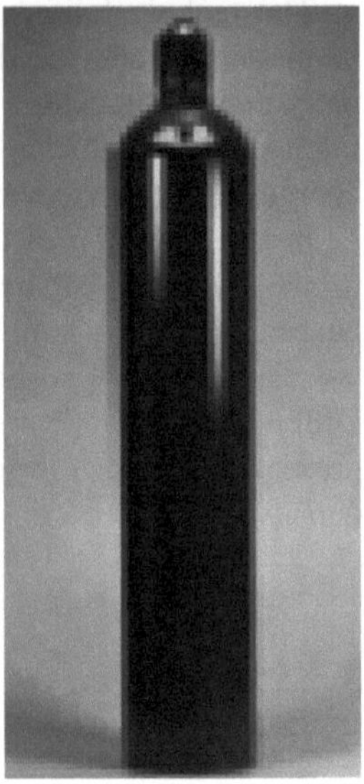

**Fig. 3.17** Standard cylinder of compressed hydrogen at 200 bar (50 l)

### 3.5.5.1 Energy Storage with Hydrogen

The main hydrogen storage systems include:

- compressed gaseous hydrogen;
- liquid hydrogen;
- metal hydrides;
- carbon nanostructures;
- crystal microspheres;
- ammonia.

#### Storage of Compressed Hydrogen

New materials allowed the manufacturing of storage tanks able to keep hydrogen at extremely high pressures, therefore decreasing weight and size (Fig. 3.17).

These aspects are fundamental for the applications in the sector of transports, since they can be a cause of severe limitations in terms of autonomy and loading capacity. The introduction of tanks with metallic or thermoplastic liners, reinforced with carbon, glass and aramid fibres, reduced the weight by 3–4 times compared to common tanks, and allowed therefore to partly exceed the restrictions linked to the use of traditional cylinders. These tanks are able to operate at pressures up to 700 bar, therefore allowing to obtain energy storage densities suitable to the use onboard of vehicles.

The compression work is given by:

$$L = \frac{1}{k-1}\frac{p_1}{\rho_1}\left(\beta^{\frac{k-1}{k}} - 1\right) = \Delta H = c_p \cdot (T_2 - T_1) \tag{3.39}$$

where $k$ is the ratio between the specific heat at constant pressure ($c_p$) and the specific heat at constant volume ($c_v$) and $\beta$ is the compression ratio

$$\beta = \frac{p_2}{p_1} \tag{3.40}$$

As already seen for other applications, in case of high compression ratios as with hydrogen compression for storage purposes, compression takes place in several interrefrigerated stages.[31] Indicating with $n$ the number of stages, the compression ratio of each single stage[32] $\beta_s$ is:

$$\beta_s = \sqrt{[n]}\beta \tag{3.41}$$

By proceeding similarly to natural gas for the calculation of $c_p$, it is possible to calculate k:

$$k = \frac{c_p}{c_v} = \frac{c_p}{c_p - R} \tag{3.42}$$

In considering isentropic compression, the following is obtained:

$$\frac{T_2}{T_1} = \left(\frac{p_2}{p_1}\right)^{k-1/k} = \beta_s^{k-1/k} \Rightarrow T_2 = T_1 \cdot \beta_s^{k-1/k} \tag{3.43}$$

At this point, the calculation of the enthalpy reaction is made (of the single stage) in the ideal case, considering isentropic compression (reversible adiabatic transformation):

$$\Delta H_{s,i} = c_p \cdot (T_2 - T_1) \tag{3.44}$$

It is possible to take irreversibility into account by considering a *thermodynamic compression efficiency* $\eta_c$ defined as follows:

$$\eta_c = \frac{\Delta H_{s,i}}{\Delta H} \Rightarrow \Delta H_s = \frac{\Delta H_{s,i}}{\eta_c} \tag{3.45}$$

By assuming, as it is allowed in first approximation calculations, that the input temperature (and therefore also the output temperature) is the same for all the stages, it clearly follows that also the enthalpy reaction is the same for all stages. Therefore, the total enthalpy reaction is

---

[31] Between the exit from one stage and the access to the next one, the fluid is cooled at constant temperature. In this way, the total compression work decreases.

[32] Considering the compression ratio as equal for the different stages.

**Table 3.11** Hydrogen compression work

| Storage pressure (bar) | Work for different production pressures in % of the energy content stored | | |
|---|---|---|---|
| | 5 (%) | 10 (%) | 20 (%) |
| 200 | 5.46 | 4.31 | 3.22 |
| 350 | 6.43 | 5.24 | 4.10 |
| 700 | 7.68 | 6.43 | 5.24 |

$$\Delta H = \Delta H_s \cdot n_s \tag{3.46}$$

Considering an efficiency of electric motors $\eta_{\mathrm{ME}}$ activating compressors, it is possible to calculate the work $L$ for hydrogen compression:

$$L = \frac{\Delta H}{\eta_{\mathrm{ME}}} \tag{3.47}$$

It is readily evident how the compression work depends, besides on the storage pressure and the efficiency of the conversion process, also on the pressure at which hydrogen is produced. For instance, many electrolysers available in the market produce hydrogen at a pressure ranging between 5 and 20 bar; reformers produce hydrogen at pressures that range between 8 and 30 bar according to their size. Table 3.11 shows the hydrogen compression work for different values of the storage pressure and of the pressure at which hydrogen is produced, considering a compression in three stages, a compression thermodynamic efficiency of 87% and an efficiency of electric motors of 93%.

Other relevant characteristics for storage systems include the following:

- *storage capacity*, that is the ratio (expressed as a percentage) between the mass of the stored vector and the total mass of the storage system (including the vector);
- *energy density in volume*, that is the energy stored per volume unit of the storage system;
- *energy density in mass*, that is the energy stored per mass unit of the storage system (including the vector).

As an example, let's consider the cylinder in Fig. 3.17 that is able to contain 50 l of hydrogen at the pressure of 200 bar. Its unladen weight amounts to 44 kg, it has an external diameter of 229.5 mm and a length of 1,435 mm.[33] Hydrogen density at 200 bar at a room temperature (25°C) is 14.05[34] kg/m$^3$, therefore the

[33] Blugas: http://www.blugas.com/prodotti-bombole-acciaio.php

[34] Calculated through the Van der Waals equation. At high pressures, since the effects of interactions between the gas particles and the volume of particles themselves are not negligible (being both considered as null in the ideal gas law) the ideal gas law is not applicable, since it originates significant errors (hydrogen density at 200 bar and 25°C calculated with the ideal gas law is 16.16 kg/m$^3$).

**Table 3.12** Storage capacity of a hydrogen cylinder compressed at 200 bar

| Hydrogen cylinder compressed at 200 bar | | |
|---|---|---|
| Storage capacity (% in weight) | 1.5 | % |
| Energy density in volume | 1,419 | kJ/litre |
| Energy density in mass | 1,884 | kJ/kg |

cylinder contains 0.702 kg of hydrogen, corresponding to an energy content of 84,240 kJ. The characteristics of said cylinder are summed up in Table 3.12.

### Storage of Liquid Hydrogen

By condensing gaseous hydrogen in a liquid form, through cryogenic techniques and the reaching of low and very low temperatures (in the case of hydrogen −253°C), it is possible to obtain storage characteristics far better than in the case of compressed hydrogen. At the liquid state hydrogen has a density amounting to 0.071 kg/l (namely 71 kg/m$^3$, against 14.66 and 23.60 kg/m$^3$ of compressed hydrogen at 200 and 350 bar, respectively) and therefore a lower heating value of 8,515 kJ/l. Comparing it to other liquid fuels, is to be considered that the lower heating value of petrol amounts to nearly 31,700 kJ/l.

Liquefaction processes use a combination of compressors, heat exchangers, expansion motors and butterfly valves to obtain the cooling desired. The easiest liquefaction process is the *Linde process* or the *Joule–Thompson expansion process*. Through this process, the gas is compressed at room temperature and then cooled in a heat exchanger before passing through a valve in which it undergoes the Joule–Thompson expansion process, producing a liquid. Once the liquid is removed, the gas goes back to the compressor passing through the exchanger, as indicated in Fig. 3.18.

These processes are based on temperature variation in the "free" expansion (meaning without exchange of heat and work, thence $\Delta H = 0$) of real gases[35]: this variation is expressed through the *Joule–Thompsoncoefficient*, a function of pressure and temperature, defined as the change of temperature compared to the change of pressure in a Joule–Thompson process:

$$\mu_{T,p} = \left(\frac{\partial T}{\partial p}\right)_H = \frac{(2 \cdot a/R \cdot T) - b}{c_p} \tag{3.48}$$

where $a$ and $b$ are the Van der Waals constants, R the universal gas constant and $c_p$ the molar specific heat at constant pressure. The temperature variation corresponding to a given pressure variation is

$$\Delta T = \mu \cdot \Delta p \tag{3.49}$$

[35] In ideal gases, in a free expansion, temperature remains constant.

**Fig. 3.18** The Linde process

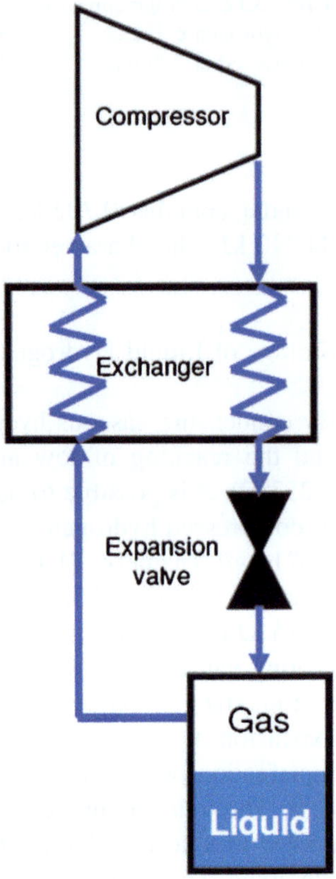


Since in the expansion $\Delta p < 0$, it is evident that in the expansion itself the gas cools if $\mu > 0$ that is to say if the cycle operates at temperatures lower than the so-called *inversion temperature*[36] of the cooled gas.

Different gases have an inversion temperature higher than room temperature (nitrogen, oxygen); others, among which hydrogen, have much lower inversion temperatures.

The inversion temperature is obtained, by definition, by considering $\mu = 0$. From Eq. (3.48) it follows:

$$\mu_{T,p} = \frac{(2 \cdot a/R \cdot T) - b}{c_p} = 0 \Leftrightarrow \frac{2 \cdot a}{R \cdot T_i} - b = 0 \Rightarrow T = \frac{2 \cdot a}{R \cdot b} \tag{3.50}$$

It is evident how hydrogen, having an inversion temperature much lower than room temperature, cannot be liquefied by using the Linde process, or better, the Linde process can be used to liquefy hydrogen only after it was cooled at

[36] For all gases a temperature exists, called *reversal temperature*, below which the Joule–Thomson coefficient is positive, whereas in case of higher temperatures it is negative.

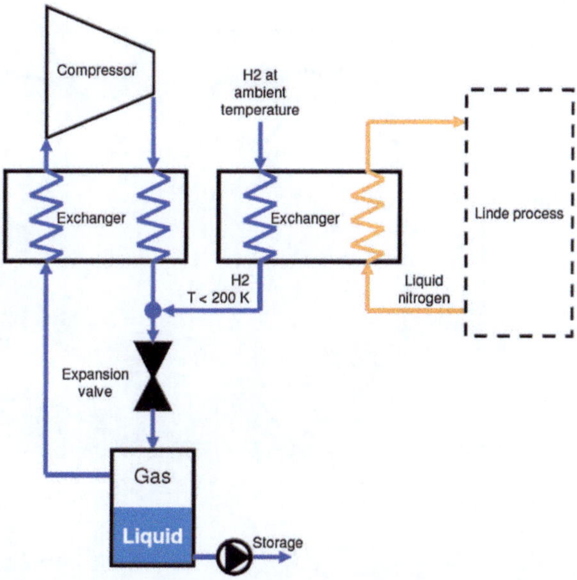


**Fig. 3.19** Outline of the hydrogen liquefaction cycle

temperatures lower than the inversion temperature. The hydrogen liquefaction process includes, then, two different phases:

- cooling in a heat exchanger up to a temperature lower than the inversion temperature;
- liquefaction through the Linde process.

This process is schematically shown in Fig. 3.19. Hydrogen is cooled with liquid nitrogen[37] (which can be obtained through the Linde process) at a temperature of nearly 78 K before passing through the expansion valve. Nitrogen is subsequently retrieved and recycled in the continuous refrigeration cycle.

The difference of free energy ΔG (therefore, the minimum theoretical liquefaction work), between gaseous hydrogen at room temperature (300 K) and liquid hydrogen at 20 K amounts to 11,640 kJ/kg (9.7% of hydrogen energy content) [2]. The real energy consumption largely varies according to the size of the liquefaction plant: it ranges from nearly 50 MJ/kg (over 40% of the liquefied hydrogen energy content, with an efficiency[38] of the system amounting to about 23%) for medium-size plants (capacity of 200 kg/h of liquid hydrogen) to nearly 30 MJ/kg (amounting to about 25% of the liquefied hydrogen energy content, with an efficiency close to 39%) for large-size plants (capacity of 12,000 kg/h of liquid hydrogen) [3].

In order to keep these temperatures, double-walled tanks were fine-tuned, with a cavity in which vacuum is created ("dewar" tanks). This technology is

[37] Other gases as helium can be used as well.

[38] Considered as the ratio between the minimum liquefaction work of hydrogen ($\Delta G$ between gaseous hydrogen at 300 K and liquid hydrogen at 20 K) and the energy actually spent.

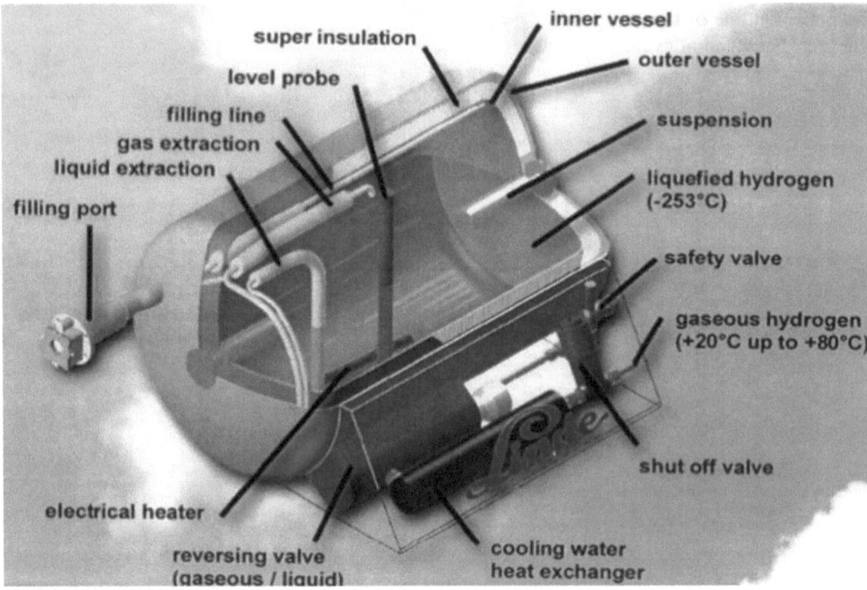


**Fig. 3.20** Liquid hydrogen tank by BMW 750hL—Linde

**Table 3.13** Storage capacity of a liquid hydrogen tank

| Liquid hydrogen tank | | |
|---|---|---|
| Storage capacity (% in weight) | 6.6 | % |
| Energy density in volume | 5,900 | kJ/litre |
| Energy density in mass | 8,868 | kJ/kg |

well-established in Germany, where BMW has been using it for over 15 years in hydrogen cars with internal combustion engines. The tank shown in (Fig. 3.20) can contain nearly 8 kg of liquid hydrogen (957.600 kJ) and its unladen weight is about 100 kg. Table 3.13 shows the storage capacity of this tank.

The tank is made up of a double-walled structure made of two stainless steel plates with a 2 mm thickness, and one vacuum super-insulating layer (thermal conductivity of $10^{-4}$ W/m K) of 30 mm thickness, which separates the external tank from the internal one. The insulation system is used to keep liquid hydrogen at a pressure of 3–5 bar and at a constant temperature of nearly –250°C in the long term. The period expected for the complete emptying of a half-full hydrogen tank is nearly nine days. But also in this phase, the car is still able to cover about 20 km (12 miles) with the remaining hydrogen stored in the tank. The accumulation in the liquid form is perhaps the technology that apparently better meets the automotive needs although showing some limits. Among the cons of liquid hydrogen there is the higher complexity of the system, not only onboard the vehicle, but also in terms of distribution and refuelling, and the higher related costs. Also the energy cost of liquefaction is considerable, since it amounts to about 30% of the fuel energy content, against a value between 4 and 7% for compressed hydrogen.

For hydrogen conservation at the liquid state, the phenomena linked to electronic spins[39] of hydrogen atoms must be considered. Hydrogen molecules are made up of two electrons and two protons; the combination between the two electronic spins leads to the formation of a molecular bond only in case of anti-parallel spins. This entails the existence of two different types of hydrogen molecules, according to their total nuclear spin:

- $I = 0$ with *anti-parallel nuclear spin*;
- $I = 1$ with *parallel nuclear spin*.

When $I = 0$ hydrogen is indicated with "*para*", whereas when $I = 1$ it is indicated with "*ortho*".

At room temperature (300 K), 25% of hydrogen is in the *para* form, whereas 75% in the *ortho* form. By cooling hydrogen, the concentration of ortho-hydrogen varies, shifting from 75% (at room temperature) to 50% at 77 K, up to 0.2% at the liquefaction temperature (21 K). The conversion reaction from *ortho* to *para* is exothermic and the conversion heat depends on temperature: at 300 K the conversion heat amounts to 270 kJ/kg and increases at 519 kJ/kg at 77 K. In case of lower temperatures, the conversion enthalpy remains rather constant and amounting to 523 kJ/kg. This value is higher than the latent heat of hydrogen vaporization (451.9 kJ/kg) at 21 K; therefore, in storing non-converted liquid hydrogen from *ortho* to *para*, the conversion enthalpy causes hydrogen evaporation. It is therefore necessary to preventively convert *ortho* hydrogen into *para*[40] [2]: this conversion can be catalysed by active surfaces (activated charcoal) and by paramagnetic species (chromium oxide and gadolinium); in this way, nuclear spins can be converted without causing the breaking of the H–H bond.

### Hydrogen Storage in Metal Hydrides

Different pure or bonded metals can be combined with hydrogen, producing *metal hydrides*. These compounds are able to trap hydrogen at relatively low temperatures. Hydrogen penetrates the crystal lattice of the metal, and occupies the interstitial spaces.

Hydrogen can be stored in the form of hydride, at low pressures, with a higher density compared to simple compression, similar (according to literature data, even higher) to liquid hydrogen. The use of this storage system, safe and efficient,

[39] Electrons own an intrinsic angular phase, associated to a magnetic phase. This property is defined as *electronicspin* and can have two different stages, characterised by a magnetic quantum number of spin, $m_s$, which can have two values only: +1/2 or −1/2.

[40] Theoretically speaking, the conversion process from *ortho* to *para* is spontaneous, but very slow: the time to halve the conversion is over one year at 77 K. This means that, considering the cooling times, this conversion de facto does not take place. Furthermore, from the energy viewpoint, it is convenient to carry out the conversion at high temperatures since the conversion enthalpy increases as temperature decreases.

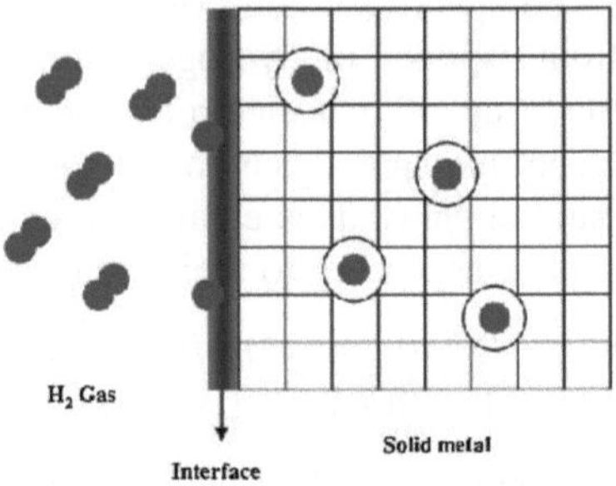


**Fig. 3.21** Hydrogen storage in metal hydrides

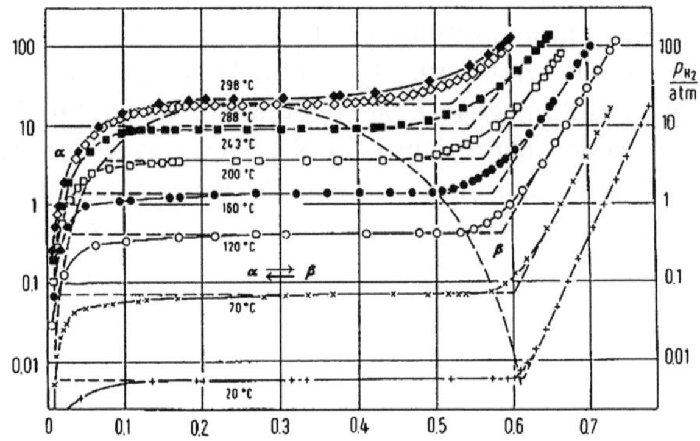


**Fig. 3.22** Palladium isotherm at different temperatures

is linked to the identification of a metal with sufficient absorption capacity operating at appropriate temperature intervals.

Potentially, the storage volume is reduced by 3–4 times compared to compression, hence allowing the use of these systems in cars, whereas specific energy also depends on the specific gravity of the base metal. The percentages in weight of hydrogen, out of the total weight, range between 1 and 12.7% (LiH). By way of comparison, in case of normal cylinders, this percentage is slightly higher than 1%, and therefore these storage systems are potentially very promising.

A weak point of this technology is represented by the total weight of these storage systems. In the sector of transports, giving the same weight, the vehicle has an autonomy three times lower than the one obtainable with liquid or compressed hydrogen with advanced tanks. There are, on the contrary, undoubted advantages in terms of convenience, compactness, storage stability and intrinsic safety.

The process of hydrogen storage in hydrides is schematically shown in Fig. 3.21: molecular hydrogen dissociates on the surface of the metal and therefore migrates (in an atomic form) to interatomic spaces. It is necessary to supply a certain quantity of energy to dissociate the molecule, but the reaction is globally exothermic.

Hydrogen absorption and release take place by varying temperature and/or pressure. Figure 3.22 shows the absorption of hydrogen in palladium hydride according to the pressure at different temperatures. The ideal situation occurs when hydrogen absorption and release take place at room temperature, driven "only" by pressure.

**Table 3.14** Characteristics of simple metal hydrides

| Metal | Hydride | Capacity (% weight) | T (°C) (@1 bar $H_2$) | PROS | CONS |
|---|---|---|---|---|---|
| Palladium | $PdH_{0,77}$ | 0.72 | 147 | Low temperatures and working pressure | High cost |
| Titanium | $TiH_{1.97}$ | 3.98 | 643 | Low cost | High working pressure |
| Vanadium | $VH_2$ | 3.81 | 12 | Low temperatures and working pressure | High cost |

**Table 3.15** Characteristics of some binary hydrides of light metals

| Hydride | Capacity (% weight) | T(°C) (@1 bar $H_2$) | PROS | CONS |
|---|---|---|---|---|
| LiH | 12.7 | 894.000 | High capacity | High working temperature |
| NaH | 4.2 | 421.000 | | High working temperature |
| $MgH_2$ | 7.6 | 287.000 | Low cost | Absorption and release kinetics |
| $CaH_2$ | 4.8 | 1.074 | | High working temperature |

The thermodynamic aspects of the formation of hydride from gaseous hydrogen are described by the LHV (*Pressure-composition isotherms*) curves of the type shown in Fig. 3.22. They are obtained as follows. At a given temperature and in the presence of low pressure hydrogen, the host metal initially dissolves hydrogen in solid solution (phase alpha), after the dissociation of molecule $H_2$ into atomic hydrogen at the surface of the material. As pressure increases, the concentration of hydrogen in the metal undergoes slight increase, and consequently the local interactions between hydrogen atoms and the atoms of the host lattice become increasingly important, until the nucleation process and the growth of the hydride beta phase begin. Until the solid solution and the hydride coexist, isotherms (hydrogen pressure at a given temperature according to hydrogen concentration in the material) show a plateau, whose length indicates the quantity of hydrogen that can be reversibly stored at that temperature with slight pressure variations. When the alpha $\Leftrightarrow$ beta transition is completed, the hydrogen pressure suddenly increases with the concentration. The two-phase region ends in a critical point $T_c$ above which there is a continuous transition between phase alpha and beta. The equilibrium pressure (position of the plateau) strongly depends on temperature, and is connected to enthalpy reactions $\Delta H$ and entropy variation $\Delta S$ according to the following Van't Hoff equation:

$$\ln(P_{\mathrm{eq}}) = \frac{\Delta H}{RT} - \frac{\Delta S}{R} \tag{3.51}$$

Different types of hydrides can be used for hydrogen storage:

- simple metal hydrides ($MH_x$);
- complex metal hydrides;
- light metal hydrides.

Table 3.16 Characteristics of some complex hydrides of light metals

| Hydride | Theoretical capacity (% weight) | Release Temperature (°C) | | Capacity (% weight) at minimum temperature |
|---|---|---|---|---|
| | | $T_{min}$ | $T_{max}$ | |
| $LiAlH_4$ | 10.6 | 100 | 150 | 6.40 |
| $NaAlH_4$ | 7.5 | 180 | 250 | 4.30 |
| $LiBH_4$ | 18.3 | 75 | 330 | 0.46 |
| $NaBH_4$ | 10.7 | 240 | 400 | 0.69 |

Table 3.17 Fundamental classes of hydrides for hydrogen storage

| Hydrides for hydrogen storage | |
|---|---|
| AB($LaNi_5H_{6.5}$,$LaNi_{4.7}Al_{0.3}$ $H_{6.5}$) | Low capacity (max 2%/w), good cycle |
| AB($FeTiH_2$) | Low capacity (max 1.9%/w), loss in cycles |
| AB2 ($ZrV_2H_{5.5}$) | Low capacity (max 3%/w), need to activate |
| A2B ($Mg_2NiH_4$) | 3.6%w, slow kinetics, high absorption *T*, need to activate |
| $MgH_2$ | 7.6%w, slow kinetics, high absorption *T*, need to activate |
| Lamellar phases | Ca Al X (X = Si), 5%w, not reversible. |
| Complex hydrides | High capacity, little reversible |

Table 3.14 shows the main characteristics of some simple metal hydrides.

Compared to simple metal hydrides, complex hydrides show a similar performance (or better performance) than palladium in hydrogen absorption and release; lower costs, but a good storage capacity. Table 3.15 shows the absorption features of some binary hydrides of light metals, whereas Table 3.16 shows the main features of some complex hydrides of light metals; they have high theoretical storage capacity, and the main problem with their use is represented by a difficult rehydrogenation.

Hydride technology (Table 3.17) is under study in the US, in Europe and in Japan, where automotive manifacturers fine-tuned special titanium alloys in the framework of their programme envisaging the introduction of fuel cell vehicles.

Interesting results were obtained with magnesium alloy, light, and with a high hydrogen storage capacity (7% in weight, Tdesorb: 300°C).

As concerns the storage capacity, the Series HBOND 1,500 tank can be considered as an example; its main features are shown in Table 3.18, from which it is possible to easily calculate the storage characteristics of this tank (Table 3.19).

As regards energy for storage purposes, please refer to the enthalpy of hydride formation: in the case in point, at 21°C.[41]

This enthalpy amounts to 30.1 kJ/mole of absorbed hydrogen, corresponding to nearly 12.5% of the energy content of the hydrogen stored.

An important role in storage is played by the dynamic behaviour of hydrides; in particular, the kinetics of adsorption and desorption must be considered. In practice, besides in a maximum value of the flow of hydrogen instantly available—

[41] The formation enthalpy varies with the temperature.

**Table 3.18** Characteristics of the hydride tank Series HBOND 1500

| Hydride tank series HBOND 1500 | |
|---|---|
| Storable hydrogen ($Nm^3$) | 1.5 |
| Height (mm) | 284 |
| Diameter (mm) | 145 |
| Weight | 13 kg (including 10 kg metal hydride) |
| Material: aluminium alloy | type ENAA-6063 |
| Internal exchanger | Tube: diameter 10 mm, length 6 m |
| Safety valve (bar) | 20 |
| Charge temperature | 20°C |
| Discharge temperature | 10–50°C |

**Table 3.19** Storage capacity of hydride tank Series HBOND 1500

| Hydride tank series HBOND 1500 | | |
|---|---|---|
| Storage capacity (% in weight) | 1.0 | % |
| Energy density in volume | 3,412 | kJ/litre |
| Energy density in mass | 1,218 | kJ/kg |

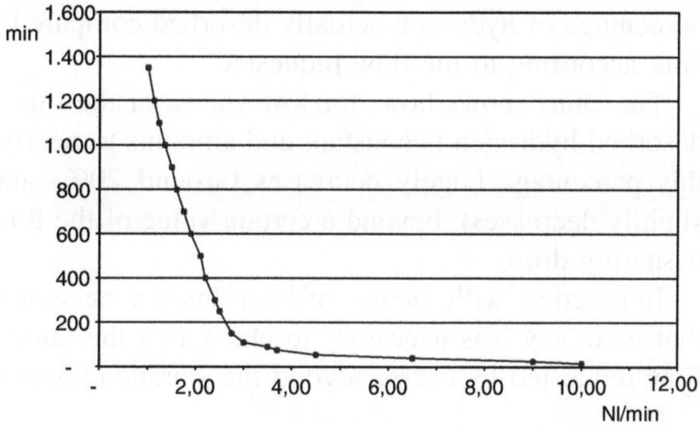


**Fig. 3.23** Maximum hydride holding time

this leads to a decrease of the hydrogen actually supplied by the hydride[42] as the flow requested increases.

In this connection, Fig. 3.23 shows the maximum maintenance time of hydride according to the flow requested.[43] From the chart it is possible to define, besides the maximum flow available (nearly 10 Nl/min), the dynamic behaviour of the hydride. Let us consider, for instance, the values of the hydrogen flow requested amounting to 1.5 and 6.5 Nl/min. In the former case there is a maximum

[42] Similarly to what happens in other accumulators.

[43] This phenomenon depends on the temperatures and conditions of thermal exchange. The diagram in the figure refers to tests made at 50°C and with an internal heat exchanger (described in Table 3.18): this mode of thermal exchange is more efficient than the one of a temperature bath, in which the thermal power is supplied through the external wall of the tank.

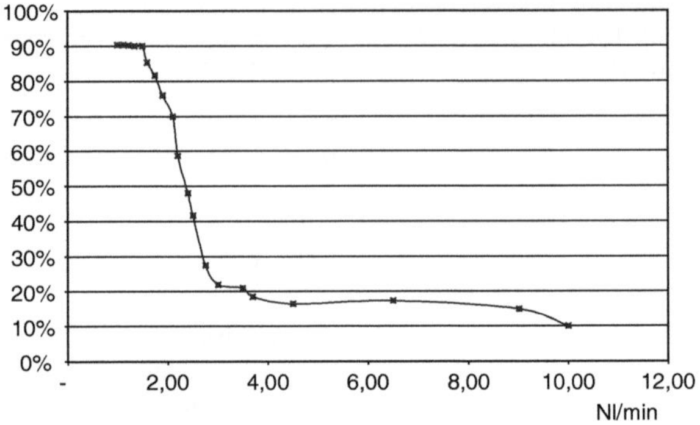


**Fig. 3.24** Percentage of hydrogen really desorbed compared to the nominal capacity of the tank

maintenance time of 900 min (meaning that the quantity of hydrogen actually desorbed amounts to 1,350 Nl), whereas in the latter case the maintenance time amounts to 40 min (the quantity of hydrogen really desorbed amounts to 260 Nl). Therefore, in real operation conditions, hydride is not able to desorb the entire quantity of hydrogen adsorbed, and the quantity of hydrogen actually desorbed varies with the flow requested, decreasing as it increases. Figure 3.24 shows the percentage of hydrogen actually desorbed compared to the normal capacity of the tank according to the flow requested.

The chart shows how, for low values of the flow requested, the percentage of desorbed hydrogen is constant and amounts to nearly 90%. As the flow increases, this percentage hugely decreases (around 20%) and is kept rather constant (it slightly decreases); beyond a certain value of the flow (maximum flow available), it sharply drops.

In practice, with the use of hydrides it is necessary to consider their behaviour that is, to say it is necessary to check that the tanks used are able to provide the flow requested by users, beyond the storage capacity of tanks themselves.

### Hydrogen Storage in Carbon Nanostructures

Carbon nanostructures (carbon nanotubes and nanofibres) might show good hydrogen adsorption capacity, with results in some cases surprising.

Researches on these materials are being carried out by several workgroups, but the results obtained, often contrasting, in many cases are not comparable since they are referred to samples of materials of different types, tested in very different pressure and temperature conditions, and measured with non-comparable testing procedures. Pressure range varies from a few bar up to some hundreds of bar, temperature ranges between 80 and 800 K, the weight adsorption percentages apparently obtained range from values lower than 1% up to an incredible 60%.

Figure 3.25 shows the different structures of carbon nanotubes: armchair (a), zigzag (b) and chiral (c). Nanotube diameters range between 0.6 and 3 nm.

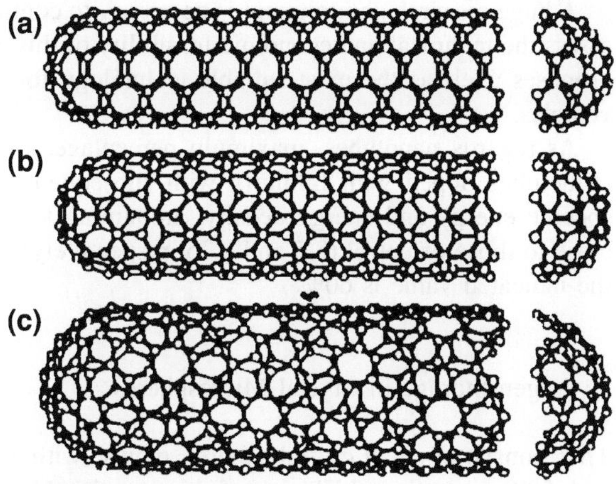


**Fig. 3.25** Different types of carbon nanotubes: armchair (**a**) zig–zag (**b**) and chiral (**c**)

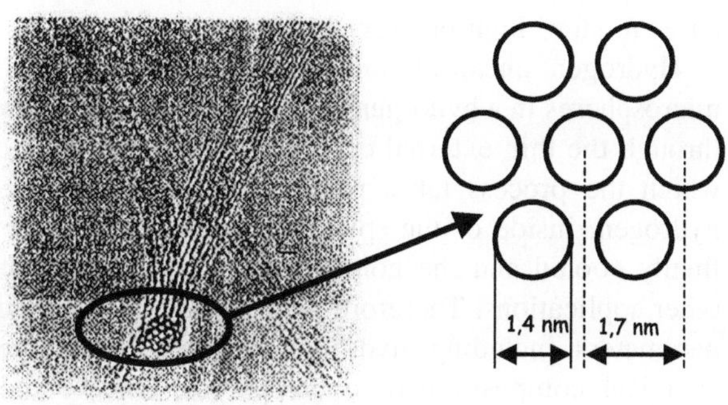


**Fig. 3.26** Possible arrangement, with relevant size, of carbon nanotubes

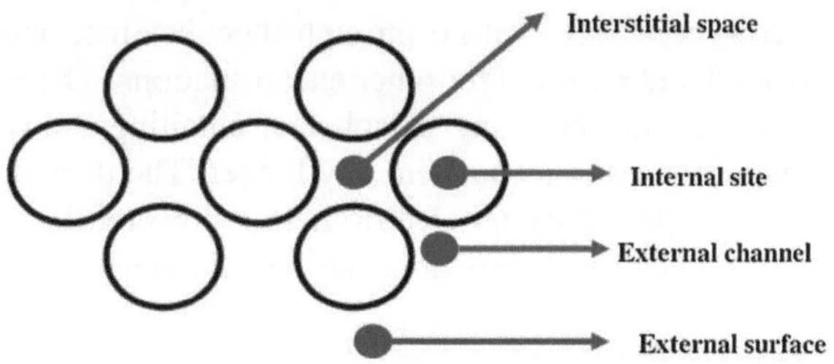


**Fig. 3.27** Hydrogen storage in a structure of carbon nanotubes

Figure 3.26 shows a possible arrangement, together with the relevant size, of carbon nanotubes. In this arrangement, hydrogen can be stored, also with high density and low pressures.

Figure 3.27 shows the "positions" that hydrogen can have inside of a carbon nanotube: hydrogen can occupy the interstitial spaces (between one tube and the other inside of the structure), the internal spaces (inside of the tube), an external channel (between one tube and the other on the external surface of the structure) and the external surface.

A huge research commitment is necessary to confirm the results obtained and to assess the technical and economic feasibility of this technology, which potentially proposes itself as the most suitable technology for hydrogen storage onboard of vehicles.

As regards nanotubes, maximum percentages of 20% in weight are reported (nanotubes doped with lithium), with subsequent very high values in terms of specific energy and energy density, amounting to 6.66 kWh/kg and 6 kWh/litre, close to those of petrol (8.5 and 8.75, respectively), and in the case of nanofibres, the indicated value is 60%.

### Hydrogen Storage in Crystal Microspheres

They consist of small crystal spheres, empty, with a diameter ranging between 25 and 500 μm and a thickness of 1 μm only. Microspheres are processed and commercially transported in the form of fluid powder. They can be used on large bases to store high pressure hydrogen.

Hydrogen encapsulation is made through the heating of a bed of empty microspheres in a hydrogen-dense environment. Hydrogen is introduced in spheres through the thin external crystal sheath, permeable, thanks to the temperatures at which the process takes place (from 200 to 400°C). This process ends when hydrogen, inside of the spheres, reaches the same external pressure. The bed is finally cooled and the non-encapsulated hydrogen is either released or kept for other applications. Therefore the process efficiency depends on the values of some parameters, including: hydrogen pressure, bed temperature and volume, size and chemical composition of microspheres. Once cooled at room temperature, the spheres withhold hydrogen, and are subsequently covered, stored in low-pressure tanks and transported in the form of thin powder. The extraction of hydrogen from microspheres takes place through their heating, and subsequently they are again covered and recycled for other encapsulations. The release of hydrogen can also be caused by the breaking of spheres, entailing; however, the disadvantage of not being allowed to reuse them any longer. The thermal energy necessary to this end can be supplied by an electric heater powered by a small hydrogen cell or by a battery, recharged through a mechanical generator.

### Hydrogen Storage in Ammonia

The use of ammonia as a hydrogen storage system clearly entails the production of ammonia starting from the hydrogen to be stored and its decomposition at the moment of hydrogen demand.

Ammonia represents a good example of hydrogen storage at the liquid non-cryogenic state, which consists in the use of liquids with high hydrogen content from which extracting this element on-demand that is, to say when users demand it. This solution has its main points of strength in the stability of liquids, partly in

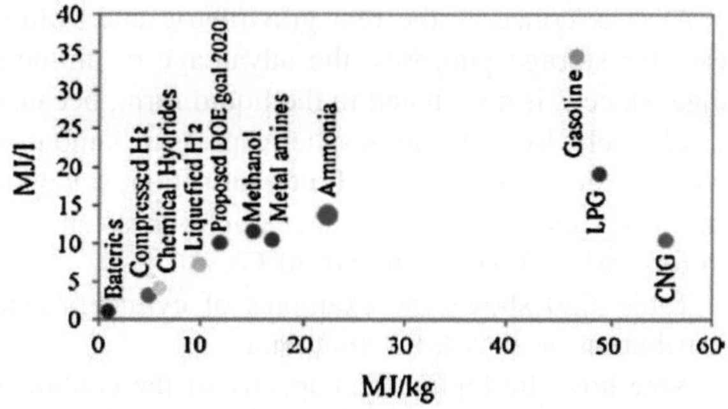


**Fig. 3.28** Storage capacity for different systems and DOE's goals for 2020

**Table 3.20** Examples of ammonia cylinders presently available in the market

| Capacity [l] | Dimensions [l] | $Nh_3$ (kg) | Cylinder weight (kg) | Storage Capacity (%) | Energy density in volume (kJ/l) | Energy density in mass (kJ/kg) |
|---|---|---|---|---|---|---|
| 1 | 1.5 | 0.4 | 1,5 | 3.7 | 2,893 | 2,224 |
| 5 | 9.1 | 2.5 | 10 | 3.5 | 2,909 | 2,112 |
| 20 | 31.1 | 10 | 27 | 4.8 | 3,401 | 2,855 |
| 40 | 54.0 | 20 | 47 | 5.1 | 3,910 | 3,153 |

the storage conditions that in general are not particularly prohibitive as concerns temperature and pressure, but mostly in the density of storable hydrogen, which can be even higher than liquid hydrogen. Other liquids are useful for the same purpose, but ammonia has such characteristics that allow it to be one of the few energy vectors that, exception made for fossil fuels, in line with the DoE's goals for 2020, as shown in Fig. 3.28 [4].

Other advantages to take into consideration include the possibility of performing the decomposition reaction without any additional reagents (which might lower the real energy density of the liquid, as it happens for instance, with methanol, whose decomposition needs water), the absence of carbon atoms in its molecule (which excludes the presence of carbon dioxide and carbon monoxide among the reaction products); the fact that it is by now a fundamental product in chemical industry (which eliminates the critical aspects relating to the marketing and distribution, typical of those cases in which there is the will to extend the use of new materials outside of laboratories). The most interesting characteristic consists therefore in the high density of the storable hydrogen (in fact, its hydrogen content accounts for 17.6% in weight), which is even higher than the one obtainable by pure hydrogen.

If gaseous ammonia and hydrogen are considered at the conditions of 25°C and 1 bar, then the densities of nearly 0.771 and 0.09 g/l are obtained, respectively. Since the molecular weight of the two species is respectively 17 and 2 g/mole, then ammonia has a hydrogen density amounting to 1.5 times the same pure hydrogen:

$$\frac{\rho_{H_2}^{NH_3}}{\rho_{H_2}} \approx \frac{0.136\,gl^{-1}}{0.09\,gl^{-1}} \approx 1,5 \tag{3.52}$$

Also as concerns the real gravimetric and volumetric density of the devices used for storage purposes, the advantage of ammonia compared to hydrogen is huge, since it is distributed in the liquid form, because its liquefaction temperature is relatively low (the atmospheric pressure amounts to −33.4°C). Furthermore, cylinders do not demand high pressurisation, since it amounts to the tension of the saturated vapour of the substance (and, for instance, varies from nearly 4 bar at 0° up to nearly 20 bar at around 50°C).

Table 3.20 shows the examples of cylinders available in the market for the distribution of anhydrous ammonia.

Note how the higher the capacity of the containers considered, the higher the percentage of weight of ammonia and hydrogen contained, obviously corresponding to a higher energy density; for commercial purposes, the cylinders presently used are the same type as those used for the storage of hydrogen at 200 bar. Ammonia is at the liquid state and the pressure inside of cylinders is the pressure of ammonia vapour saturated at the temperature of the cylinder: nearly 4 bar at 0°C up to nearly 20 bar at around 50°C. It is evident that the cylinders used are oversized and, therefore, the use of ad hoc cylinders would allow to reduce volume and weight (given the same ammonia contained) and to increase the storage capacity and energy density in volume and in mass.

As already said, the use of ammonia as a hydrogen storage system clearly entails the production of ammonia itself, starting from the hydrogen to be stored and its decomposition at the moment of hydrogen demand. The synthesis and decomposition reactions of ammonia are the following:

$$\text{Synthesis} \quad 3H_2 + N_2 \rightarrow 2\,NH_3$$
$$\text{Decomposition} \quad 2NH_3 \rightarrow 3H_2 + N_2$$

In the synthesis phase, the energy necessary is the one needed to take gases to the working conditions (pressure and temperature at which the synthesis reaction takes place), the reaction enthalpy and the energy to liquefy ammonia. It is then necessary to add the energy spending for nitrogen production.

During the decomposition phase, the following must be considered: the energy necessary to take liquid ammonia to the gaseous phase, the energy to take gas to the decomposition temperature, the decomposition enthalpy, and an energy spending that will depend on the conditions of hydrogen use (pressure and temperature).

As a matter of fact, energy waste strongly depends on the peculiar characteristics of each production and utilisation plant, and therefore it can be more realistic to evaluate energy spending according to a "black-box" scheme starting from the consumption of the system itself.

As concerns the production process, at an industrial level it is presently evaluated mainly through the Bosch-Haber process or similar ones (Fauser, Casale, Claude, NEC, Mont-Cenis) that over the time turned out to be more convenient than other processes as the distillation of hard coal or the hydrolysis of calcium cyanamide $CaCN_2$. In practice, the procedure entails the direct synthesis of

**Table 3.21** Pressure and temperature conditions as well as reaction enthalpy for the processes of synthesis and decomposition of ammonia

| Reaction | T (°C) | P (bar) | ΔH ($kJ\ mol^{-1}$) |
|---|---|---|---|
| $3\ H_2 + N_2 \rightarrow 2\ NH_3$ | 400–600 | 200–300 | −92.4 |
| $2\ NH_3 \rightarrow 3\ H_2 + N_2$ | 700–800 | 9–36 | 66.5 |

hydrogen and nitrogen at the gaseous state, with different pressure and temperature conditions according to the processes used (which de facto differ for the synthesis plants), but in most of cases range between 200 and 300 bar and between 400 and 600°C. An exothermic reaction occurs, but it shows a very high activation energy (of nearly 230–420 $kJ\ mol^{-1}$ [5]), for which reason it is necessary to use catalysts made of nickel–alumina, iron, ruthenium, without which it would be necessary to work at much higher temperatures.

As regards the decomposition process, it is generally considered as the inverse process of synthesis, although the working conditions are not identical in the two reactions. Decomposition is an endothermic reaction that needs higher temperatures than synthesis, of nearly 700–800°C and prefers low pressures [6]. However, it must be underlined that also in this case the use of catalysts allows a huge reduction of the working temperature and improves the percentage of ammonia produced. Based on data in the literature, catalysts can reduce working conditions at pressures and temperatures between 9 and 36 bar and 400–600°C. Table 3.21 shows the working conditions of the synthesis and decomposition process, as well as the relevant enthalpy variations.

For the calculation of the energy spending, it is possible—as an example—to consider a facility powered through a 50 MW wind farm: 45 MW are used for hydrogen production through water electrolysis, 2 MW for air liquefaction and nitrogen production, 3 MW for the Haber–Bosch synthesis process, with a daily production of 24 tons of $H_2$ (and 196 tons of $O_2$), 112 tons of nitrogen and 136 tons of ammonia [7]. To analyse the cost to store hydrogen in ammonia, it is necessary to start from hydrogen availability and consider the energy spending for the production of nitrogen plus the one needed for the synthesis process. In consideration of the above, the following energy spending applies:

Nitrogen synthesis 37.0 kJ/mole of $N_2$

Ammonia synthesis 32.4 kJ/mole of $NH_3$

Total production of ammonia 50.9 kJ/mole of $NH_3$

Considering that a mole of ammonia ($NH_3$) contains 1.5 mol of hydrogen ($H_2$) the energy spending referred to the mass unit of the hydrogen contained in ammonia amounts to nearly 16,967 kJ/kg, corresponding to 14.1% of the energy content of the hydrogen stored.

As concerns the decomposition process, since a system of preferential use has not been identified yet, then the field of testing is still open.

**Table 3.22** Reaction of ammonia decomposition inside of a SOFC

| Ammonia Decomposition inside of a SOFC | |
|---|---|
| Anode | $2NH_3 + \text{calorie} \rightarrow 3H_2 + N_2$ |
| | $3H_2 \rightarrow 6H^+ + 6e_2^-$ |
| Cathode | $\frac{3}{2}O^2 + 6H^+ \rightarrow 3O^-$ |
| | $3O^- + 6H^+ \rightarrow 3H_2O + \text{calorie}$ |
| Global reaction | $2NH_3 + \frac{3}{2}O_2 \rightarrow N_2 + 3H_2O$ |

As an example, consider the use of a Solid Oxide Fuel Cell (SOFC), powered by ammonia on the anode and by air on the cathode, inside of which the decomposition reaction of ammonia into nitrogen and water takes place (Table 3.22) [8]:

The energy spending for the ammonia decomposition phase amounts to 11.25 kJ/mol$H_2$ that is, to say 5,625 kJ/kg of hydrogen (accounting for 4.4% of hydrogen energy content). The value shown does not consider the possible recovery of the heat produced at the anode that would significantly increase the overall efficency.

On the basis of the examples made, the overall energy spending (synthesis and decomposition of ammonia) amounts to 22,592 kJ/kg of hydrogen, corresponding to 18.8% of the energy content of hydrogen itself.

A further possibility found in the literature is the use of ammonia at the solid state (Solid State Ammonia Storage - SSAS), meaning by this not the solidification of pure ammonia, but rather the production of ammonia-based chemical compounds that, at room temperature and pressure, appear at the solid state, similarly to hydrogen storage through complex hydrides.

This solution is considered as suitable in terms of size and safety, although in the literature data relating to production and use are not yet available.

*Other Methods for Hydrogen Storage*

In the framework of chemical processes, a system was tested that allowed to keep hydrogen in small spheres of sodium, potassium or lithium compounds. If released in water, they combine by releasing hydrogen and originating compounds, in turn recyclable, as sodium hydroxide, potassium hydroxide, etc. The US company Power Ball tested this system using spheres covered with a plastic sheath that allows hydrogen release when demanded, through a process that divides spheres one at the time. Another technique, presently still in an experimental stage, assumes the use of powdered iron and water. At high temperatures, their reaction produces rust and hydrogen. This process is inexpensive and has a 4.5% efficiency. The only need is iron retrieval through the elimination of oxygen in order to allow its recycling. The disadvantage of this technique consists in the huge quantity of iron needed. Furthermore, systems similar to metal hydrides are under study, with the use of particular hydrocarbons or chemical products (methanol, liquid methane and ammonia) as metal replacement, as well as the use of sodium borohydride ($NaBH_4$ powders.)

**Table 3.23** Energy spending for hydrogen transportation

| Transportation System | Energy spending (%/1,000 km) | |
|---|---|---|
| | Compressed (200 bar) | Liquefied (−163°C) |
| Road | 31.4 | 8.1 |
| Railway | 15.5 | 4.0 |
| Sea | 9.0 | 2.3 |

*Hydrogen Transportation and Distribution*

As regards hydrogen transportation in non-dedicated networks, considerations similar to the considerations made in Sect. 3.5.1, and Table 3.2 apply.

The lower heating value of hydrogen amounts to 120,000 kJ/kg; considering Rp values amounting to 0.07 for hydrogen compressed at 200 bar[44] and 0.27 for liquid hydrogen at −353°C[45] the values shown in Table 3.23[46] are obtained.

For the transportation in dedicated networks (hydrogen pipelines[47]) it is possible to use most of the tested technologies used for the transportation of natural gas in gas pipelines, with some key adjustments in materials used.

For the calculation of energy spending, the same procedures used for natural gas apply. By analogy, the same conditions in terms of diameter of pipelines are considered: 1 m, pressure 75 bar. Given the same energy transported, the velocity of hydrogen shall be three times as much as higher than natural gas, that is to say 30 m/s. With these hypotheses, it is possible to obtain Re = 19,708,520, from which $\lambda = 0.0086$ and $J = 0.39465$. Considering a section of 1,000 km it is possible to obtain $L_u = 1{,}402{,}200 + 450 + 3{,}781{,}538 = 5{,}184{,}188$ J/kg.

With similar considerations as those seen for natural gas on concentrated losses and the performance of the pumping stations, it is possible to obtain:

$$L_{\mathrm{eff},D} = \frac{\Delta H}{\eta_{\mathrm{SP}}} + \frac{J \cdot g \cdot 1.2}{\eta_{\mathrm{SP}}} D \tag{3.53}$$

For a section of 1,000 km it is possible to obtain $L_{u,\mathrm{eff}} = 13.185.470$ J/kg, corresponding to nearly 11% of the energy content of the gas transported.

For uniformity purposes, the values of energy spending for hydrogen transportation are set for a distance of 1,000 km, similarly to what was done for oil, coal and natural gas. In this connection, it is suitable to underline that, while the fossils mentioned are originally primary sources and the distances of transportation are linked to the location of the resource and of the demand—and, de facto, these vectors really have to be transported along thousands of kilometres—hydrogen is

---

[44] The density of hydrogen compressed at 200 bar is 14.2 kg/m$^3$.

[45] The density of liquefied hydrogen at −253°C is 71 kg/m$^3$.

[46] The values shown refer to transportation only and do not consider the energy necessary for compression and liquefaction.

[47] All over the world (USA, Japan, France, Germany, Italy, etc.) some thousands of kilometres of pipelines are nowadays used for the transportation of hydrogen.

Fig. 3.29 Liquid (a) and gaseous (b) hydrogen distribution unit

not a primary source and, therefore, it has to be "produced". Obviously, this production can occur on-site where needed and, therefore, the distances of transportation that can be considered are those relating to a distribution at an industrial district or city district level, amounting to a maximum of about one hundred kilometres.

Also as concerns distribution, hydrogen can be distributed both in a liquid and in a gaseous form (Fig. 3.29). Today there are already some experimental hydrogen filling stations for both types. Figure 3.29 shows the filler for the supply of liquid and compressed gaseous hydrogen.

## 3.5.6 Synthetic Fuels as Energy Vectors

Synthetic fuels are obtained through the processing of biomasses (in that case, we refer to *biofuels* or *renewable fuels*) or fossil fuels as natural gas and coal.

The main synthetic fuels include:

- Biodiesel
- Ethanol
- Methanol
- Syngas
- Biogas

### 3.5.6.1 Biodiesel

Biodiesel is obtained starting from agricultural products such as soya, rape and sunflower, but also from animal fat, original or waste vegetable oil and fat, such as waste edible fat and oil (exhausted frying oil). The seeds of the crops mentioned above represent the basic material: they are pressed to obtain a vegetable oil, which then reacts with methanol in the presence of alkaline catalyst, to form methyl ester and, as a secondary product, raw glycerine. There can be different production procedures but, in all the cases, through this chemical procedure called transesterification, it is possible to obtain a blend of fatty acid methyl ester, which is called biodiesel.

#### 3.5.6.2 Ethanol

Ethanol is a *short-chain alcohol*, also known as *ethyl alcohol* or *spirit of wine*; its chemical formula is $CH_3CH_2OH$.

At room temperature, it appears as a colourless liquid with a typical smell. It is basically volatile and extremely inflammable.

It is completely soluble in many organic solvents—for instance chloroform—and in water, with which it forms in a proportion of 95:5 a low-boiling-point azeotrope that makes it impossible to obtain by simple distillation an ethanol with a higher purity.

Pure ethanol at 100%, or *absolute* or *anhydrous ethanol* can be obtained by removing water from the azeotrope through the addition of benzene and subsequent fractional distillation, or by using metallic magnesium that, when added to the water/ethanol azetrope, quantitatively reacts with water and originates hydrogen, which is separated by the solution through degassing, and magnesium hydroxide through the distillation of absolute ethanol. In this way, the remaining water is completely eliminated.

Bioethanol is ethanol produced through a fermentation process of biomasses that is, to say farm products rich in sugar (glucides) such as cereals, sugary crops, starchy products and marcs.

In the energy sector, bioethanol can be used as a component for petrol or for the preparation of ethyl tertiary buthyl ether (ETBE), a high-octane derivative. It can be used blended in petrol with percentages up to 20% without modifying the engine, or even in higher quantity or even pure in so called Flex Engines.

The production of bioethanol occurs through the anaerobic fermentation of glucose; the theoretical reaction is the following:

$$C_6H_{12}O_6 \rightarrow 2CH_3CH_2OH + 2CO_2$$

On the basis of this reaction, from 1 kg of glucose it is possible to obtain 0.51 kg of ethanol and 0.49 kg of carbon dioxide. As a matter of fact, what is obtained is an aqueous solution with an ethanol content ranging between 5 and 15% according to the type of biomass and to the processes used.

Alcohol retrieval from the aqueous solution occurs through distillation, which is the most energy-consuming phase of the entire process.

Bioethanol can also be produced starting from synthetic gases (syngas[48]), gases rich in hydrogen also containing CO. Bioethanol, using syngas alone, can be produced according to the reaction:

---

48 See Sect. 3.5.6.4.

$$2CO + 4H_2 \leftrightarrow C_2H_5OH + H_2O$$

Another possibility is using methanol with syngas; in this case the reaction is:

$$CH_3OH + CO + 2H_2 \leftrightarrow C_2H_5OH + H_2O$$

#### 3.5.6.3 Methanol

Methanol is the simplest among alcohols, also known as *methyl alcohol* or *wood spirit*, and its chemical formula is $CH_3OH$.

At room temperature it appears as a colourless liquid with a typical smell. It is highly volatile and extremely inflammable. The methanol flame is invisible.

It is completely soluble in many organic solvents, such as chloroform and water.

Methanol is toxic and causes damage to the optic nerve and to the retina. A lethal dose for a human being ranges from 0.3 to 1 g/kg of body weight.

Methanol can be produced from natural gas, coal and biomasses.

From the energy viewpoint, the balance of methane transformation into methanol is negative, although the same transformation is convenient as far as the transportation of natural gas in the form of methanol is concerned.

Given the lower heating value as well as the toxicity characteristics of methanol against ethanol, the latter is usually indicated as preferable as fuel.

#### 3.5.6.4 Syngas

The syngas deriving from gasification processes is mainly made up of carbon monoxide, carbon dioxide, hydrogen, water vapour and methane, to which adding, in case of gasification with air, nitrogen in the following percentages:

- $H_2$: 9–60%
- CO: 14–50%
- $N_2$: 0–50%
- $CO_2$: 9–20%
- $H_2O$: 5–30%
- $CH_4$: 1–7%

The lower heating value ranges between 5,400 and 11,300 kJ/Nm$^3$.

Besides these substances, *there are always compounds*, mostly unwanted, as *particulate* (solid particles contained in gas), *sulphur compounds* ($H_2S$ and COS) and *nitrogen compounds* ($NH_3$ and HCN), *halogenated compounds* (HCl), tar (a complex of oxygenated compounds produced during the pyrolysis and that undergo transformations in the subsequent phases of the gasification process), *alkali* (KOH, KCl), *phenols* and *others*.

The sulphur and nitrogen present in the biomass, in fact, are mainly converted into $H_2S$ and in $NH_3$, but in the biomass used in gasification processes, contents in

S and N are extremely low (<0.3%). Biomass is made up of over 95% of carbon, oxygen and hydrogen, in addition to percentages of sulphur, nitrogen and other elements depending on the type of soil and water, as well as on the type of fertilisers and parasiticides used.

#### 3.5.6.5 Biogas

Biogas is a gas mixture whose composition varies according to the conditions at which it is obtained. Its calorific value is lower than methane and ranges between 18.81 and 27.17 MJ/Nm$^3$. In any case, this mixture is mainly made up of methane, carbon oxide and dioxide, nitrogen, hydrogen, besides lower percentages of hydrogen sulphide and other gases. As a rule, the percentages are the following:

- $CH_4$: 50–80%
- $CO_2$: 30–40%
- $H_2$: 5–10%
- $N_2$: 1–5%

#### 3.5.6.6 Energy Efficiency Considerations for Synthetic Fuels Production

*Energy efficiency* means the ratio between the useful energy and the energy spent in the production process. It is important to underline that useful energy ($E_u$) is, besides the energy content of the fuel produced, also the energy deriving from the possible energy use of other products of the supply chain. The energy spent ($E_s$), using biomass as primary resource includes, in general, the energy used for farming, the energy spent for the production of the biomass used, starting from the cropped plant, and the energy of the processes carried out to obtain biofuels:

$$\mathrm{RE} = \frac{E_u}{E_s} \tag{3.54}$$

When energy efficiency is lower than 1, reference is made to *negative energy efficiency*: this means that the energy spent for fuel production is higher than the energy content of the fuel itself. In case of energy efficiency higher than one, reference is made to *positive energy efficiency*.

It is important to make some considerations on the meaning of energy efficiency so defined from the viewpoint of the sustainability of energy systems.

Besides the above, sustainability must be considered; in this sense, it is necessary to underline that fuel can be obtained from a renewable source. In this case, although the value of energy efficiency, defined as seen above, was lower than 1 (negative energy efficiency) there would be in any case the advantage of producing a fuel (that would replace fossil fuels in end use) without any consumption of non-renewable energy. Therefore, it is probably more appropriate to distinguish the kind of energy spent in producing the energy vector and to better relate energy efficiency to the consumption of *non-renewable energy*, and say *that a negative energy efficiency is obtained if the consumption of non-renewable energy for its production exceeds the energy content of the fuel produced*: this means that it appears as necessary to consider as energy spent (3.54) only non-renewable energy.

The calculation of energy efficiency demands a detailed analysis of all the processes involved (from farming to the entire fuel cycle) and the knowledge of all the sub-processes involved in order to correctly quantify the useful energy produced and the energy spent in the processes.

For instance, biodiesel can be produced (always through the transesterification of biomasses) starting from the cropping of sunflowers: this process provides biodiesel, as a single product that can be used from the energy viewpoint. In this case, the energy spent for production purposes also has to include, besides the energy for the production process itself, also the one relating to the cropping of sunflowers. Another system to produce biodiesel includes the use of pomace (a sub-product of olive-oil production process) and this process allows to supply, besides biodiesel, also pellets and syngas. In this case, the energy spent is only the one dealing with the production process itself, and while calculating useful energy, the energy contained in pellets and/or in syngases must be considered as well.

In the case of production of bioethanol from sugar cane, finally, all the energy spent (bioethanol production process and cropping of sugar cane) shall be ascribed to the production of bioethanol. Furthermore, in calculating useful energy, one must consider that from other parts of the same plant it is possible to produce syngas.

Apart from the energy efficiency of the single processes (that in any case play a fundamental role as far as energy efficiency is concerned), it is evident how it is possible to obtain very different energy efficiencies according to all the processes and sub-processes somehow related to fuel production process. Obviously, as already mentioned, energy efficiency is also influenced by the technologies used: for instance, biomasses suitable for thermochemical processes (defatted pomace in the case of production of biodiesel from pomace and bagasse and leaves in the case of production of bioethanol from sugarcane) can be used in different ways: *directly* (or upon pre-processing obtaining pellets) as renewable solid fuels, or, *indirectly*, producing syngas through gasification.

## *3.5.7 Heat-Transfer Fluids as Energy Vectors*

### 3.5.7.1 Heat Transportation

Energy can be transported and stored in the form of heat. As regards transport, the heat energy vector is only suitable for short distances, since in case of long distances energy losses would be excessive. An example of heat transport is represented by the systems for room heating: in this case, the distance covered by heat depends on the type of system used. Options include the independent heating

**Fig. 3.30** Operation scheme of a heat accumulator

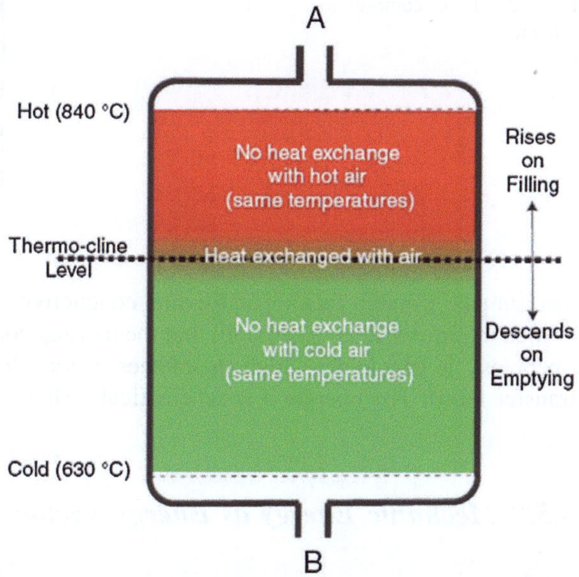


system, with minimum distances, limited to heat distribution in the single apartment, up to the case of district heating, in which the heat produced through a central heating system at the district (or small city) level, is distributed to all the users in a district for distances reaching some kilometres, passing through the condominium heating systems. In all these cases, the heat-transfer fluid is water: temperature amounts to nearly 70°C for the use in radiators, whereas in case of district heating, superheated water at nearly 120°C[49] is used. In general, on the basis of specific needs, a gas (hot air, $CO_2$, dry water vapour, etc.) is preferibly used as heat-transfer fluid.

#### 3.5.7.2 Heat Storage

For storage purposes, water is often used (electric boilers, water heaters) or alternatively other fluids with suitable physical characteristics; when heat must be stored at high temperatures, solid vectors are used as in the case of heat storage in solar plants for the production of electricity or hydrogen (still in an experimental phase). In this case, the storage system is schematically made up of a thermally insulated container of a suitable size, filled with small spheres (pebbles) or other solid objects of an appropriate shape, compactly distributed (packed bed), so that

[49] In this case, the fluid used for (primary) transport is not the same that reaches the radiator (secondary); the former releases heat to the latter in a heat exchanger; therefore, it is necessary to have a fluid at a temperature markedly higher than the end use on, in order to allow heat exchange.

Fig. 3.31 Fluid compression scheme

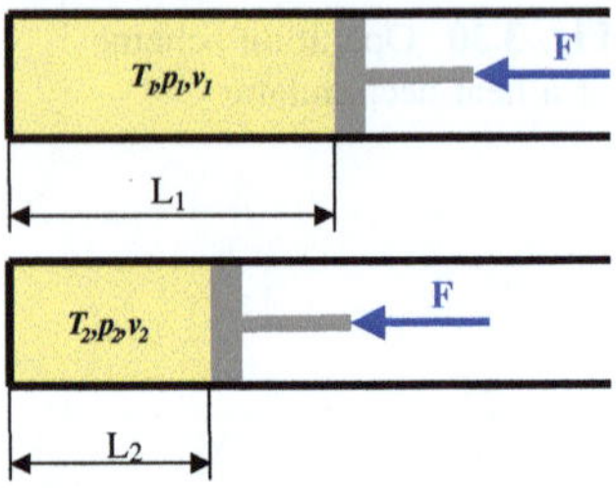


gas can pass through them. The thermal conductivity of the whole set is very low (which minimises heat dispersion), but the thermal contact between the gas and the "pebbles" is excellent (which guarantees a good heat exchange with the heat-transfer fluid). The operation is schematically shown in Fig. 3.30 [9].

## 3.5.8 Mechanic Energy as Energy Vector

The storage and transport of mechanic energy can mainly take place in the ways described below:

- pressurised fluids (transport and storage);
- flywheels (storage alone);
- springs (storage alone);
- shafts and mechanical transmission systems (transportation alone).

### 3.5.8.1 Pressurised Fluids or Pressure-Transfer Fluids

One way for storing mechanic energy envisages the compression of elastic fluids. Referring to Fig. 3.31, the work made by force $F$ on the fluid is stored in the compressed fluid of the force itself.

As it is well-known, in closed systems the reversible work of fluid compression is:

$$L = -\int_1^2 p dV \tag{3.55}$$

If the compression transformation can be considered as adiabatic,[50] this work can be expressed according to the initial state and the final state of the fluid:

[50] Since heat transmission is a "slow" phenomenon, quick transformations can be considered with a good approximation as adiabatic.

$$L = \frac{1}{k-1}\frac{p_1}{\rho_1}\left[\left(\frac{\rho_2}{\rho_1}\right)^{k-1} - 1\right] = \frac{1}{k-1}\frac{p_1}{\rho_1}\left[\left(\frac{p_2}{p_1}\right)^{k-1/k} - 1\right] \tag{3.56}$$

where $k$ is the ratio between the specific heats at constant pressure ($c_p$) and at constant volume ($c_v$):

Therefore, once the initial conditions of the fluid and the mechanic energy to be stored are known ($L_{12}$), the final conditions of the fluid itself can be calculated:

$$p_2 = \left(L\frac{k-1}{k}\frac{\rho_1}{p_1} + 1\right)^{k/k-1} \cdot p_1 \tag{3.57}$$

From adiabatic equations it is also possible to see that:

$$\frac{T_2}{T_1} = \left(\frac{p_2}{p_1}\right)^{k-1/k} \Rightarrow T_2 = T_1\left(\frac{p_2}{p_1}\right)^{k-1/k} \tag{3.58}$$

And from the equation of state:

$$\frac{p_2}{\rho_2} = R \cdot T_2 \Rightarrow \rho_2 = \frac{p_2}{R \cdot T_2} \tag{3.59}$$

Obviously, in allowing the fluid to expand from the new conditions to the previous conditions, the work spent in compression is given back after deducting the compression and expansion losses:

$$L_e = \frac{1}{k-1}\frac{p_2}{\rho_2}\left[\left(\frac{p_1}{p_2}\right)^{k-1/k} - 1\right] = L \tag{3.60}$$

These losses can be taken into account through the *compression* ($\eta_c$) and expansion ($\eta_e$) efficiency: the storage efficiency ($\eta_s$) is therefore the outcome of the two efficiencies

$$\eta_s = \eta_c \cdot \eta_e \tag{3.61}$$

If the compressed fluid is stored for longertime, it is necessary to consider that, due to the unavoidable thermal exchanges with the outer part, the temperature of the fluid decreases (going back, at most, to the initial value $T_1$) and, therefore, the "energy content" of the fluid decreases, as well as the energy stored. In fact, in considering the borderline case mentioned above ($T_3 = T_1 < T_2$), since heat exchange is an isochoric process (at a constant volume), it follows that:

$$\frac{p_3}{T_3} = \frac{p_3}{T_1} = \frac{p_2}{T_2} \Rightarrow p_3 = \frac{T_1}{T_2}p_2 < p_2 \tag{3.62}$$

And the reversible work obtainable in the expansion is therefore:

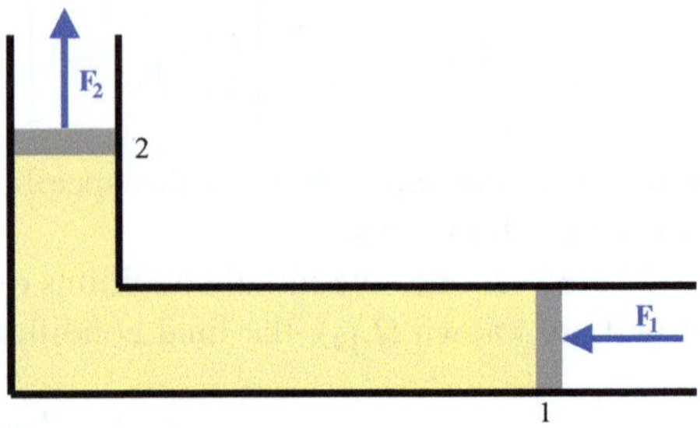


**Fig. 3.32** Transportation of mechanic energy with pressurised fluids

$$L'_e = \frac{1}{k-1}\frac{p_3}{\rho_2}\left[\left(\frac{p_3}{p_1}\right)^{k-1/k} - 1\right] < L_e \tag{3.63}$$

According to the same principle, it is possible to transport mechanic energy from one point to another through pressurised fluids. In fact, in a container with pressurised fluid, the fluid itself exercises the same pressure in all the points and in all directions. As regards Fig. 3.32, by exercising force $F_1$ in section 1 the work $L$ made by such force can be given back by the fluid (always, obviously, minus losses) in section 2. If the fluid is a liquid (incompressible), then reference will be made to *hydraulic drives*; if it is a gas (compressible) then the definition of *pneumatic drives* shall apply.

Having defined $A_1$ and $A_2$ the surfaces of sections 1 and 2 and $s_1$ and $s_2$ the movements of said surfaces, for energy conservation there will be, after deducting losses:

$$\left.\begin{aligned} A_1 \cdot s_1 &= A_2 \cdot s_2 \\ F_1 &= p \cdot A_1 \\ F_2 &= p \cdot A_2 \end{aligned}\right\} \frac{F_1}{p}s_1 = \frac{F_2}{p}s_2 \Rightarrow F_1 \cdot s_1 = F_2 \cdot s_2 \tag{3.64}$$

If $A_1 = A_2$, then $s_1 = s_2$ and $F_1 = F_2$ and there is only transportation of mechanic energy, from sections 1 to 2. Conversely if $A_1 \neq A_2$ then $s_1 \neq s_2$ and $F_1 \neq F_2$, in addition to transportation there is also a *transformation of mechanic energy*, since characteristics change.

#### 3.5.8.2 Flywheels

Flywheels are quite commonly used to lower the degree of periodical irregularity by alternative storing and releasing *kinetic energy*. Having defined $\omega$ angular velocity (rad/s), the kinetic energy $E$ (J) stored by a flywheel with a moment of inertia $I$ (kg m$^2$) is:

$$E = \frac{1}{2} I \cdot \omega^2 \tag{3.65}$$

The *moment of inertia* $I$ (which expresses the capacity to store energy) of a mass cylinder $m$ and radius $R$ is:

$$I = \frac{m \cdot R^2}{2} \tag{3.66}$$

The *mechanic energy* spent to bring the flywheel from an angular velocity $\omega_1$ to $\omega_2 > \omega_1$ by applying a torque C is stored by the flywheel, in the form of a moment of inertia, and is "returned" in the moment in which the torque stops (or decreases).

#### 3.5.8.3 Springs

Springs can store mechanic energy in the form of *elastic potential energy*. The elastic potential energy $E_{p,\text{el}}$ stored by deforming a spring, having defined $P$ the applied force and $f$ the arrow, is:

$$E_{p,\text{el}} = \frac{1}{2} P \cdot f \tag{3.67}$$

After deducting losses, the mechanic energy spent for deformation is stored in the form of elastic potential energy and is "returned" by taking the spring back to its free configuration (unchanged).

#### 3.5.8.4 Shafts and Mechanical Transmissions

Shafts and the different types of mechanical transmissions (cogwheels, belts, chains, crank mechanisms, etc.) are mechanisms that allow the transportation of mechanic energy for very short distances. The "energy spending" of this transportation is linked to the friction losses typical of mechanical transmissions. In many cases, besides transportation, there is also a transformation of mechanic energy, changing the characteristics of motion (torque, angular velocity) or the type of motion (for instance from, reciprocating to rotating).

### *3.5.9 Radiant Energy as Energy Vector*

As already said, energy transportation—and not storage—can take place in the form of electromagnetic radiations: for instance, solar energy reaches the earth and the bodies exchange energy by radiation. As concerns the solar energy reaching the earth, dispersions (that is to say the energy "cost" of transportation) are due to the crossing of the atmosphere and depend on the length of the atmosphere crossed (therefore on the relative position earth-sun) and on weather conditions.

## 3.6 The Era of Energy Vectors

The future of energy depends on the energy vectors that human intelligence will be able to develop and to integrate in its development systems. After the era of oil, it is not realistic to wait for the era of another primary source; the next era is *the era of vectors*. It is an era characterised by the capacity of energy systems to use different primary sources in the different places and for different local socio-economic conditions, but also to be able to standardise applications, energy conversion and transformation systems, in order to identify the most flexible energy vector, the one that shows the highest production potentials and the lowest utilisation problems.

The success of each energy vector is in fact linked to its appropriateness, adaptability and integrability with the local and global energy system. The most important characteristics for the penetration of energy vectors in the modern energy system, besides their suitability for storage and transportation, is the possibility of their production from different energy resources and the possibility of their end use with the lowest possible level of waste and the minimum impact on the environment. In both cases, with the highest possible efficiency.

The integration of vectors in the energy system, from the primary source to end use, always entails costs in terms of energy spending, conversion and transformation. These costs can be fully repaid, and can even lead to advantages when at least one of the following conditions applies:

1. *Acceptability of the energy spending*: if the energy spent for the production of the vector wouldn't be otherwise used for environmental, technological or socio-economic reasons;
2. *Transportation or storage needs*: if energy transportation or storage wouldn't be possible without the passage through the conversion or transformation in the vector identified for environmental, socio-economic or technological reasons;
3. *Restrictions for end-use*: if the energy end use wouldn't be possible without the passage through the conversion or transformation in the vector identified for environmental, socio-economic or technological reasons.

In the analysis of modern energy systems, and even more by considering future ones, it is possible to realise that at least one of the conditions mentioned above is, and will increasingly be, present.

In case of option 1 (acceptability of energy spending), it is possible to easily note that it always occurred when the source is available in places or times different from those relating to the use (this is the case of fossil fuels, but also of energy production centralised in large or very large size plants with a level of production scarcely adjustable as in the case of nuclear power), or its exact availability cannot be foreseen (and this is the case of many renewable energies).

In the case of option 2 (transportation or storage needs), it is evident how the real availability of huge quantities of energy is increasingly linked to the characteristics of the vector selected, in terms of availability, energy spending and transportation and storage.

In the case of option 3 (restrictions for end-use), the characteristics of environmental compatibility and standardisation of utilisation technologies clearly indicate how energy vectors with specific characteristics are increasingly requested by users, with particular and growing attention paid to the possibility of the vector of being used in technological solutions able to provide the useful effect desired with minimum or zero polluting emissions and with the widest range of possible uses (electricity in this case is the glaring example).

Therefore, if the need to integrate energy vectors in the global energy system is clearly destined to grow—since the three conditions provided have more and more frequently occurred at the same time in most of real situations—the new and important challenge against the world of energy with the growing era of energy vectors is not that much linked to "***whether***" integrating one or several vectors at any level of analysis or design of an energy system, but rather "***which***" vector is suitable to choose to perform such an integration. The choice of the vector, merely from the energy viewpoint, would only be possible with the consideration of the energy spending needed for its production, transportation, storage and end use. This pathway always has to be followed since it is evident that the good planning of the energy system minimises its waste and therefore maximises the energy that can be really used by end users compared to the one available at the moment of accessing the primary resource in order to make it available.

The only analysis of the efficiency of the chain of transformations and conversions that can provide the ***energy*** and ***power*** demanded is however already insufficient today to meet the real users' needs. Some characteristics of energy vectors and the possibility of their exploitation in technologies able to guarantee the qualitative levels requested and the meeting of the increasingly strict conditions will be more and more important for the technologies and commercial the increasingly success of a vector compared to another one. It is therefore important to take into consideration these characteristics, and according to them to select those vectors that allow to meet them at the lowest energy cost.

### 3.6.1 Possible Production from Different Primary Sources

The age of energy vectors begins as a response to the growing scarcity of non renewable primary energy sources due to the problems linked to their natural availability, but also for the unforeseeable economic–political conditions of many areas of our planet. The energy vectors destined to be the protagonists of the new era *must therefore guarantee the possibility of production from several different sources*, also allowing the passage from one source to another over the time, as the availability conditions change.

Among the vectors analysed in this chapter, certainly *electricity, synthetic fuels* (either liquid or gaseous), *heat-transfer fluids*, and *hydrogen* show the highest potential as concerns their possible production starting from different sources. These four energy vectors shall therefore be taken in due consideration while conceiving present and future energy systems able to provide the necessary flexibility of the primary source. The achievement of the results in the field of technological research and development for the production, transportation, storage and the most efficient end use of these vectors has a strategic value.

### 3.6.2 Environmental Compatibility During Use

The environmental compatibility of energy solutions is a condition increasingly influencing the choices of policy makers, businessman, managers and, most of all, consumers and regulators. The emission of substances that are harmful for human beings and for the environment due to energy conversion processes is less and less tolerated at any level of the global system, and this is absolutely clear at the end-use level.

Environmental compatibility while using the energy vector with which the energy is made available at the point of demand is therefore a characteristic to be considered as a necessary condition for the economic success and the social acceptability of any solution. In closed environments, for portable applications, and increasingly also for all urban use, the useful effect in terms of mechanical, thermal, electric and light energy has to be provided at zero emissions. The same is happening for the markets of tourism or as concerns locations with a high environmental and architectural interest, involving wider and wider target customers and that can be somehow associated to mass market.

The capacity of a vector of being used in zero emission energy conversion systems is therefore a crucial characteristic for the selection of the vectors to be focused upon. The concept itself of "zero emissions" is becoming more and more important: the potential possibility for a neutral balance in terms of carbon dioxide emissions, relevant for all emission reduction considerations and actions concerning climate change, is certainly very much interesting and allows the success of all vectors that can be produced from renewable resources and nuclear power, that is to say, once again, electricity, synthetic fuels (liquid or gaseous), heat-transfer fluids and hydrogen.

However, the most evident characteristic that can be immediately detected by end users is the absence *of any type of polluting emissions at the point of use, not only* $CO_2$. This limits further the group of vectors that show the highest possibilities of being proposed for zero emission applications, with a large evidence relating to electricity, heat-transfer fluids and hydrogen.

### 3.6.3 Conversion Efficiency

The remarkable economic, energy and technological expenditure necessary for the production, transportation and storage of vectors with all the characteristics mentioned above—that is to say electricity, heat-transfer fluids, hydrogen and synthesis fuels—makes it particularly important to achieve higher efficiency levels in the different conversion and transformation processes are necessary to access the needed quantity and quality of energy at the end use.

The efficiency of conversion systems, especially small ones used by end users, therefore becomes an even more important characteristic for an energy system that still resorts to abundant sources and that does not need many and complex conversion processes to make the final energy vector available (especially if it is of a fossil origin).

The availability of high performance technologies is therefore another crucial characteristic for the selection of the right energy vector in a complex system. The characteristic requested confirms electricity as a main candidate to keep and increase its primary role in the landscape of future vectors, thanks to the high energy performance of the technologies that allow its use and of the transportation systems, as well as the indirect storage systems.

In analysing the level of energy performance of the technologies used, also hydrogen confirms its very high potential. Fuel cells, in fact, since not linked to thermodynamic performance thresholds, can allow levels of global efficiency (electric + thermal) very close to the unit.

## References


1. ENEA, Rapporto energia e ambiente (2010); also: European Environment Agency (2008), Energy and Environment Report 2008, Rep 6/2008, ISBN 978-92-9167-980-5.
2. Zuttel A (2004) Hydrogen storage methods. Springer, Heidelberg
3. Innocenti L, Il consumo energetico nella produzione, stoccaggio e distribuzione dellı idrogeno, http://didattica.dma.unifi.it/website/pub/Energetica/EMT1_05_matdid/Idrogeno.pdf
4. Zamfirescu C, Dincer I (2008) Using ammonia as a sustainable fuel. J Power Sources 185(1):459–465
5. Modak JM (2002) Haber Process for ammonia synthesis. Resonance, 7(9):69–77 DOI: 10.1007/BF02836187
6. Holladay JD, et al (2009) An overview of hydrogen production technologies. Catalysis Today. 139(4):244–260
7. Gruhn S, Sackett B (2008) Wind to NH3: Freedom fertilizer - Green fertilizer and fuel for the future, the 5th Ammonia Fuel Conference; Minneapolis
8. Zamfirescu C, Dincar I (2007) Ammonia as a green fuel and hydrogen sources for vehicular applications, Elsevier
9. Maccari A, Vignolini M (2001) Progetto di massima di un impianto pilota per la produzione di 2000 $m^3$/giorno di idrogeno solare basato sul processo UT-3. ENEA Report, Rome

# Chapter 4
# Energy Conversion and Transformation Plants

## 4.1 Overview

The energy necessary for end-use has different forms (mechanic, luminous, electric, thermal); it can be of different origin, and there are different processes that make it available.

The primary sources directly available in nature are not always conveniently and efficiently usable in the processes from which the most suitable form of energy for end-use can be obtained. For this reason, in almost all cases, primary sources undergo *conversion* or *transformation processes*, whose outcome is indicated by the so-called *secondary sources*.

*Conversion* processes are characterised by a *passage from one energy form to another* (for instance from electric power to mechanic power, as it happens in electric motors), whereas in *transformation* processes the *form of energy remains the same although with different characteristics* (as it happens, for instance, in electric transformers, which vary in terms of current intensity and voltage, keeping—net of the losses—the same power; or in heat exchangers in which only energy flow characteristics, like temperature, vary, but not the form of energy).

## 4.2 Energy Conversion Plants

As already mentioned, energy production processes include the set of single energy transformation and conversion processes that allow to transform/convert the resource used in the desired energy form. According to the technology adopted, these processes can be higher or smaller in number and of different types. For the production of electric power, for instance, conversion plants can be classified according to the number and type of conversion and/or transformation processes used (Fig. 4.1).

F. Orecchini and V. Naso, *Energy Systems in the Era of Energy Vectors*,
Green Energy and Technology, DOI: 10.1007/978-0-85729-244-5_4,

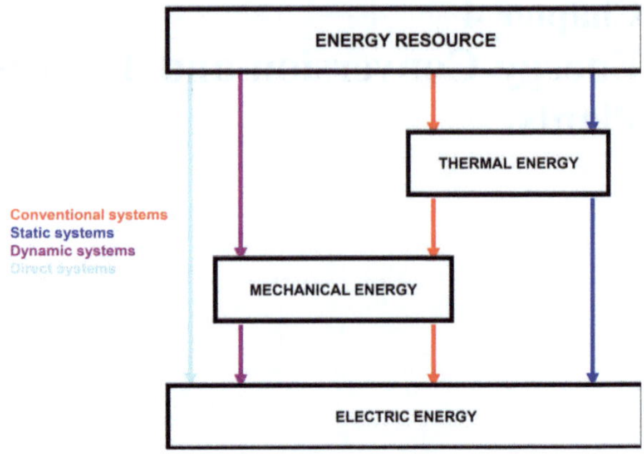


**Fig. 4.1** Electric power production processes

In the so-called *conventional systems* (thermo-electric power plants) the chemical energy of fuels is converted into heat through a combustion process; such heat is converted into mechanical energy, through a thermodynamic[1] cycle, which is in turn converted into electric power by on electric generator.

In the so-called *dynamic systems* the resource used is available in the form of mechanic energy (i.e. water kinetic energy in hydroelectric power plants and wind kinetic energy in wind power plants), *transformed* to obtain mechanical power on a shaft, which can in turn be converted into electric power by a generator.

In *static systems* (owing their name to the fact that they have no moving components), as magneto-hydrodynamic, thermoelectric and thermionic, the chemical energy of the fuel is converted into heat[2] that is directly transformed (without the intermediate passage in the form of mechanical energy) into electric power through the electromagnetic, thermoelectric and thermionic conversion processes.

Finally, *direct systems* directly convert the resource exploited in electric power: these types of systems are the *photovoltaic plants*, which directly convert solar radiations into electric power through a photoelectric process, and the fuel cells, that convert the chemical energy of a fuel (generally hydrogen) directly into electric power through an electrochemical process.

All the systems mentioned above can also envisage other transformation processes such as *mechanical variation systems* (either reduction or increasing), for instance introduced in order to obtain a suitable number of revolutions of the rotor of the generator starting from the number of revolutions of the turbine shaft, and *systems for the transformation of the voltage and frequency* of the electric power produced in order to obtain the suitable utilisation characteristics.

This paragraph contains the description of the most common conversion plants used for the production of the main energy vectors: heat, electricity and hydrogen.

---

[1] This introduces a thermodynamic limit to the conversion process efficiency, the so called *Carnot efficiency*.

[2] These plants can also be powered with solar, nuclear, or other types of energy.

Table 4.1 summarizes the systems under consideration with the relevant conversion.

### *4.2.1 Mechanical Energy-to-Electric Power Conversion Plants*

#### 4.2.1.1 Hydroelectric Plants

*Plant Cycle and Layout*

The cycle for the production of electric power through hydropower plants can be summarizes as follows: due to the effect of the solar energy, the water of seas, rivers and wet surfaces evaporates. The masses of damp air, transported by winds, encounter favourable condensation conditions and originate rainfalls. The water that falls on mountains has a potential mechanic energy that generally disperses through friction by flowing downstream; by limiting this waste by means of suitable pipes, such energy, instead of being dissipated in the form of heat, is transformed (unless unavoidable losses occur) into usable kinetic energy; through a hydraulic turbine, then, the kinetic energy of water is transformed into mechanic work.

Basically there are two plant layouts: *run-of-river plants*, that use the flow rate of the river by following its natural variation, and *reservoir plants*, which allow to fill reservoirs with volumes of water hence making them available according to energy needs, clearly within the volume limits of the reservoir itself. The works that are necessary for the construction of a hydroelectric power plant include:

- *Weirs*: normally, in reservoir plants, intake works consist of dams, whereas in run-of-river works there are simple weirs. Dams are taller since, besides intercepting the water flow, also create a reservoir that is useful to regulate flow rates. Weirs, on the contrary, are works of moderate height that generally limit the level of water upstream within the line of the riverbed.
- *Intake works*: which allow to transport water from the weir to the plant. They are made up of an intake work (endowed with grids and interception devices) followed by a diverting channel that can be either free-surface or a pressure pipeline. Free-surface channels generally have a trapezoidal section and can be dug into the ground and endowed with a concrete coating; pressure pipelines, on the contrary, have a round section and are made with steel pipes.
- *Adduction and transportation work at the plant*: these pipes begin from the forebay reservoir or from the surge tank and bring water to the machines of the plant; they are heavily inclined and made of round steel plate. They always have head and foot valves that allow to prevent the passage of water when needed.
- *Disconnection work*: at the end of the diversion channel is located a *forebay reservoir* if the diversion is an open channel, or a *surge tank* in case of a pressure derivation. The disconnection aims at reducing the effects of the sharp changes in the flow rate caused by the regulation of the power to be produced,

**Table 4.1** Some energy conversion plants

| Conversion | | Plant | Conversion and transformation |
|---|---|---|---|
| From | To | | |
| Mechanical | Electric | Hydroelectric power plants | Transformation of the potential energy of water into kinetic energy<br>Transformation of kinetic energy into mechanical power on a shaft<br>Conversion from mechanic energy into electric power |
| | | Wind power plants | Transformation of the kinetic energy of the air into mechanic power on a shaft<br>Conversion from mechanic energy into electric power |
| Radiant | Electric | Photovoltaic plants | Direct conversion |
| Chemical | Electric | Thermoelectric plants (gas, steam and combined gas-steam plants) | Conversion of the chemical energy of a fuel into thermal energy (combustion)<br>Conversion of thermal energy into mechanic energy (thermodynamic cycle)<br>Conversion of mechanic energy into electric energy |
| | | Fuel cells | Direct conversion |
| Chemical | Mechanic | Internal combustion engines | Conversion of the chemical energy of a fuel into thermal energy (combustion)<br>Conversion of thermal energy into mechanic energy (thermodynamic cycle) |
| Chemical | Electric | MHD (Magneto Hydro Dynamics) | Conversion of the chemical energy of the fuel into thermal energy of a plasma conversion of thermal and mechanical energies og the plasma into electric one |
| Radiant | Thermal | Flat-plate collectors | Direct conversion |
| | | Concentrators | Direct conversion |
| Thermal | Chemical | Chemical reactors[a] | Direct conversion |
| Electric | Chemical | Electrolysers | Direct conversion |
| Radiant | Chemical | Photoelectrolysis | Conversion of radiant energy into electric energy<br>Conversion of electric energy into chemical energy |
| Electric | Thermal | Compression heat pumps | Conversion of electric energy into mechanic energy (refrigerant fluid compression)<br>Conversion of mechanic energy into thermal energy (heat flow) |
| | | Peltier effect heat pumps | Direct conversion |
| Electric | Radiant | Incandescent lamps | Conversion of electric energy into thermal energy (Joule effect)<br>Conversion of thermal energy into luminous radiant energy |
| | | Discharge lamps | Direct conversion |

[a] "Direct conversion" is referred to the only water thermolysis chemical plant that uses heat for the scission reaction. According to the heat production procedures, the plant as a whole operates a transformation of chemical energy (in case heat is produced from the combustion of fossils) or a conversion from radiant energy to chemical energy in case heat is produced in concentration solar plants.

**Table 4.2** Criteria for the construction of hydroelectric power plant

| Function of the work | Reservoir plant | Run-of-river plant |
|---|---|---|
| Weir | Dam | Weir |
| Intake | Pressure | Free-surface |
| Transportation | Pressure tunnel | Open-air or tunnel |
| Disconnection | Surge tank | Forebay reservoir |
| Adduction at the plant | Pressure pipeline | Pressure pipeline |
| Energy production | Open-air or cavern | Outdoor |
| Drain | Outdoor or tunnel | Outdoor |

introducing at the end of the diversion channel a tank that can either temporarily absorb the overflow or provide additional overflow in case of need. These works also aim at reducing the effects of pressure variations (*water hammers*) in water channels because of the manoeuvres on the machines of the plant.

- *Plants*: they can be open-air with the building above the ground, sunken or in a vertical shaft, underground or as caverns. They include the turbine, the electric transformer and all the auxiliary components.
- *Outlet works*: they are made up of an open channel or a pressure channel that gives back the water flow used in the watercourse, hence bringing water from the plant to the watercourse.

The criterion at the basis of the construction of these works varies according to the type of plant made as shown in Table 4.2.

*A Brief Mention on Hydraulic Machines*

The type of turbine to be installed in a hydroelectric plant depends on the head available and on the flow rate. Figure 4.2 shows the diagram relating to the use of the three most widespread turbines: *Pelton*, *Francis* and the *bulb-type turbine* or *Kaplan*. Following the principle of similitude, the two parameters, head and flow rate, define the *characteristic velocity* or the *specific speed*, a parameter that characterises "hydraulically similar" machines. This parameter can be expressed with the following formula[3]:

$$N_s = \frac{n\sqrt{N}}{H_m^{5/4}} \tag{4.1}$$

where $N_s$ is the *characteristic velocity*, n the *nominal speed of the machine*, *N* the *nominal power* and $H_m$ the *hydraulic head.*

[3] It can also be expressed according to mass flow rate *Q*, instead of power *N*, that is to say:

$$N_{sq} = \frac{n\sqrt{Q}}{H_m^{3/4}}$$

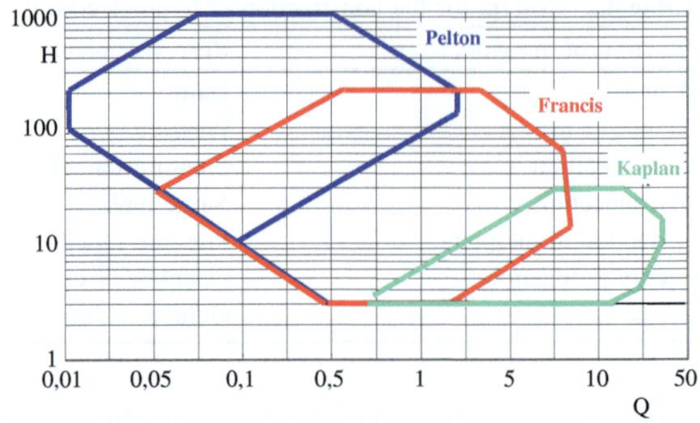


**Fig. 4.2** Graph showing the utilisation of Pelton, Francis and Kaplan turbines according to the flow rate and the hydraulic head

**Table 4.3** Characteristic number for the different types of turbines

| Turbine | $N_{sq}$ |
|---|---|
| Pelton | 5–22 |
| Francis | 20–100 |
| Kaplan | 100–300 |

The characteristic velocity has an important practical meaning since, given the same hydraulic head and power, it is proportional to the rotation speed of the machine, and since the weight of the machine (and thence its cost) decreases as speed increases, machines are developed with the highest possible characteristic velocity. Table 4.3 shows the indicative values of the characteristic velocity of the most commonly used hydraulic turbines[4] [1].

*The Net Head*

The *net head* of a hydraulic turbine is the *difference of hydraulic loads*, i.e. *of the mechanic energy owned by the liquid, between two conventional sections immediately upstream and downstream the machine*. These sections are chosen with operational criteria for the different types of turbines.

Indicating with $H_i$ and $H_o$ the hydraulic loads, with $z_i$ and $z_o$ (in m) the heights (as compared to an arbitrary reference), with $p_i$ and $p_o$ (in kg/m$^2$) pressures, and with $V_i$ and $V_o$ (in m/s) the speeds corresponding to the input and output sections of the machine mentioned above, the net head expressed in meters is:

$$H = H_i - H_o = \left(z_i + \frac{p_i}{\gamma} + \frac{V_i^2}{2g}\right) - \left(z_o + \frac{p_o}{\gamma} + \frac{V_o^2}{2g}\right) \tag{4.2}$$

where $\gamma$ is the specific gravity of water (in kg/m$^3$).

The reference chart is shown in Fig. 4.3: $z$ is the geometrical height, $p/\gamma$ is the static head and $V^2/2$ g is the velocity head.

[4] The values shown are obtained by expressing $n$ in revolutions/min., $Q$ in m$^3$ and $H_m$ in m.

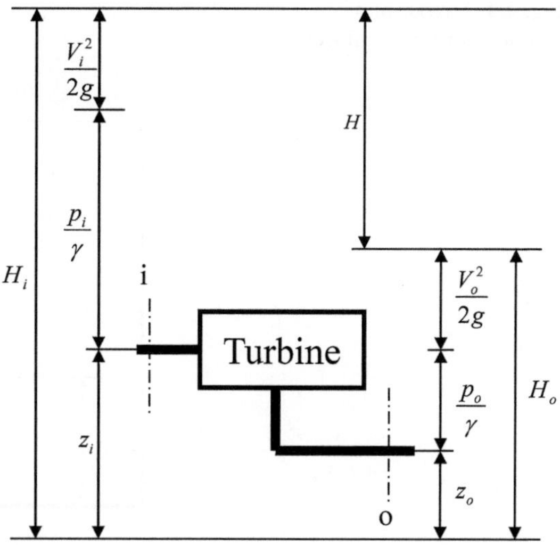


**Fig. 4.3** Layout of a hydroelectric plant

In case of impulse turbines[5] (Pelton) the impeller completely rotates free in the air, hit by one or several jets, therefore the outlet pressure is the atmospheric pressure, namely $p_o = p_a$. Furthermore, the outlet velocity is to be considerated as null, considering calculations in the corresponding energy (kinetic) dissipated in the turbine (the turbine is charged with a loss that as a matter of fact is not related to it). Therefore in case of impulse turbines, the following applies:

$$H_o = z_o + \frac{p_a}{\gamma} \tag{4.3}$$

As an outlet section, the height of the touching point of the jet axis at the average diameter of the wheel is considered; in case of horizontal axis two-jet turbines, the average of the heights of the touching points defined above is considered. In case of multi-jet turbines, the height of the plane containing them is taken into consideration. As an inlet section, the connection flange between the pressure pipeline and the machine is considered. By replacing (4.3) in (4.2), it follows that the net head for impulse turbines is:

$$H = \frac{V_i^2}{2g} + \frac{p_i}{\gamma} + z_i - z_o - \frac{p_a}{\gamma} \tag{4.4}$$

In reaction turbines (Francis and Kaplan), the outlet pressure is different from atmospheric pressure due to the presence of the diffusion-aspiring pipe. This allows not to lose the net head corresponding to the height of the outlet section of the machine at the discharge canal. As concerns Fig. 4.4 by applying the

[5] In general, action (*or impulse*) turbines are those in which there is no pressure difference in the operating period between inlet and outlet from the impeller, whereas in reaction turbine the outlet pressure is lower than the one at the inlet of the impeller.

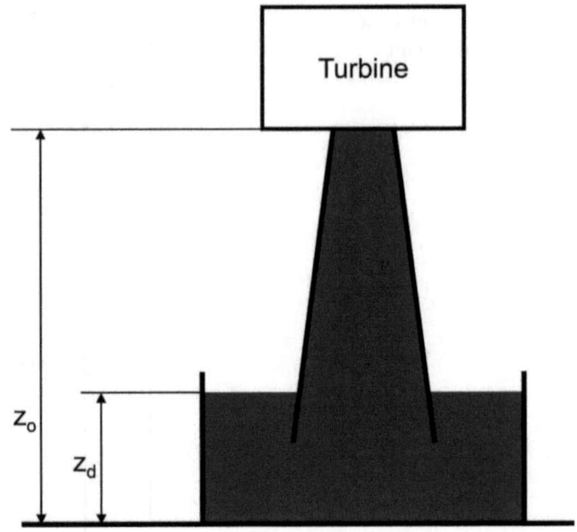


**Fig. 4.4** Layout of a hydraulic turbines diffuser

Bernoulli's principle (with no losses) between the outlet section of the machine (*o*) and the discharge section (d), it follows that:

$$\frac{V_o^2}{2g} + \frac{p_o}{\gamma} + z_o = \frac{V_d^2}{2g} + \frac{p_d}{\gamma} + z_d \tag{4.5}$$

At the discharge, velocity can be considered as null and pressure is the atmospheric one, therefore (4.5) it becomes:

$$\frac{V_o^2}{2g} + \frac{p_o}{\gamma} + z_o = \frac{p_a}{\gamma} + z_d \tag{4.6}$$

From which, bearing in mind (4.2), it is possible to write the equation of the net head for reaction turbines:

$$H = \frac{V_i^2}{2g} + \frac{p_i}{\gamma} + z_i - z_d - \frac{p_a}{\gamma} \tag{4.7}$$

To determine the expression of the hydraulic head according to the height of the intake reservoir, of the discharge canal and of the machine, as regards Fig. 4.5 by applying the Bernoulli equation between the intake section (in) and the inlet section to the machine (*i*), it is possible to obtain:

$$\frac{V_{in}^2}{2g} + \frac{p_{in}}{\gamma} + z_{in} = \frac{V_i^2}{2g} + \frac{p_i}{\gamma} + z_i \tag{4.8}$$

In the intake section, velocity can be considered as null and pressure is the atmospheric one, therefore (4.8) becomes:

$$\frac{p_a}{\gamma} + z_{in} = \frac{V_i^2}{2g} + \frac{p_i}{\gamma} + z_i \tag{4.9}$$

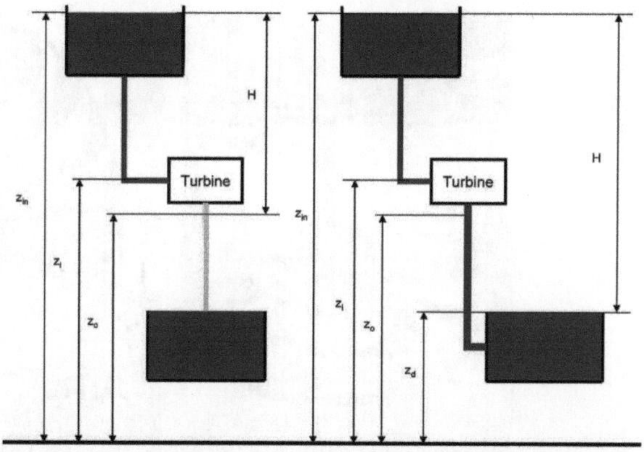


**Fig. 4.5** Layout of a hydroelectric plant with impulse and reaction turbine

**Fig. 4.6** Pelton turbines

By replacing (4.9) in (4.4) and in (4.7) the expressions of the net head for impulse turbines (Himp) and reaction turbines (Hreact) are obtained:

$$H = z_{in} - z_o \tag{4.10}$$

$$H = z_{in} - z_d \tag{4.11}$$

It is therefore possible to see how in impulse turbines the net head corresponding to the height of the outlet section from the machine is lost compared to the discharge canal.

*Field of Application and Efficiency of Hydraulic Turbines*

It is now possible to state that the Pelton turbines are suitable for the highest net heads (impulse turbines, Fig. 4.6), the Francis reaction turbines are used for intermediate heads (Fig. 4.7), whereas for small heads the Kaplan reaction turbines are used (Fig. 4.8). The fields of application, in general, are those shown in Table 4.4 [1].

In case of not-too-high heads, it is economically convenient to use the Francis reaction turbines, even when it would be possible to use the Pelton ones. With the

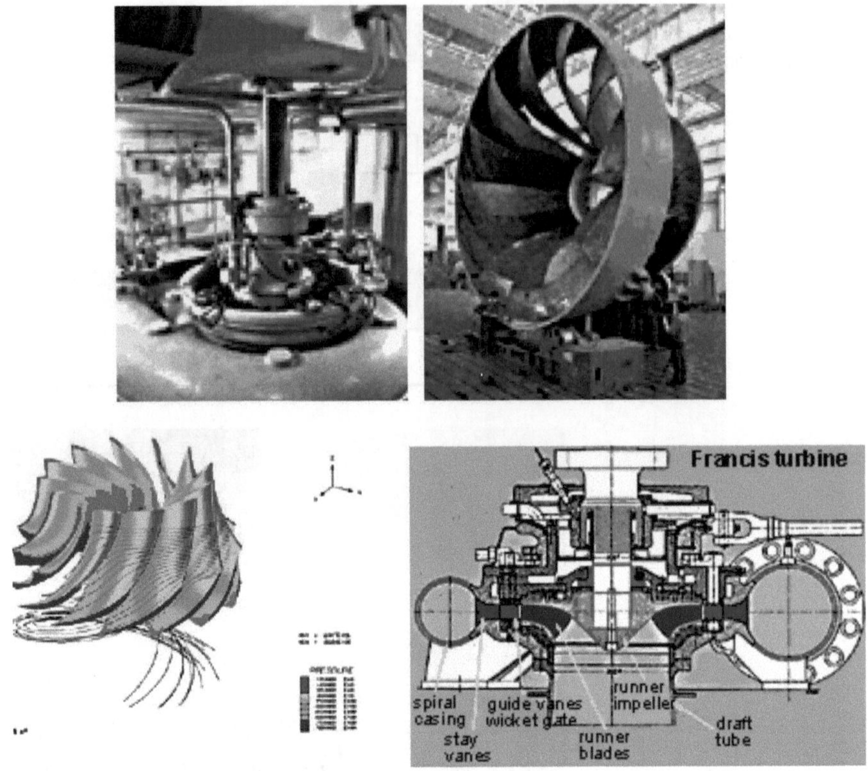


**Fig. 4.7** Francis turbines

continuous fine-tuning of the fluid-dynamic design of blades and with the experience acquired in the solution of problems related to high heads, as well as the use of high discharge heads, there is a continuous trend by Francis turbines to trespass the Pelton turbines' field of application. As the head increases, however, the economic convenience of the Francis turbines decreases, due to the worsening of seal and cavitation problems.

In case of small heads with large flow rates, Kaplan turbines are used: with them, the adjustable blades allow not only to keep a good efficiency in terms of adjustment, but also a good operation in case of variations of the net head—including considerable ones.[6] For this reason, despite of the construction complexity and, therefore, of high costs, the Kaplan turbines tend to trespass the Francis turbines' field of application.

As concerns efficiency, Pelton and Kaplan turbines are endowed with an adjustment system (variation of the jet section with Pelton, variation of the propeller-blade angle with Kaplan) that allows to keep a good efficiency for wide flow

[6] For small heads (i.e. 10 m), head variations, although limited in absolute terms, entail variations that from a percentage viewpoint are relevant for the head itself.

**Fig. 4.8** Kaplan turbines

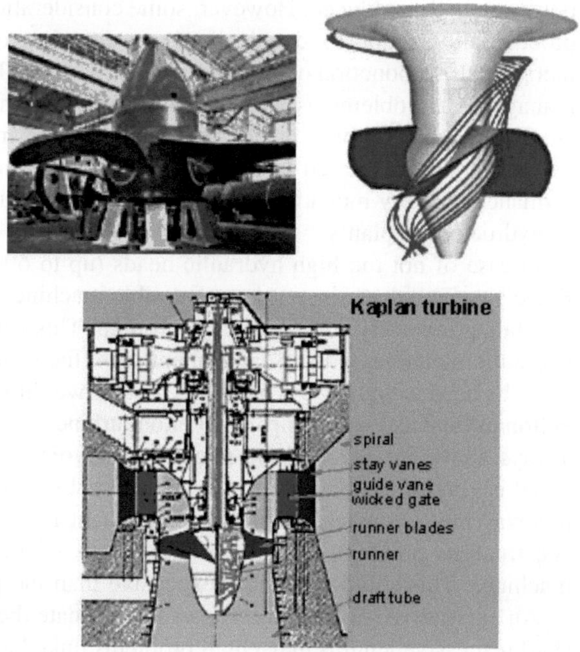


**Table 4.4** Fields of application of hydraulic turbines

| Turbine | Hm (m) |
|---|---|
| Pelton | >500 |
| Francis | 70–500 |
| Kaplan | <70 |

rate variations (given the same hydraulic head and power); the Francis turbine has no adjustment systems and its efficiency markedly decreases for flow rate variations—including limited ones.

*Pumped-storage Plants*

One particular feature of hydraulic plants is that demand of they allow the construction of *pumping plants*, which, in the situations of low electricity, use electric power to pump upstream the water.

These devices are typical of pondage installations in which a pump (Francis turbines are reversible, they can work both as a turbine and as a pump, and therefore in some cases there is one machine only) takes water from the basin downstream back to the basin upstream, consuming electric power (usually in the night lower demand). The alternator, being a reversible engine, while pumping works as an electric motor. It is obvious, while considering efficiency, that the electric power consumed in the pumping phase is higher than the one produced during generation, and therefore the entire process wastes part of the electric power globally produced. However, some considerations are needed. The pumping process allows not to drastically intervene, interrupting the water flow and therefore the production of electricity in some cases. This interruption would entail management problems as well as energy waste higher than the energy spending entailed by the pumping plant. And, what is more important, this process makes the conversion requested to thermoelectric plants more efficient, since their performance are very much influenced by sudden variations of loads than in the case of hydroelectric plants.

In case of not too high hydraulic heads (up to 600 m) it is possible to use the Francis turbine that, since it is a reversible machine, during the pumping process, in rotating inversely, can also act as a pump. This solution, which only includes a hydraulic machine, is more convenient from the economic viewpoint.

If the head is higher, it is necessary to use two different hydraulic machines: one Pelton turbine and one pump. The Pelton turbine, working as an impulse turbine, is always above the level of the afterbay reservoir, whereas the pump, in order to avoid cavitation problems, must be placed below the level of the lower basin; in this way the two machines are located distant from one another, and this entails construction problems linked to the length of the shaft connecting the two machines. This solution is more expensive than the previous one.

An alternative solution that allows to eliminate the problem of the length of the shaft (excessive lengths may entail problems linked to shaft vibrations, at the point of making the work practicably impossible), consists in the creation of two separate plants in overlapping caves, one at a higher height with Pelton-alternator groups, the other one at a lower height with pump-electric motor groups. The cost of these plants heavily exceeds the one of single-cave plants.

Another solution, successfully tested and implemented consists in making a multistage pump working as a turbine by making it turn in the opposite direction. The adoption of particular profiles of the impeller allows to obtain a good efficiency both in terms of generation and pumping. This type of machines allows to adopt binary groups (i.e. groups with one hydraulic machine only) for high heads (600–1,300 m) and high power (larger then 150 MW) [2].

One limitation for multistage reversible groups is the impossibility of adjusting the power supplied since the hydraulic machine has no diffuser and works “all or nothing”.

*General Classification of Hydroelectric Plants*

The power of hydraulic plants varies from some tens to thousands of kW. Obviously there are different construction criteria for the different sizes; the following categories of plants can be identified:

- Micro-hydraulic up to 100 kW.
- Mini-hydraulic from 100 kW to 1 MW.
- Small-hydraulic from 1 to 3 MW.
- Hydraulic over 3 MW.

**Fig. 4.9** Main components of a wind turbine

Larger hydroelectric power plants (small-hydraulic and, most of all, hydraulic) need a long construction time and remarkable initial investments, especially due to the civil engineering works that are necessary. The construction costs for this type of plant is very high, whereas the cost for the production of the electric power supplied is among the lowest ones in the different types of plants.

The potential energy-electric power conversion efficiency amounts to nearly 80% and is mainly determined by the efficiency of turbines (about 90%), by energy waste in pipelines (1–2%) and by the efficiency of the alternator (around 90%).

The utilisation depends on the type of plant: in run-of-river plants, the variation of the power supplied follows the natural course of the flow rate available and the number of hours of operation per year is, on average, 5,000. In basin plants, on the contrary, it is possible to adjust the power supplied according to the demand and reach the highest number of hours of operation per year (nearly 5,800).

#### 4.2.1.2 Wind Power Plants

Overview

Wind power plants convert the kinetic energy of moving air masses into mechanic energy of blade rotation. The set of blades represents *the rotor* of the wind turbine, the *nacelle* contains all the other elements that are necessary for energy conversion, and the tower acts as a support (Fig. 4.9).

In general, the mechanic energy linked to blade rotation can be exploited in two different ways:

- As mechanic energy, for instance to operate a pump (windpumps).
- Transforming it into electric power (aero-generators, also commonly called wind turbines).

Having considered the remarkable and growing importance that the latter machines have taken in world market, further on in this chapter reference is exclusively made to wind power plants destined to the production of electric power.

Apart from this first distinction, wind turbines are subject to several classifications, that are shown below.

*Classification of Wind Turbines*

A first classification divides wind machines on the basis of the position of the rotation axis of blades; it is possible to find *horizontal axis turbines* and *vertical* axis turbines. Vertical axis turbines, as the Darrieus turbine, have had little fortune until today, despite of some evident advantages as the possibility of working at very low wind regimes and the automatic alignment with the instantaneous direction of the wind. Their main limit is represented by a lower conversion efficiency compared to horizontal axis turbines.

As concerns the position of the rotor and of the nacelle, wind turbines can be placed *upwind* or *downwind*.

The almost universally adopted solution is *upwind*, in which moving air "encounters" the rotor and subsequently the nacelle and the tower. This position allows to minimise the disturbance of the tower and the nacelle against the wind before it encounters the blades.

The *downwind* configuration is exclusively used in particular cases, for some small-sized yaw turbines, since the nacelle passively aligns to the direction of the wind. This choice, however, shows the huge disadvantage of the interference of the tower and the nacelle with the air flow, entailing the following consequences:

- Increase of the oscillating forces acting on the machine;
- Lower conversion efficiency;
- Increase of acoustic emissions.

Wind turbines can be employed to feed isolated individual users, or to simultaneously supply electric power to the electricity grid. When a wind turbine has to be connected simultaneously to the electric network (which is the most frequent case nowadays), it is necessary that the energy produced has strict quality features in terms of voltage and frequency. Assuming to generate electric power through an alternator connected to the rotor of the turbine, it must be considered that the rotation speed of the alternator depends on the frequency through:

$$n = \frac{60 \cdot f}{p} \tag{4.12}$$

where $n$ is the rotation speed of the alternator (revolutions per minute), $f$ is the mains frequency[7] (Hz) and $p$ is the number of polar couples of the generator.

The number of polar couples is limited, due to their size and to the cost of the generator, to a maximum of 3. Therefore, in case of a frequency of 50Hz it is possible to obtain the rotation velocity of 3,000 rpm (1 polar couple), but also 1,500 rpm (2 polar couples) and 1,000 rpm (3 polar couples). If we consider, for instance, a rotation speed of 1,000 rpm and a turbine with a diameter of 40 m, the peripheral speed of the tip of the blade amounts to over 2,000 m/s. It is evident that such a speed would entail an overstress for the blades. It is therefore necessary to differentiate the rotation speed of blades from the speed of the generator.

Many manufacturers solve this problem by inserting a reduction gear with a ratio of about 1:50. This is a configuration with a direct connection to the mains and entails a constant rotation speed for the rotor of the wind turbine. The consequence of this option is a non-suitable behaviour whenever conditions differ from those of the design.

An alternative option is to *use axial flux permanent magnet slow generators.* A generator with 30 poles, for instance, rotates with a speed of 200 rpm. This rotation frequency entails a peripheral speed that is still acceptable for small-sized turbines (some tens of kW), due to the small standard radius. For instance, a turbine of 10–20 kW, with a diameter of 8 m, would have a speed at the blade tip of nearly 84 m/s corresponding to a rotation frequency of 200 rpm. Thanks to this configuration, it is possible to make a direct coupling between rotor and generator, avoiding the presence of the reduction gear, hence with a huge saving in terms of weight, cost and energy losses.

The most frequently adopted configuration to avoid the use of the reduction gear, however, is the one of the indirect connection to the grid. In such a configuration, a synchronous generator (alternator), directly connected to the rotor, produces electric power in multifrequency alternating current (since the rotation speed varies); this current is transformed into direct current by a *rectifier*, and subsequently transformed again into constant-frequency alternating current through an *inverter*.

The disadvantages of this configuration are linked to the additional cost of transformation devices. The advantages include the possibility of avoiding the presence of the reducion gear and to obtain a variable rotation speed for blades, with the subsequent optimisation of efficiency for a wide range of wind speed.

The *tip speed ratio*, $\lambda$, is defined as the ratio between the maximum peripheral speed of the blade (that is to say the speed at the tip of the blade, $v_{p)}$, and wind speed ($v$):

$$\lambda = \frac{v_p}{v} = \frac{\omega \cdot r}{v} \tag{4.13}$$

[7] 50 or 60 Hz (see Sect. 3.5.4.1).

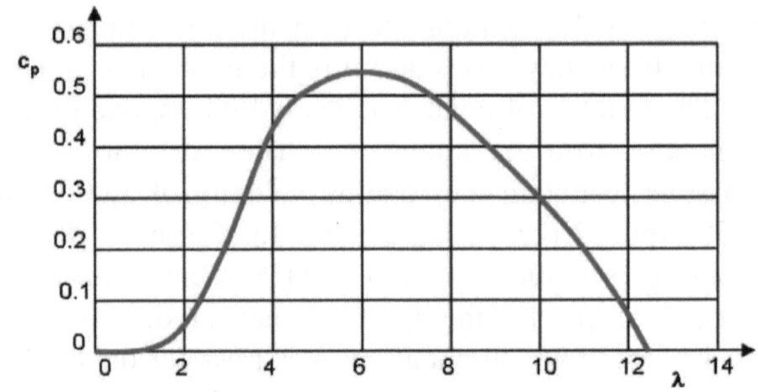


**Fig. 4.10** Power coefficient as $\lambda$ varies

Furthermore, the *power coefficient*, $c_p$ is defined as the ratio between the energy transferred to the rotor and the one contained in the wind current. Please note that $c_p$ is proportional to the efficiency of the machine, although it takes on higher values since it considers neither the mechanical loss in the other components (for instance the reduction gear), nor the electric ones in the generator.

The power coefficient of a wind turbine varies with $\lambda$, reaching a peak whenever a given value applies. Figure 4.10 shows the qualitative trend of the power coefficient according to $\lambda$.

*Fixed rotation speed turbines*, for which $\lambda$ necessarily varies as wind speed changes, operate at optimal $\lambda$ only for a particular value of wind speed. *Variable speed turbines*, on the contrary, operate at optimal $\lambda$ in a wider field of wind speed values, since they are able to adapt their rotation frequency to the instantaneous velocity of the wind, therefore keeping the value of $\lambda$ constant. The result is that the variable rotation speed allows an increase of average efficiency. The choice of the optimal number of blades for a wind turbine is linked to the $\lambda$ ratio. The turbines with a high number of blades have a low value of the optimal ratio of peripheral speed ($\lambda = 1$) and are therefore characterised by a low number of revolutions, given the same wind velocity. They show, however, a remarkable starting torque, and this is the reason why they are used for water pumping or for other purposes in which mechanic energy is directly necessary. The high torque demands a great strength for the rotor and therefore a high number of blades.

Conversely, the turbines with a lower number of blades are used for electric power generation and are characterised by a high rotation speed and a high value of $\lambda$ ($5 \div 8$).

Nowadays, turbines with $\lambda > 8$ are not manufactured, since the level of noise, approximately proportional to the peripheral speed of the blade, would reach inacceptable levels.

The large-size turbines that are presently used for electric power generation are almost exclusively manufactured with three blades. The need for blades with a large diameter, in fact, entails high peripheral speeds at the tip of blades, given the same wind velocity. This means, in practice, a high value of $\lambda$. Three-blade turbines are then preferred, since they operate at the highest efficiency with high values of $\lambda$. The recent trend to manufacture turbines with two or even one blade is linked to the cost reduction that could be obtained. All the large-size turbines

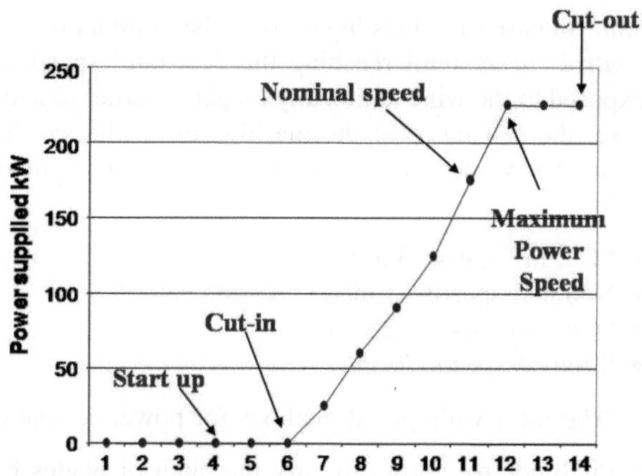


**Fig. 4.11** Power supplied according to wind speed

belonging to the first generation were endowed with two-blade rotors. Nowadays a different approach is followed, according to which a three-blade rotor is preferred, given its more uniform weight distribution in the area swept out. This circumstance entails a higher uniformity of the stress on the structure and also a lower visual impact during the rotation of blades.

According to the present regulations for the certification of wind turbines, the resistance of components must be tested with less frequent and strong winds (between 50 and 70 m/s). It is readily evident that it is not convenient to design a machine able to operate at such values of wind speed, since the additional cost would be huge, while the increase of additional energy production negligible, having considered the low frequency of very strong winds. For this reason, designers prefer to adopt measures that reduce the power absorbed by a wind turbine as the wind speed increases, and put it "out of use" in case such velocity becomes excessive. According to international regulations, this operation includes two braking systems, one for operation and one for emergency. Namely, a *power regulation* is carried out. This operation consists of dispersing excess energy, besides a given value of wind velocity, in order to prevent damages to the plant.

Usually, there are *four running speeds* relating to wind turbines:

- Standstill rotor
- Partial load operation
- Adjustable nominal power operation
- Stop.

In order to characterise the operation of wind turbines, a series of values for the speed of the wind hitting the rotor are defined (Fig. 4.11).

Until the wind reaches a *minimum speed*, called "*cut-in*", the energy transferred to the blades is not sufficient to overcome the inertia of the rotor and of the generator, and the turbine cannot then be started up. Subsequently, the *nominal speed* is defined, as the one at which corresponds the rated power of the machine. Between the "cut–in" and the nominal speed, the machine operates at a *partial load*. In case of values higher than the nominal one, the turbine works in a *power control speed*, until reaching the "cut-out" speed, at which the machine is not exposed to the wind action any longer, in order to avoid structural damages. In any case, the resistance of the machine up to the *survival speed* is guaranteed. The indicative values of the speeds described for a modern turbine are summed up below:

- "Cut-in" speed: 3.5 m/s
- Nominal speed: 13 m/s
- "Cut-out" speed: 25 m/s
- Survival speed: 70 m/s.

The most widespread methods for power regulation are the following:

- Pitch control allows to vary the pitch of blades by rotating them around their longitudinal axis, so as to "take them out" of the wind. Thanks to this control, in case of speed values exceeding the nominal one, the power remains constantly as the rated power. The pitch variation allows a mild control of power, although demanding a blade rotation mechanism as well as an active control of output power that controls the implementation of the pitch.
- *Stall* is based on the homonymous phenomenon. Stall consists in the detachment of the fluid vein from the surface of the blade as the wind speed increases. Following this detachment, the lift force decreases due to the rotation of blades, and the excess energy is dispersed in the form of vortexes. Compared to the control for pitch variation, the "passive" use of the stall phenomenon requires neither a mechanism of blade rotation around their longitudinal axis, nor a control mechanism. The stall, on the other hand, needs a high-level aerodynamic design and also leads to the development of vibrations on the wind machine.
- *Yaw*: the rotor is "taken out" by the wind through a rotation around the vertical axis.
- *Tilt variation*: the rotor is "taken out" by the wind through a rotation around the horizontal axis.

Normally, high-power turbines are controlled by stall or pitch variation, whereas small-sized ones by yaw or by tilt variation.

As a matter of fact, the yaw mechanism is used in all turbines to point the nacelle towards the direction of the prevailing wind, as this direction varies in time. This variation can be carried out passively, by means of a weathercock, or through an active control operated by an anemometer. The signal of the anemometer records the wind direction and transmits this information to the yaw system. The use of weathervanes is allowed for rotors with a diameter up to 10 m, because in case of larger turbines it would be too large to be economically convenient.

The power produced by wind turbines, as already underlined, depends on wind speed. The *nominal power* is the one that corresponds to the nominal speed for a given machine. The energy actually produced by a machine in a given period of time, normally 1 year, cannot be simply calculated by multiplying the nominal power by the number of operations hours, since speed varies in time.

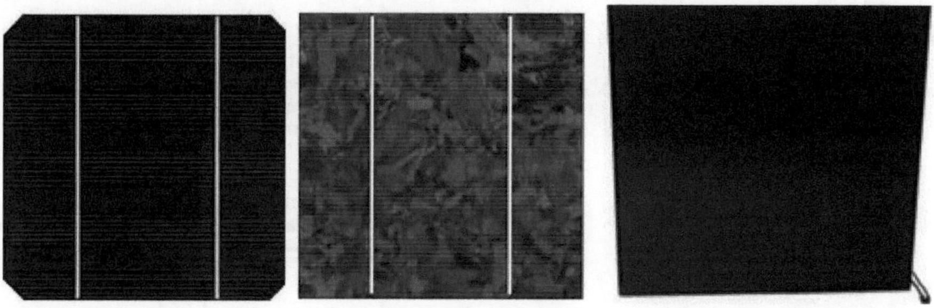

**Fig. 4.12** Photovoltaic cells in single-crystalline, polycrystalline and amorphous crystalline silicon

The producibility of a wind turbine is therefore strongly related to the *speed distribution of the site* and can be expressed in *equivalent hours*, meaning the hours of operation at the nominal power that usually reach a maximum number of 3,000 per year. From an economic viewpoint, the use of wind turbines to generate electric power becomes convenient in those sites in which the equivalent hours of operation per year range at least between 1,500 and 2,000.

## *4.2.2 Radiant Energy-to-Electric Power Conversion Plants*

These plants are based on the "direct" conversion of solar radiant energy into electricity as a consequence of the *photovoltaic effect.*

### 4.2.2.1 Photovoltaic Plants

The Photovoltaic Cell

The element at the basis of photovoltaic devices is the *photovoltaic cell* (Fig. 4.12). It is available is different sizes and can be made of different materials. The most widespread cells are in crystalline silicon (monocrystalline and polycrystalline) and in amorphous crystalline.[9] A typical form is squared, with a surface of nearly 100 $cm^2$, although larger cells can be manufactured.

The curve describing the trend of the current of the cell according to its voltage, called *characteristic curve*, shows the behaviour of the cell both in the dark and in the light (Fig. 4.13).

By applying a potential difference in the dark, there is a flow of current only if the value exceeds a given threshold.

[9] Other materials (for instance organic materials) are being studied.

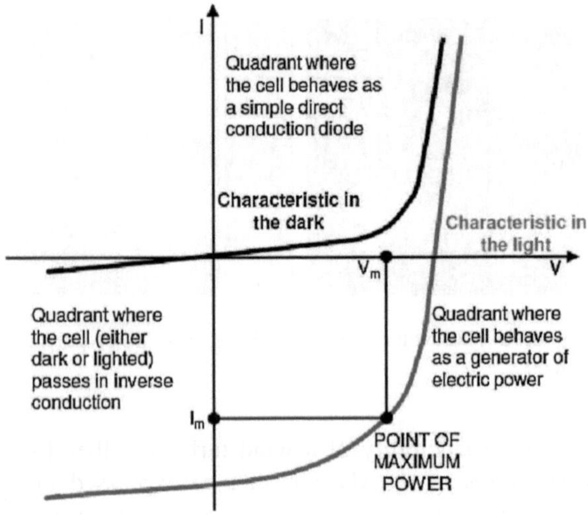


**Fig. 4.13** Typical curve *I–V* of a photovoltaic cell

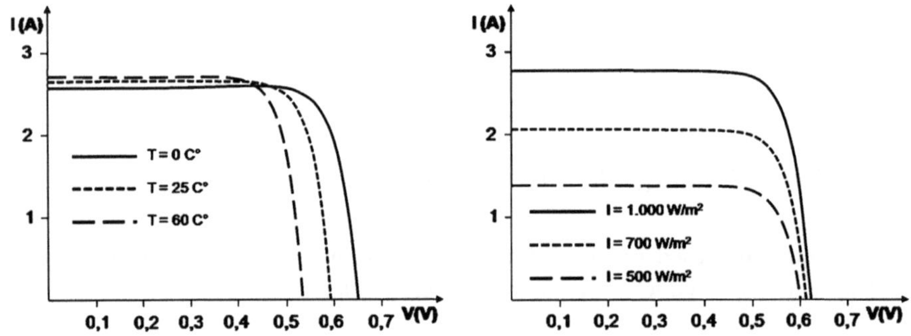


**Fig. 4.14** Variation of curve *I–V* with the temperature of the cell and with the solar radiation intensity

On the contrary, by exposing the cell to the light electric power is generated in any case; if the junction is in short circuit ($V = 0$) a maximum current is generated ($I_{sc}$), whereas in case of open-circuit (I=0), a maximum voltage is determined on the two sides ($V_{oc}$). In the case of standard silicon cells ($10 \times 10$ cm), $I_{cc}$ is nearly 3 A, whereas $V_{oc}$ amounts to about 0.6 V. Having considered that the power of a generator is the product of its voltage by the current generated, in the two cases described above the power will be null, and from the characteristic curve it is possible to identify the point in which the system provides the maximum power.

The characteristic curve of a cell depends, besides its *intrinsic properties* (the material), also on two important parameters: the *intensity of luminous radiation* and the *temperature of the cell itself* (Fig. 4.14). It is evident, therefore, that at the maximum power it takes different values as $I$ and $V$ vary; the corresponding values of current and voltage are called *current* ($I_{pmax}$) and *maximum power voltage* ($V_{pmax}$).

Temperature increase cause a decay in the electric performance of the photovoltaic device. In a silicon cell, for instance, whilst the short circuit current undergoes a negligible variation, the voltage of the open circuit decreases as temperature increases (nearly 2 mV/°C). The heating of a photovoltaic cell entails, therefore, an overall decrease of its performance.

Since photovoltaic system are powered by a variable source and since, as just analysed, their performance are linked to the characteristics of solar radiation. In order to quantify the performance of a photovoltaic cell (and more generally of whatever photovoltaic device), it is necessary to refer to precise conditions, known as *standard conditions*:

- *Radiation intensity* 1.000 W/m$^2$
- *Cell temperature* 25°C
- *Solar spectrum* AM1, 1.5 (correspong to a solar zenith angle 4819).

The rated data of the power generated by a photovoltaic cell is expressed in *Watt peak* (Wp); the *peak power* is the one generated in the standard conditions mentioned above. The *conversion*, efficiency is the ratio between the electric power *P*, provided by the cell and the radiation power that strikes the frontal surface, expressed as a product of the density of incident radiation *I* (W/m$^2$) and the area of the surface exposed *A* (m$^2$):

$$\eta = \frac{P}{I \cdot A} \tag{4.14}$$

The *nominal conversion* efficiency is calculated in standard conditions: therefore the density of incident radiation *I* is 1,000 W/m$^2$ and by indicating with $P_n$ the peak power of the cell:

$$\eta_n = \frac{P_n}{1{,}000 \cdot A} \tag{4.15}$$

The nominal efficiency of monocrystalline silicon commercial cells also reaches 18%; those in polycrystalline silicon do not exceed 14%, whereas those in amorphous silicon reach nearly 8%.

The power generated in operation is almost always different from the maximum nominal power. Two important factors influence, in fact, the output of a photovoltaic generator: *radiation intensity* and *temperature* of the photovoltaic cell (Fig. 4.14).

The first factor influences the short circuit current and subsequently the current generated at maximum power. One can approximately say that the short circuit current proportionally decreases as radiation decreases.

The second factor, on the contrary, acts on the open-circuit voltage and therefore also on maximum power. In this case as well it is possible to assume that the maximum power voltage undergoes a linear decrease with temperature increase.

The result is that the operating power decreases if compared to the reference value (maximum power or peak power).

*The Applications of Photovoltaic Technology*

There are many possible applications of photovoltaic technology and range from the power devices of very few hundreds of Watt (mostly pocket calculators and watches) up to plants of several MWp.

In the past, having considered the prohibitive costs that characterised these systems, mainly *small-size applications* were developed since they did not require burdensome investments. However, as the cost of modules decreased, larger and larger applications gained ground until reaching the 90s with *photovoltaic plants*. They showed that high quantities of electric energy can be easily produced with photovoltaic technologies, and that the only limitation to their development is cost. If prices decrease to levels that allow to compete with electricity production from traditional sources, there will be no more barriers to its spreading. The constant decrease of the cost of modules led the sector's operators to perform a careful evaluation of those costs that are not directly depending on the photovoltaic module. Thanks to this new approach, the attention of scholars and designers shifted from large size power plants (high costs for bearing structures) to smaller integrated systems in buildings, which represent the most promising solution for photovoltaic systems.

*The Photovoltaic Plant*

Starting from *finished cells* it is possible to constitute *modules*, by adapting the voltage and the current of the group of cells to users' needs. The voltage of a single cell, in fact, is on average low for all materials (around 0.5 V), whereas, by connecting cells in series, the desired potential difference is obtained, which is the sum of the various potentials. Several panels, electrically connected in series, constitute a *string* whose voltage will be the sum of the voltage of the different panels forming the string itself, whereas the current will be the same as in a single panel. Several strings, electrically connected in parallel, make up the *photovoltaic field*: the voltage of the field is the same as one of the strings, and the current is the sum of the current of the stack.

Another component of photovoltaic plant is the *maximum power point tracker* (MPPT). The MPPT allows working current and voltage to be the maximum power ones at any operating (incident solar radiation, temperature) and load condition.

The other components of plants vary according to the following types of plants:

- *Stand-alone systems*
- *Grid-connected systems*.

Stand-alone systems envisage a *storage system* that guarantees current supply even in the hours of lower lighting or of dark. The elements making up a stand-alone system include the *photovoltaic field*, the *charge controller*, the *battery* and, in case the equipment must be powered through alternating current, an *inverter*.[10]

[10] Device converting the inlet direct current into alternating one.

The storage system is normally *electrochemical (electric accumulators)*; the choice is between lead–acid batteries, the most widespread ones, nickel–cadmium and nickel–metal hydride batteries, lithium batteries which are more expensive.

The charge controller acts as a protection of accumulators. In fact, according to the type of batteries selected, some precautions must be made on the voltage at which the charge and discharge phases are carried out. In case the threshold voltage is exceeded, the charge controller interrupts the current, and it allows current to flow again only when voltage decreases below the threshold value.

This type of plant is used in those cases in which the electric grid is absent or hardly accessible (i.e. remote houses, mountain refuges, pumping systems, telecommunications systems, repeater stations, mark buoy, etc.).

The systems connected to the grid release the excess of energy generated in the hours of sunshine and take it in the hours in which the generator is not sufficient to meet the overall demand. They necessary include an inverter that transforms the direct current produced by a photovoltaic generator into an alternating current as similar as possible to the one supplied by the grid. However, the type of periodical function that represents the current trend over time have a perfectly sinusoidal trend, differently from the inverters in stand-alone applications, which almost always can provide a square wave signal (low-quality of the signal) without creating any problems to the equipment they supply. In this case it is necessary to obtain an output signal with a high degree of purity. It is also necessary that the current injected in the grid is "in-phase" with the grid itself. Quite often, gride-connect system—for either simplicity, or feed-in tariff advantages-supply all the produced electricity to the grid even when integrated in buildings.

*Energy Produced by a Photovoltaic System*

Since the nominal power (peak power) of a photovoltaic plant is only a reference value, it is fundamental to evaluate the actual quantity of energy that the plant itself is able to actually produce in a given period of time.

Besides the nominal characteristics of the plant itself (number and efficiency of panels), such quantity of energy depends also on other parameters such as:

- *The insolation conditions* in the installation site: the incident solar energy (kWh/$m^2$ year);
- *The position of modules in space*: tilt, azimuth and eventual shadings;
- *Balance of system (BOS) efficiency*: the efficiency of all the other systems of the plant (inverters or accumulators), the losses in connection cables, etc..

Once the site characteristics (latitude, altitude and horizontal radiation) and the position of panels (tilt, azimuth and eventual shadings) are known, it is possible to calculate the incident specific energy $H_p$ (kWh/$m^2$), on panels (Fig. 2.9): once the useful surface of panels $A$ ($m^2$) is known, the total incident solar energy on panels is calculated. The electric power (in direct current) $E_{dc}$ (kWh) produced by the photovoltaic field is:

$$E_{dc} = H_p \cdot A \cdot \eta_p \tag{4.16}$$

Such energy shall be converted into alternating current and then transferred to users, and in these operations part of the energy is lost. The BOS efficiency measures the amount of said losses[11]. The *useful energy* $E_u$ produced by the plant is, therefore:

$$E_u = E_{dc} \cdot \mathrm{BOS} = H_p \cdot A \cdot \eta_p \cdot \mathrm{BOS} \tag{4.17}$$

By referring to the surface unit and assuming $\eta_p = 16\%$ and BOS $= 90\%$, (4.17) becomes:

$$E_u = E_{cc} \cdot BOS = H_p \cdot 0.144 \tag{4.18}$$

that is to say the useful energy accounts for 14.4% of the incident solar radiation. In considering, as an example, $H_p = 1.800$ kWh/m$^2$ year[12] it follows that with 1 m$^2$ of panels it is possible to produce nearly 260 kWh/year.

Equation 4.17 can be written according to the nominal power of the plant $P_n$ (kW). By definition of the nominal efficiency of the panel[13]:

$$\eta_p = \frac{P_n}{A} \tag{4.19}$$

By replacing (4.19) in (4.17) it follows that:

$$E_u = H_p \cdot P_n \cdot \mathrm{BOS} \tag{4.20}$$

As regards the nominal power unit, the useful energy is a fraction (equal to BOS) of the specific incident solar energy. The product of $H_p$ by BOS represents the equivalent hours for the photovoltaic panel. On the basis of the hypotheses a 1 kWp plant produces nearly 1.600 kWh/year.

### *Costs and Sizing General Criteria*

The sizing of a photovoltaic system can follow different approaches, according to whether a stand-alone or a grid-connected plant is taken into consideration.

A stand-alone system, in fact, must in any case guarantee to users the meeting of the consumptions envisaged. The entire set of the photovoltaic system and of

---

[11] As a matter of fact, the definition of BOS efficiency often includes other factors of loss that are not linked to other parts of the system, but still to the module and in particular to some of its peculiar operating conditions. These are the *losses due to the effect of temperature, losses due to reflection on the frames of modules*, etc. In the case of isolated systems, the battery charge and discharge process entails an additional loss, which can be estimated as amounting to nearly 10% of energy.

[12] Solar radiation in Rome with a tilt of 30°, azimuth equal to 0 and without shading.

[13] Standard conditions provide for an incident radiation of 1 kW/m$^2$.

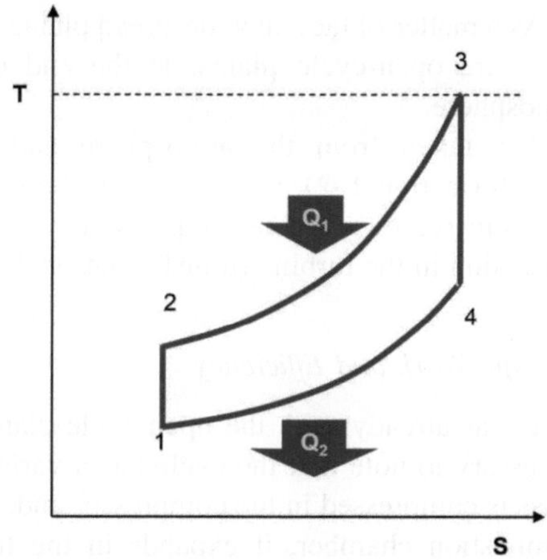


**Fig. 4.15** The ideal Joule cycle

storage batteries must therefore be able to meet the entire electric power needs, and sizing is consequently strictly related to consumption.

In case of grid-connected systems, on the contrary, there is not any new constraint, since users can be directly supplied through the electric grid, which acts as an almost unlimited storage capacity. The starting points for sizing can therefore be different from consumption and, in particular, they can be linked to the surface available or to the initial investment budget, as well to feed-in tarrif rules.

The cost of photovoltaic modules is obviously different according to the size of the plant to be developed, and is continuously evolving. Photovoltaic modules are presently sold "on a power basis", meaning that, for instance, several polycrystalline modules are sold at the same price as a minor quantity of single-crystalline modules provided that the nominal power of the two groups is the same.

Prices are continuously evolving, also in consideration of the incentives in force, and the present trend indicates a progressive reduction of prices.

### *4.2.3 Chemical Energy-to-Electric Power Conversion Plants*

#### 4.2.3.1 Gas Turbine Plants

Thermodynamic Cycle

The basic reference cycle is the Joule cycle (Fig. 4.15[14]) made up of two adiabatic processes (compression and expansion) and two isobaric processes (heating and cooling).

[14] Considering air, a perfect gas is $H = c_p{\cdot}T$ and being $c_p$ constant $H$ and $T$ only differ for a scale factor, therefore the cycle has the same form on planes *H–S* and *T–S*.

As a matter of fact, in widespread plants the cooling process does not exist since these are open-cycle plants; at the end of expansion, air is released into the atmosphere.

Air, taken from the atmosphere and compressed in an axial compressor (transformation 1–2), is then sent to the combustion chamber where its temperature is increased at a constant pressure (transformation 2–3); this is followed by the expansion in the turbine (transformation 3–4) and the release into the atmosphere.

*Specific Work and Efficiency*

Since, as already said, the open-cycle plant is an *internal combustion* plant, it is necessary to note that the cycle has a variable mass, since a given quantity of air mass is compressed in the compressor and, together with the fuel introduced in the combustion chamber, it expands in the turbine. Having defined $\alpha$ as the ratio between the air mass and the fuel mass, and LHV the lower heating value of the fuel, it follows that:

$$Q_1 = \frac{\mathrm{LHV}}{\alpha + 1} = c_p(T_3 - T_2) \Rightarrow \alpha = \frac{H_i}{T_3 - T_2} - 1 \quad (4.21)$$

Let's consider, as an example, the following values of the characteristic parameters:

LHV = 10,000 kcal/kg (gasoil)

$T_3 = 1{,}300$ K

$T_2 = 549$ K ($T_1 = 27°$C, $\beta = p_2/p_1 = 9$)

$c_p = 0.26$ (mean value between $T_3$ and $T_2$).

From (4.21) it is possible to obtain $\alpha = 53.8$ which means that to each kg of air react with 0.0186 kg of fuel; therefore the constant mass can be approximately accepted.

Here it is recalled that compression 1–2 (1–2′ in the real case) and expansion 3–4 (3–4′ in the real case) originate, respectively, compression work $L_c$ and expansion work $L_t$, with regard to Fig. 4.16. The efficiency of the ideal cycle[15] is:

$$\begin{aligned}\eta_l &= \frac{L_t - L_c}{Q_1} = \frac{c_p(T_3 - T_4) - c_p(T_4 - T_1)}{c_p(T_3 - T_1)} \\ &= 1 - \frac{T_4 - T_1}{T_3 - T_2} = 1 - \frac{T_1 T_4/T_1 - 1}{T_2 T_3/T_2 - 1}\end{aligned} \quad (4.22)$$

Taking into consideration the status equations and processes, the result is:

$$\frac{T_1}{T_2} = \left(\frac{P_2}{P_1}\right)^{\frac{k-1}{k}}; \quad \frac{T_4}{T_1} = \frac{v_4}{v_1}, \frac{T_3}{T_2} = \frac{v_3}{v_2}; \frac{v_3}{v_2} = \frac{v_4}{v_1} \Rightarrow \frac{T_4}{T_1} = \frac{T_3}{T_2} \quad (4.23)$$

[15] In the ideal cycle, the reversible isentropic compression and expansion transformations are considered.

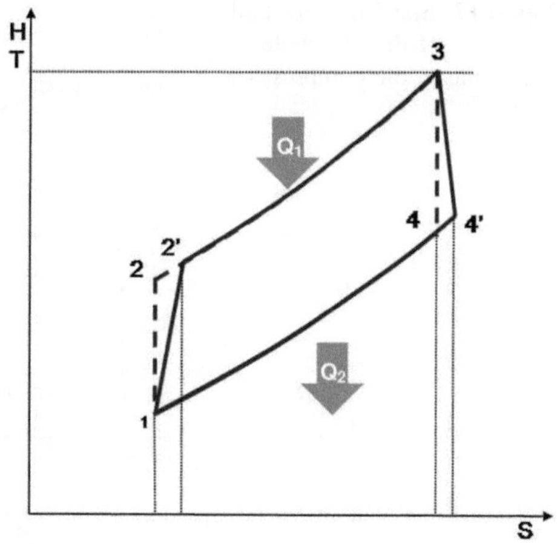


**Fig. 4.16** Ideal (*dashed line*) and real (*continous line*) Joule cycle

Subsequently, since the compression ratio $\beta$ and the parameter $\varepsilon$ are defined as follows:

$$\beta = \frac{p_2}{p_1} \tag{4.24}$$

$$\frac{k-1}{k} = \varepsilon \tag{4.25}$$

By replacing (4.23), (4.24) and (4.25) in (4.22), the ideal efficiency $\eta_i$ is obtained:

$$\eta_i = 1 - \frac{1}{\beta^{\varepsilon}} \tag{4.26}$$

Indicating with $\eta_c$ the *efficiency of the compressor* and with $\eta_t$ the *efficiency* of the *turbine*, the real specific work of cycle $L_r$ is:

$$L_r = L_{tr} - L_{cr} = c_p\eta_t(T_3 - T_4) - \frac{c_p}{\eta_c}(T_2 - T_1) \tag{4.27}$$

After adimensionalising, dividing by $c_pT_3$ ($T_3$ is fixed at a maximum value for technological reasons), and considering

$$\tau = \frac{T_3}{T_1} \tag{4.28}$$

through a procedure analogous to the previous one, it is possible to obtain:

$$\frac{L_r}{c_pT_1} = \eta_t\left(\tau - \frac{\tau}{\beta^{\varepsilon}}\right) - \frac{1}{\eta_c}(\beta^{\varepsilon} - 1) \tag{4.29}$$

The trends in efficiency and in the specific work according to $\beta$ are shown in Fig. 4.17. The respective maximum values are obtained for different values of the

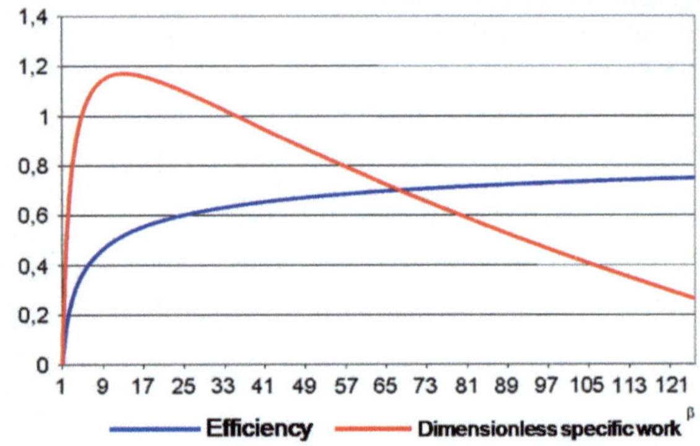


**Fig. 4.17** Specific work and efficiency of the gas cycle according to the compression ratio

compression ratio. Assuming $T_3 = 1.300$ K, $T_1 = 300$ K, $k = 1.38$, $\eta_c = 0.88$ and $\eta_t = 0.9$, the maximum specific work is obtained with $\beta = 8.4$ ($L = 1.0231$ kJ/kg and $\eta = 0.3454$), whereas the maximum efficiency is obtained with $\beta = 20$ ($\eta = 0.388$ and $L = 859.4$ kJ/kg). Considering that the trend of the efficiency curve near the maximum level is flatter than the specific work, intermediate values of the compression ratio are adopted, although closer to the optimum of the specific work: $\beta = 10 \div 12$ (with $\beta = 12$ it is possible to obtain $\eta = 0.3716$ and $L = 996.3$ kJ/kg).

As already mentioned, the maximum temperature of the cycle is imposed by technological factors that concern the resistance of the blades of the turbine to high temperatures: mechanical strength, creep, corrosion. By using cooled blades, the maximum temperature presently achievable is 1,300–1,350°C. With $T_3 = 1{,}350$°C the maximum specific work is obtained with $\beta = 12.9$ ($L = 2{,}147.4$ kJ/kg and $\eta = 0.4088$) whereas the maximum efficiency is obtained with $\beta = 38$ ($\eta = 0.4605$ and $L = 1{,}777.8$ kJ/kg).

*Thermal Regeneration*

With the values of maximum temperature of cycle $T_3$ and of the compression ratio $\beta$ mentioned above, it is possible to obtain a temperature $T_2$, $\approx$ 320°C at the end of the compression and $T_4' \approx 500$°C at the end of the expansion. Having considered $T_4' > T_2'$ thermal regeneration is therefore possible. It consists in using the heat of flue gases at the discharge of the turbine (corresponding to area CF4′D in Fig. 4.18) to preheat the air before accessing the combustion chamber. Taking into consideration the thermal gradients that are necessary for thermal exchange, it is not possible to perform a "complete" regeneration; the heat that is actually usable to preheat the air is the one corresponding to area C′F′4′D in Fig. 4.18 that will take the air before accessing the combustion chamber at temperature TP′ (in the ideal case of complete regeneration, the temperature TP would be reached). In this connection the *degree of regeneration R* is defined:

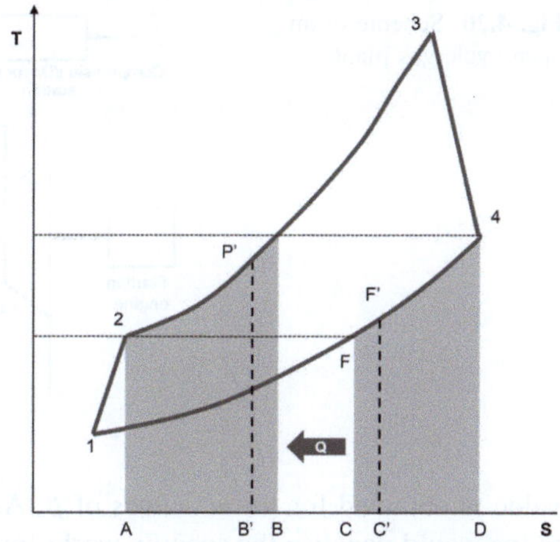


**Fig. 4.18** Joule cycle with thermal regeneration

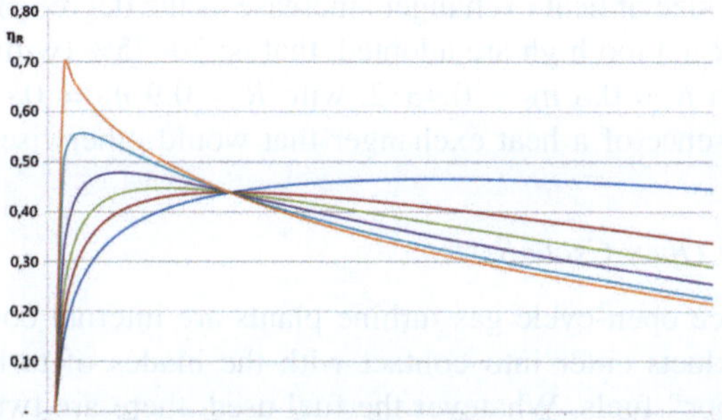


**Fig. 4.19** Gas cycle efficiency regenerated according to the compression ratio, on the basis of some values of the degree of regeneration

$$R = \frac{c_c\left(T_4' - T_F'\right)}{c_p\left(T_4' - T_F\right)} \tag{4.30}$$

The quantities of heat $Q_1$ and $Q_2$ that the fluid exchanges with the outside become:

$$Q_{1R} = c_p\left(T_3 - T_4'\right) + (1 - R)c_p\left(T_4' - T_2'\right) \tag{4.31}$$

$$Q_{2R} = c_p\left(T_2' - T_1\right) + (1 - R)c_p\left(T_4' - T_2'\right) \tag{4.32}$$

Consequently, the efficiency, according to the characteristic parameters and to the degree of regeneration, is:

$$\eta_R = \frac{1}{\eta_c\eta_t\left(1 - \frac{1}{\beta^\varepsilon}\right) - (1 - R)\left[\eta_t\left(1 - \frac{1}{\beta^\varepsilon}\right) - 1 + \frac{1}{\tau}\left(\frac{\beta^\varepsilon - 1}{\eta_c} + 1\right)\right]} \eta_c\eta_t\left(1 - \frac{1}{\beta^\varepsilon}\right) - \frac{1}{\tau}\left(\beta^\varepsilon - 1\right) \tag{4.33}$$

The efficiency trend is plotted in Fig. 4.19 against $\beta$ by varying $R$. As the degree of regeneration increases, also the maximum value of efficiency increases, and that

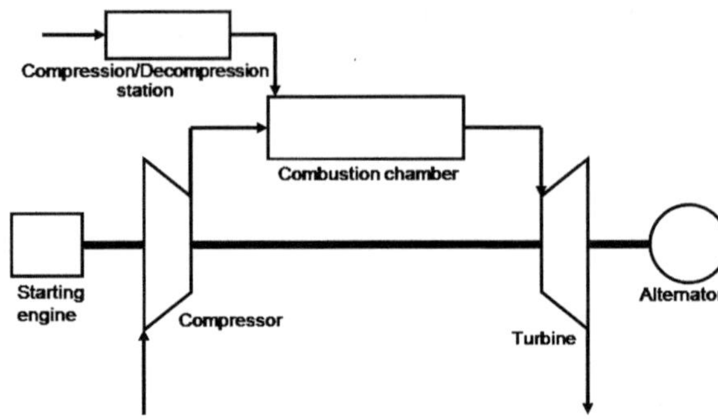


**Fig. 4.20** Scheme of an open-cycle gas plant

value is obtained for lower values of $\beta$. As a matter of fact too low compression ratios would penalise the specific work; for this reason, also considering that costs and size of heat exchangers increase as the degree of regeneration increases, values of $R$ not too high are adopted, that is: 70–75% (with $\beta = 8$; $R = 0.7$ $\eta_R = 0.4189$, with $R = 0.8$ $\eta_R = 0.4332$, with $R = 0.9$ $\eta_R = 0.4486$). Regeneration entails the presence of a heat exchanger that would otherwise be absent in gas plants.

*The Open-Cycle System*

Since open-cycle gas turbine plants are internal combustion systems, combustion products enter into contact with the blades of turbines; this prevents the use of "poor" fuels. Whatever the fuel used, there are two types of limits of tolerability of "contaminants": ***solid particles*** that exceed some sizes and cause *erosion*, and ***chemical contaminants*** (vanadium, sodium, potassium, lead, nickel, calcium) that cause *corrosion*. The problem of corrosion and erosion of turbine blades remains in any case among the most severe ones for these systems.

One of the attractive peculiarities of these plants is their relatively simple construction. In its simplest form, the plan of the system is shown in Fig. 4.20.

The turbine, the compressor and the alternator are assembled on the same shaft: the compressor and the alternator are operated by the turbine. In order to be injected in the combustion chamber, the fuel must be at a given pressure, normally amounting to nearly 20 bar.[16] The pressure at which gas is supplied varies according to the type of gas distribution network; in any case, there is usually no guarantee of a constant pressure of supply.

In the decompression station, gas decompression usually takes place (normally the supply pressure is nearly 35 bar) or, when the supply pressure drops below

[16] At the usual values of the compression ratio, nearly $\beta = 12$, the air is in a combustion chamber at the pressure of nearly 12 bar. In those cases in which high compression ratio values are adopted, $\beta = 30$, obviously a higher pressure for natural gas, nearly 35 bar, is necessary.

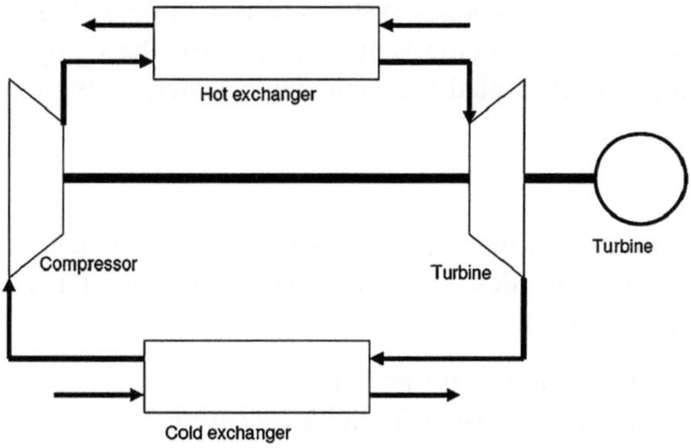


**Fig. 4.21** Scheme of a closed-cycle gas plant

some given values, a compression is performed, hence guaranteeing a constant pressure at the entry point of the combustion chamber.

At the starting of the plant, the turbine is boosted by a suitable starting engine (usually a Diesel one) that feeds it until fuel supply begins, and the ignition with suitable sparking plugs follows. Subsequently, the fuel capacity is increased according to a well-defined time law, in order to avoid thermal overloads. However, any start-up causes a stress to the plant that for the effects of ageing, costs several hours of operation at full capacity.

*The Closed-Cycle System*

As already said, in general the plant is an *open-cycle system*. It is however possible, as in the case of gas power plants, to implement the *closed-cycle system* (Fig. 4.21). This entails undeniable advantages as far as the cycle efficiency is concerned:

- By regulating the pressure at the entry of the compressor, and therefore the flow rate, it is possible to have a *wide variation of the load at a rather constant efficiency*;
- There is the possibility of *choosing the working fluid* with convenient physical characteristics;
- It is possible to work, given the same compression ratio, with pressurised fluid, at a higher average pressure of the cycle, and therefore with a *higher specific work* given the same other electric parameters;
- The system would be an external combustion plant, and therefore the fluid enters the turbine "clean", sharply reducing the problems of blade corrosion, and whatever type of fuel could be used.

On the other hand, in the closed-cycle plant, two *heat exchangers* are necessary—one hot and one cold—that were not present in the open-cycle, and therefore the plant is *far more complex and more expensive*. Furthermore, a certain availability of refrigerated water for the cool exchanger is needed, necessary in the open-cycle solution. Considering that the system simplicity is among the main strengths of turbogas systems, it is easy to understand why almost all of them are open-cycle. For the same reason, also the convenience of thermal regeneration must be evaluated for each single system at the moment of construction.

### *Further Improvements*

Further measures to improve the performance of turbogas plants have been studied. Besides the already-mentioned thermal regeneration, there are two main improvements introduced:

- *Intercooled compression*
- *Fractioned expansion with reheating.*

### *Power Regulation*

The regulation of the power supplied occurs by varying the initial expansion temperature $T_3$ acting on the fuel flow rate.[17]

Having considered that the conditions of the fluid vary, and since the rotation speed must remain constant (at least in case of electric energy production, to mantain grid frequency synchronous speed), the efficiency of the turbine, and therefore of the cycle, heavily decreases while departing from the conditions provided for in the plan. However, taking also into account that around 2/3 of the power supplied by the turbine are absorbed by the compressor and nearly 1/3 by the alternator, and that the fluid conditions at the beginning and at the end of the compression do not vary (therefore the power absorbed by the compressor remains unchanged), the decrease in the power supplied by the turbine is highly lower than the reduction of electric power that is intended to be obtained.[18] This allows rather wide regulations of the electric power supplied without excessively damaging efficiency.

A plant solution that allows regulations of the power supplied with limited decrease of efficiency is represented by *two-axle plants*. In these plants there are two different shafts: one moves the alternator, the other moves the compressor. Obviously, a turbine must be mounted on each shaft, and therefore *expansion* must be *divided* into two machines with power ratios produced approximately as 2/3 (for the compressor) and 1/3 (for the alternator).

In this way the rotation speed of the compressor is not related to the one of the alternator, and it is possible to assign to the compressor, as the load varies, the speed that allows the maximum efficiency for that load value. The two turbines can operate either in series or in parallel.

---

[17] From (4.21) it is readily evident that by decreasing the fuel flow rate, $T_3$ decreases.

[18] For instance, to halve the electric power supplied, it is necessary to bring the power supplied by the turbine at nearly 83% of the nominal value.

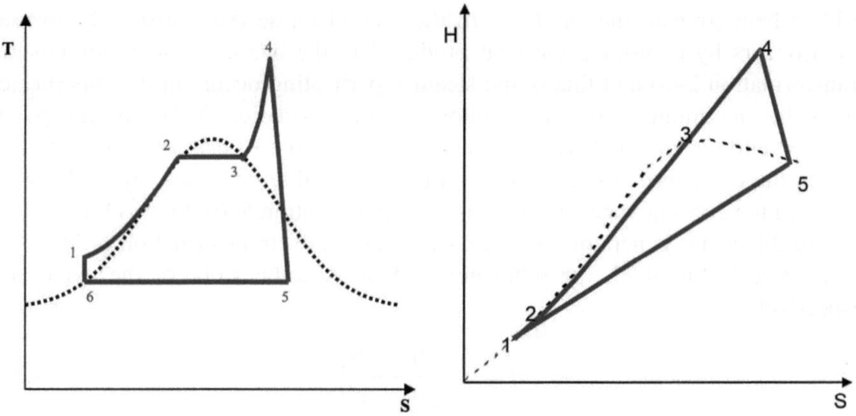


**Fig. 4.22** Hirn cycle on the enthropic (*T–S*) and on the enthalpic level (*H–S*)

*General Characteristics*

The considerable technological development of gas turbines in the last decades, with the introduction of the several improvements mentioned mentioned above, entailed huge increases in the efficiency of gas plants: from 25–27% during the 1970s to 32–35% of modern plants. A peculiar characteristic of these plants is their significant *flexibility; very short transients* (from the still position, the plant reaches the maximum steady-state condition in a few tens of minutes) allow to quickly adapt the power supplied to the demand. These characteristics make it suitable to cover the top side of the load diagram, that is to say to provide the power requested by the grid in the periods of highest consumption. In a plant there are several turbogas groups; the power supplied is regulated by acting on the number of groups operated simultaneously besides regulating the supply of each of them in an interval that ranges from the operation at full capacity to 50% of the nominal power. The hours of operation per year are 4,000–4,500. Compared to steam plants, gas plants generally have a lower power, from less than hundred of $kW_e$, to some hundreds to $MW_e$.

#### 4.2.3.2 Steam Power Plants

*Thermodynamic Cycle*

The basic reference thermodynamic cycle is the Hirn Cycle (Fig. 4.22) envisaging two isothermobaric (water evaporation (2–3) and vapour condensation (3–6)), one isobaric (steam superheating) and two adiabatic (water compression (6–1) and steam expansion (4–5)) transformations (3–4).

Water evaporation occurs in the steam generator. In a first phase (economiser) water is brought to temperature $T_2$; this process is obtained by recovering the residual heat (transformation 1–2). In the second phase (vapouriser) the evaporation occurs by exploiting the heat produced in the burner through combustion (transformation 2–3) and finally the steam superheating occurs in the superheater up to the maximum temperature allowed (transformation 3–4). At this point, saturated and superheated steam expands in the turbine (transformation 4–5) and reaches the condenser. Here, at constant pressure and temperature, by exchanging heat with the cooling water, it condenses (transformation 5–6). Pumps bring water back to the steam generator and increase its pressure (transformation 6–1).

As regards Fig. 4.22, the efficiency and the specific work of the cycle are, respectively:

$$\eta = 1 - \frac{H_5 - H_1}{H_4 - H_1} \tag{4.34}$$

$$L = H_4 - H_5 \tag{4.35}$$

In a steam power plant, in order to improve efficiency, it is convenient that condensation takes place at the lowest possible pressure. The reason for the importance of a vacuum pushed in the condenser is that work $\int dp/\rho$ is higher, given the same $\Delta p$, with low values of density $\rho$, namely with low pressures. Being steam saturated, pressure depends on temperature; therefore the pressure inside of the condenser is determined by the temperature of the refrigerant: normally working temperature is around 35°C and therefore 0.05 bar is the working pressure. In the condenser, besides condensation, steam outgassing occurs through extractors connected to the coldest areas. The choice of outgassing in the condenser is due to the fact that at the conditions therein existing, air solubility in water and air density is more than double than dry saturated steam, and therefore the quantity of steam inevitably removed together with air will be limited.

Once the conditions at the condenser are set, the cycle performance depends on the fluid conditions at the outlet of the steam generator. The search for optimal conditions is carried out by setting in advance the superheating temperature (maximum temperature of the cycle): this is set as a compromise between thermodynamic needs (Carnot effect) that demand values as high as possible, whereas technical–economic reasons impose some limitations. In consideration of these elements, the superheating temperature is typically set at 550–600°C. At these temperatures, the pressures corresponding to optimal conditions exceed the critical pressure of water (218 bar); as a matter of fact, although supercritical plants exist (in the case of large-size groups), in the vast majority of cases the outlet pressure of the steam generator is 100–180 bar. This is because the *efficiency gain* $\Delta\eta/\eta$ decreases as pressure increases, and it is therefore necessary to evaluate the convenience in pushing maximum pressure beyond some values to the detriment of costs, safety and easiness of management.

From the steam conditions mentioned above, an increase in the specific volume of over 1,000 times is reached during the duty cycle. For this reason turbines are divided into three bodies (high, medium, and low pressure ones) so that a limited

**Fig. 4.23** Hirn cycle with thermal regeneration

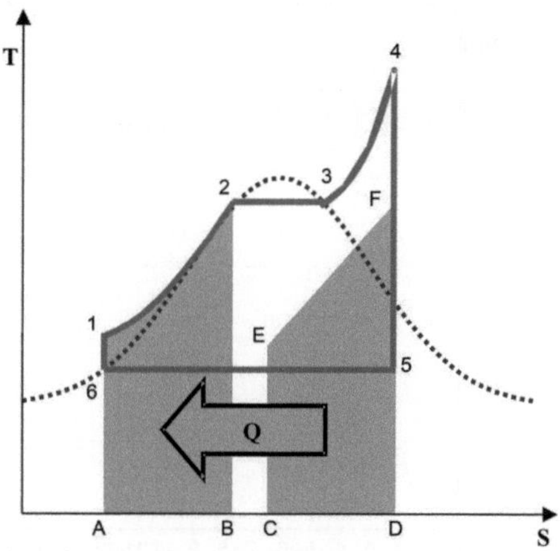


increase of the specific volume occurs in each of them. The low-pressure bodies are often divided into several elements (usually two) placed in parallel.

The cycle described until now is the reference basic cycle, to which some typical changes are made in order to improve its efficiency, mainly thermal regeneration and repeated superheating.

*Thermal Regeneration*

Thermal regeneration consists of using the heat, Q, avaible during the steam-condensation before the end of expansion (CEFD area in Fig. 4.23), in order to preheat the water coming from the condenser before entering the steam generator (area A12B).

Thermal regeneration through heat subtraction, however, is not convenient since the steam fraction at the end of expansion (point F) would be too low, endangering the integrity of the turbine or, at least, its efficiency.

The same regenerative effect can be *obtained by removing steam from expansion rather than heat* and mixing such steam with the water to be preheated. In this way the steam removed directly gives its thermal content to the liquid during the heating phase, whereas the steam in the turbine continues its expansion undisturbed. In fact, practical thermal regeneration takes place through a given number of bled-steam during expansion: the number and the position of this bleeding is fixed according to technical–economic optimisation criteria.

The regeneration degree $R$ is defined as the ratio between enthalpy variation by mass unit conferred to the liquid through regeneration $(h_{x1} - h_0)$ and the total one $(h_1 - h_0)$:

$$R = \frac{h_{x_1} - h_0}{h_1 - h_0} = \frac{h_{x_1} - h_0}{\lambda} \tag{4.36}$$

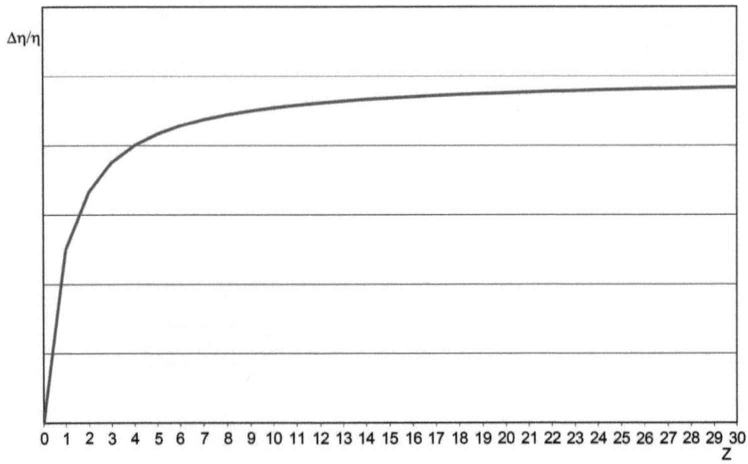


**Fig. 4.24** Efficiency gain according to the number of bleedings

where:

$$\lambda = h_1 - h_0 \tag{4.37}$$

Indicating with $z$ the number of steam bleedings, the maximum efficiency value is obtained for:

$$R = \frac{z}{z+1} \tag{4.38}$$

With these assumptions, the efficiency increase ($\Delta\eta/\eta$) is plotted against $z$ in Fig. 4.24 As $z$ increases, the efficiency gain increases to a lower rate; besides a given value it is not convenient anymore to add other bleedings, hence further complicating the plant without a valuable increment of efficiency increase. The value of $z$ for large installed powers is 8–10.

*Repeated Superheating (Reheating)*

In the past, the low superheating temperatures as well as the other high pressures for the optimisation of specific work and efficiency, led to wet steam, with too low steam fractions at the end of the expansion phase and the presence of water was an erosion caus for the turbine blades. To prevent this, the procedure of repeated superheating was developed. After expansion 40–30 (Fig. 4.25), steam is superheated at a constant pressure (transformation 30–40) up to a temperature equal to or slightly lower than $T_4$. Although developed to solve the problems related to a too low steam fraction at the end of expansion, the double superheating has advantages also from the merely thermodynamic viewpoint: by increasing the portion of heat released at high temperatures, on the basis of the multiple-source theory, overall efficiency increases. For this reason, although present superheating

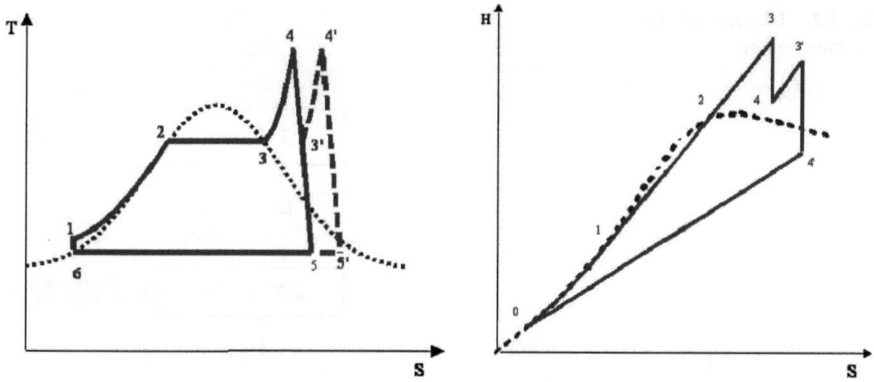


**Fig. 4.25** Steam cycle with double superheating on the anthropic and enthalpic level

**Fig. 4.26** Diagram of a multifuel steam plant with double superheating

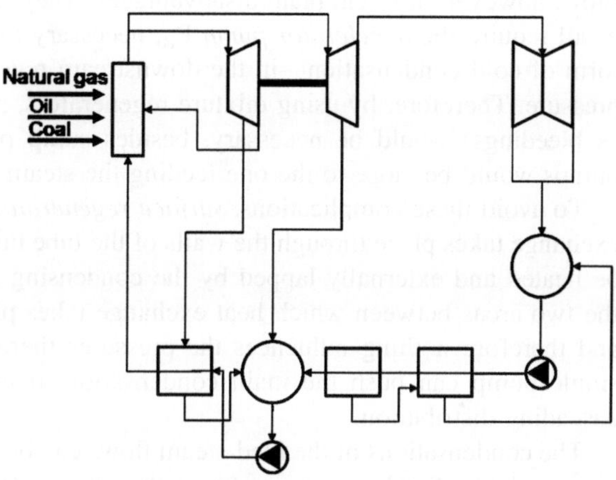


temperatures do not entail any low-fraction problems, the double superheating remains a very widespread practice.

*The Power Plant*

Figure 4.26 shows the diagram of a steam power plant with double superheating and bled-steam (with mixing *S*).

In steam plants it is possible to use whatever type of fuel, including nuclear one; the most commonly used are *fuel oil*, *coal* and *natural gas*. Modern plants are *multifuel*, that is to say they have a boiler whose combustion systems allow to use different fuels without distinction. This solution offers a huge flexibility that allows to vary the fuel used according to market needs or to product availability.

Heat exchange between bled-steam and the liquid to reheat can take place in *mixture regenerators* or in *surface regenerators*.

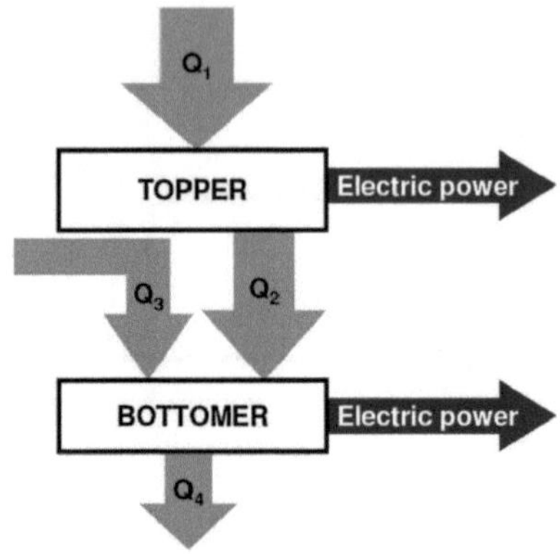


**Fig. 4.27** Diagram of the combined plant

Mixture regenerators, although efficient from the thermal exchange viewpoint, show, however, different plant disadvantages. They are bulky and heavy, but most of all require the *accelerator pump* $P_R$, necessary to introduce the liquid—in the form of cold condensation—in the downstream regenerator, which is at a higher pressure. Therefore, by using mixture regenerators, $z$ accelerator pumps (as many as bleedings) would be necessary, besides sump pumps; the capacity of these pumps would be close to the one feeding the steam generator.

To avoid these complications, *surface regenerators* are used, in which thermal exchange takes place through the walls of the tube bundles containing the water to be heated and externally lapped by the condensing steam. In these regenerators, the two areas between which heat exchange takes place are entirely independent and therefore nothing influences the pressures therein existing. It follows that a single pump can push the main condensation to the various regenerators, in a cascading distribution.

The condensations of the bled-steam flows can be sent from each regenerator to the one immediately upstream (through a flow-rate regulation valve), or to the regenerator downstream (at a higher pressure) through a pump that should only process the flow-rate of condensation relating to the bled-steam, that is to say a very limited capacity, differently from the case of mixture regenerators. The solution normally adopted is the one that provides for condensation to be sent to the upstream regenerator.

Despite of the advantages of surface regenerators, a mixture regenerator is always included in plants. The fact that it entails the insertion of a pump is not a problem, since it takes on the function of feed pump, in order to allow the feed pump and the sump pump to share the task of taking the liquid from the very low pressure in the condenser to the very high pressure in the boiler.

Furthermore, the mixture regenerator, *if operating in balanced conditions*, brilliantly performs the task of degasser, which is very important in steam plants, especially the high-power ones, and that condensers alone cannot satisfactorily perform. For this reason, the pressure in the only mixture regenerator (degasser) in the plant, performs its role in terms of balancing the condensation temperature of the bled-steam. Such pressure must exceed atmospheric value to favour the spontaneous outburst of bases from the water-steam system.

*The Cooling Circuit of the Condenser*

If a sufficient water flow is available, the circuit is open. With this type of circuit, water—after the thermal exchange—is reintroduced in the water body of origin. Conversely, if the continuous water flow available is not sufficient, then the closed circuit is used, where the cooling water, after the thermal exchange, is cooled in suitable *cooling towers*.

In the cooling towers, a close contact is established between water and an airflow; the latter absorbs heat both by convention and through the increase of its fraction (the outlet air fraction is close to one unit), following the evaporation of a small portion of cooled water.

Water is injected at a certain height of the tower and dropped on a stacking of elements in plastic, wood or zinc-coated plate, in order to be dispersed in drops or in thin layers, increasing in this way the surface of contact with the air, and subsequently be collected in a tank of cold water. On the contrary, the air comes from the bottom, covers the tower up to the top, meets water as a counter flow and outgoes from the top.

*Air circulation can be either natural* or *forced.*

In natural-draught towers, the air flows due to a chimney effect, which implies huge heights and large sections; the upper dimpling of the tower causes a decrease in speed and in air temperature and, therefore, the condensation of a part of water contained in it, with subsequent drop and recovery. Air speed in general does not exceed 3 m/s and "rain density" is about 0.001 $m^3/m^2$ s.

In forced-draught towers, the air is set in motion by fans; this allows to have far smaller towers, although with higher construction and operating cost due to the energy consumption of fans and system maintenance. In this case air speed can exceed 10 m/s whereas "rain density" amounts to nearly 0.003 $m^3/m^2s$.

As already said, a fraction of refrigerated water evaporates: water consumption $G$ due to evaporation, having defined $x_2$ and $x_1$ the values of the input and outlet air fraction, respectively, and $G_a$ the airflow rate, is given by

$$G = G_a(x_2 - x_1) \tag{4.39}$$

and generally amounts to a very limited percentage of the flow of the treated water. However, the flow of the restoration water must be higher than the one consumed due to evaporation, because the subsequent recycles would unacceptably increase salt concentration in water, and it is therefore necessary to operate some draining that must then be offset.

The flow of water necessary for steam condensation is remarkable. By considering, for instance, the following conditions at the beginning and at the end of the expansion:

- Input steam in the turbine $p = 150$ bar and $T = 540°C$
- Steam to the condenser $p = 0.05$ bar and $T = 33°C$
- Expansion efficiency in turbines: 0.86.

It is possible to calculate that, in order to produce 1 kWh, nearly 3 kg of steam are necessary, that is to say to provide a power of 1 kW a steam flow of nearly 3 kg/h is necessary.

In a 1,200 MW power plant, typically made up of four groups of 300 MW the steam flow for one group amounts to 250 kg/s. The balancing equation of heat exchange between cooling water and steam, where $m_a$ is the flow rate of water, $c$ its specific heat, assuming $\Delta T = 15°C$ (to increase higher temperature excessively large exchanging surfaces would be necessary) and $\Delta H_c = 2.440$ kJ/kg the steam condensation heat is:

$$m_v \Delta H_c = m_a c \Delta T \tag{4.40}$$

Therefore:

$$m_a = \frac{m_v \Delta H_c}{c \cdot \Delta T} \cong 9.720\,\text{kg/s} \tag{4.41}$$

to which corresponds a volume flow rate of nearly 10 $m^3$/s and therefore an overall flow rate for the four groups of 40 $m^3$/s.

*General Characteristics*

Steam power plants show rather high efficiencies which in the best possible cases reach 46%; load regulation is possible only within given limits, beyond which efficiency drastically decreases and transients are deleterious for the plant, since they negatively impact its lifetime. And in any case the thermal energy of the large fluid mass involved would not allow the quick response that is conversely possible in turbogas power plants. For these reasons, steam plants are used for the production of basic electric power, that is to say cover the lower part of the load diagram, in large-size plants often (exceeding 1,000 MW), and have a high number of hours of full load equivalent operation of 6,000–6,500 h/year, and even beyond.

A plant is usually made up of several groups, whose size ranges between 300 and 660 MW.

### 4.2.3.3 Combined Cycle Power Plants

The efficiency of thermodynamic cycles increases as the ratio between maximum and minimum operating temperature of the cycle increases. In the cycles of steam and gas thermoelectric plants, there are thresholds for these temperatures: in steam plants the maximum temperature of the cycle is limited by the relatively low value of the critical temperature of water (374.1°C), whereas in gas plants this limit concerns the temperature at the exhaust of the turbine that is imposed, after setting

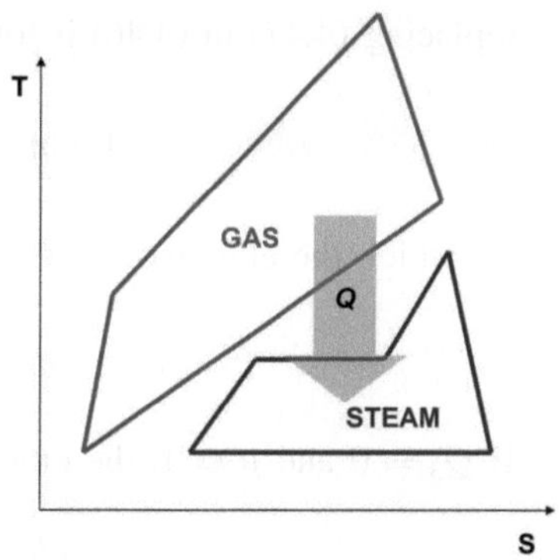


**Fig. 4.28** Gas–steam combined cycle

the initial expansion temperature and the compression ratio, by room temperature. Since the two limits are complementary, it is possible to combine two cycles so that the exhaust heat of the gas cycle represents the source of heat, or part of it, for the steam cycle. From the thermodynamic point of view, the gas plant is overlapped (*topper*) to the steam plant (*bottomer*); the maximum and minimum temperatures of the cycle "deriving" from the combination of the two simple cycles, are, respectively, the maximum temperature of the gas cycle, exceeding 1,000°C, and the minimum temperature of the steam cycle, nearly 30°C (Fig. 4.28).

The efficiency of the combined cycle depends on the efficiencies of both cycles and is, in any case, higher than each of them. Figure 4.28, having defined $\eta_c$, $\eta_g$ and $\eta_v$ the efficiencies of the combined cycle, of the gas cycle and of the steam cycle respectively, it follows that:

$$\eta_c = 1 - \frac{Q_4}{Q_1 + Q_3} \Rightarrow 1 - \eta_c = \frac{Q_4}{Q_1 + Q_3} \tag{4.42}$$

$$\eta_v = 1 - \frac{Q_4}{Q_3 + Q_2} \Rightarrow 1 - \eta_v = \frac{Q_4}{Q_3 + Q_2} \tag{4.43}$$

$$\eta_g = 1 - \frac{Q_2}{Q_1} \Rightarrow 1 - \eta_g = \frac{Q_2}{Q_1} \tag{4.44}$$

Assuming:

$$\mu = \frac{Q_1 + Q_3}{Q_1}, \tag{4.45}$$

replacing (4.45), (4.44) and (4.43) in (4.42) it is possible to obtain:

$$1 - \eta_c = \frac{Q_4}{\mu \cdot Q_1} = \frac{Q_4}{Q_2 + Q_3}\frac{Q_2 + Q_3}{\mu \cdot Q_1} = (1 - \eta_v)\frac{Q_2 + Q_3}{\mu \cdot Q_1} \tag{4.46}$$

Being:

$$\frac{1}{\mu}\left(\frac{Q_2 + Q_3}{Q_1}\right) = \frac{1}{\mu}\left(1 - \eta_g + \frac{Q_3}{Q_1}\right) = 1 - \frac{\eta_g}{\mu}, \tag{4.47}$$

replacing (4.47) in (4.46) it follows:

$$1 - \eta_c = (1 - \eta_v)\left(1 - \frac{\eta_g}{\mu}\right), \tag{4.48}$$

from which the efficiency of the combined plant is derived:

$$\eta_c = \eta_v + \frac{\eta_g}{\mu}(1 - \eta_v) \tag{4.49}$$

If $Q_3 = 0$ and $\mu = 1$, the efficiency of the combined cycle becomes:

$$\eta_c = \eta_v + \eta_g - \eta_v \eta_g \tag{4.50}$$

Equation 4.50, assuming constant $\eta_v$ and $\eta_g$, is the maximum value obtainable in (4.49).

Two fundamental solutions are possible for gas-steam combined plants: *with* or *without the afterburning* of gas turbine exhaust gases.

The efficiency of combined cycles depends on the efficiency of component cycles according to (4.49); in the last few years, the development of gas turbine technologies allowed huge increases in the efficiency of combined plants, presently reaching values exceeding 55%.

The high efficiency, as well as the use of a "clean" fuel as methane, allow to reduce $NO_x$, $SO_x$, CO, and $CO_2$ emissions, as well as particulate emission abatement compared to conventional solutions.

#### 4.2.3.4 Fuel Cells

The energy conversion systems analysed above, involving the conversion of chemical energy into electric power, perform a thermodynamic conversion. However, there are converters based on electrochemical principles often more efficient and effective than the former; they are described in this section.

*Overview*

The origin of fuel cells date back to 1839, the year in which the British William Grove generated electricity from a cell containing phosphoric acid, delimited by two electrodes reached by hydrogen and oxygen, respectively.

The basic principle of operation is the following: the cell is made up of two electrodes of porous material, separated by an electrolyte. Electrodes act as catalytic sites for the breakdown of reagents into ions; the electrolyte conducts ions preventing the passage of electricity, conveyed to an external circuit on the tips of the two electrodes.

Fuel cells are electrochemical systems that directly convert the chemical energy of a fuel (usually hydrogen) into electric power, according to an isothermal and

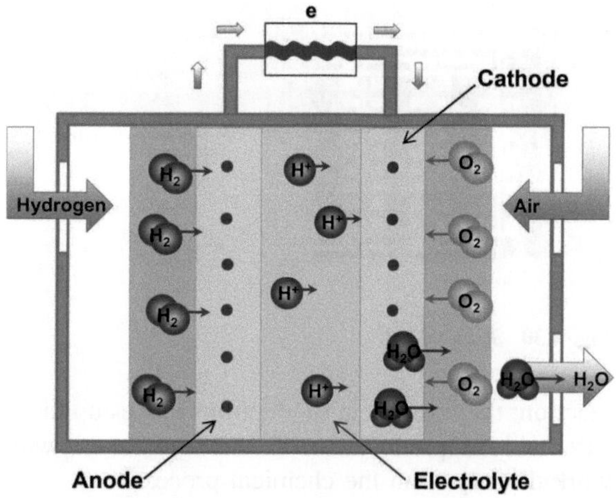


**Fig. 4.29** Diagram of fuel cells

isobaric process. Therefore, they eliminate the irreversibility associated to combustion, without considering the passage through the stage of thermal energy, and allow to depart from the maximum efficiency threshold of thermal machines, represented by the Carnot efficiency.

A distinctive feature of the operation of a fuel cell, compared to the conventional conversion of chemical energy through an oxidation reaction, is the fact that in this device both fuel and oxidiser are combined in the form of ions not directly but *through an electrolytic path*. In this case reagents are external to the cell (differently from what happens in a normal electric battery); on the other hand, neither electrodes nor the electrolyte are consumed in the complex of reaction.

In theory, any oxidation and reduction can be at the basis of a fuel cell. It is fundamental, however, that a ion reaction occurs, and that the transportation of electric charges occur rapidly.

*Operating Principle of Fuel Cells*

The cell is made up of two electrodes (anode and cathode) in a porous material, separated by an electrolyte (Fig. 4.29). Electrodes act as catalytic sites for reactions, breaking down and consuming the fuel and the oxidiser. Always within the electrodes, the product of the reaction between fuel and oxidiser is formed, and a difference in voltage arises which can be exploited by an external circuit. The electrolyte has the task of conducting the ions produced by a reaction and consumed by the other, closing the electric circuit. The cathode stores a superficial layer of negative charges, and they attract the positive ions from the electrolyte. producing an electric double layer. At the same time, the loss of electrons by the anode produces a layer of positive charges, which attracts the negative ions from the electrolyte. The double layers increase their concentration until the potentials can prevent further reactions between the electrolyte and the fuel gases. By closing the external circuit, electrons move towards the anode to participate in the

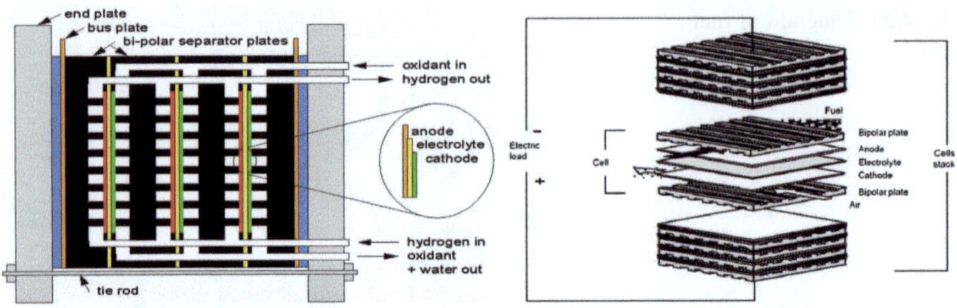


**Fig. 4.30** Stack of cells

reaction; the movement of electrons makes up the electric current that, through an external load, produces work. Therefore, in this way it is possible to obtain electric work directly from the chemical process.

Electrodes act as catalytic sites for the cell reactions that mainly consume hydrogen and oxygen, with the production of water and passage of electric current in the external circuit. The electrochemical transformation is accompanied by heat production.

| | |
|---|---|
| Anodic reaction | $H_2 \Rightarrow 2H^+ + 2e^-$ |
| Cathode reaction | $\frac{1}{2}O_2 + 2H^+ + 2e^- \Rightarrow H_2O$ |
| Total reaction | $H_2 + \frac{1}{2}O_2 \Rightarrow H_2O$ |

*The Stack*

The exposed surfaces must have a sufficient area to obtain current strengths suitable to the application needs. It is possible to have the desired surfaces, by silling the cells. The stacking of cells obtained in this way forms the so-called fuel cell stack (Fig. 4.30), which represents the basis of the typical technological use of fuel cells.

*The Thermodynamics of Fuel Cells*

In the analysis of the phenomena related to the operation of fuel cells, the standard charge is the one allowed by a mole of electrons; it is called "*Faraday constant*" (denoted as *F*). Having considered that in a mole there are $6.023 \times 10^{23}$ electrons and that one electron has the charge of $1.602 \times 10^{-19}$ C, it follows: $F = 96{,}485$[19] C/mol.

*The Faraday's Law* concerning electrolysis can be summed up as follows: *in the operation of the battery, one charge of* 96,485 C *deposits or consumes one*

[19] The exact value (National Institute of Standards and Technology—2008) is $F = 96{,}485.3399$ C/mol.

*gram equivalent of substance on each of the electrodes*. The weights of the substances that intervene in the reactions of electrons are linked to the charge transported ($q$) according to the equation:

$$q = n \cdot z \cdot F \tag{4.51}$$

where $n$ is the number of grams-atom or moles that react and $z$ is the number of electrons transported by each ion.

In considering the first Law of thermodynamics in its form:

$$\Delta U = Q - L_{\text{tot}} \tag{4.52}$$

in the case of fuel cells, $L_{\text{tot}}$ is referred both to the electric work $L_e$ and to the expansion mechanic work, $L_m$. The maximum value of the electric work amounts to the product of *f.e.m.* of the cell, $E$, times the charge that goes through it:

$$L_e = E \cdot q \tag{4.53}$$

On the other hand, using (4.52), (4.53) can be written as follows:

$$L_e = n \cdot z \cdot F \cdot E \tag{4.54}$$

And the total work, therefore:

$$L_{\text{tot}} = p \cdot \Delta v + n \cdot z \cdot F \cdot E \tag{4.55}$$

Having considered (4.55), (4.52) becomes:

$$\Delta U = Q - p \cdot \Delta v - n \cdot z \cdot F \cdot E \tag{4.56}$$

Thus, introducing the second law of Thermodynamics:

$$\Delta U = T \cdot \Delta S - p \cdot \Delta v - n \cdot z \cdot F \cdot E \tag{4.57}$$

The reactions in the fuel cell take place at constant temperature and pressure, therefore the variation of Gibbs' free energy[20] ($G$) and of the enthalpy[21] ($H$) are the following:

$$\Delta G = \Delta H - T \cdot \Delta S \tag{4.58}$$

$$\Delta H = \Delta U + p \cdot \Delta v \tag{4.59}$$

By replacing (4.59) in (4.58) it follows that

$$(\Delta G)_{p,T} = \Delta U + p \cdot \Delta v - T \cdot \Delta S \tag{4.60}$$

So, having considered (4.57), it as possible to obtain:

$$(\Delta G)_{p,T} = -n \cdot z \cdot F \cdot E \tag{4.61}$$

---

[20] $G = H - TS$.

[21] $H = U + pv$.

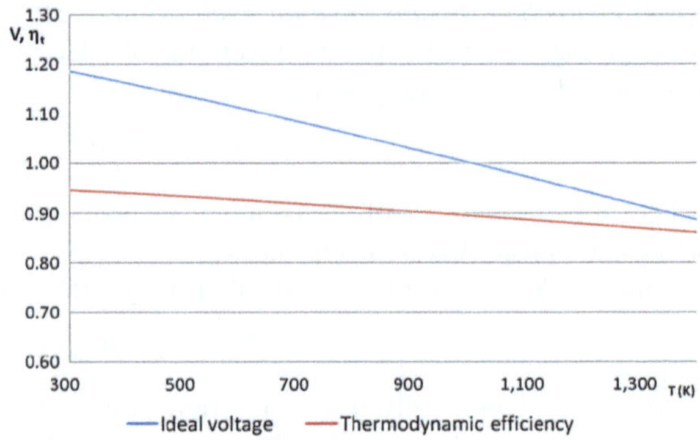


**Fig. 4.31** Variation of the ideal voltage ($V$) and of thermodynamic efficiency according to temperature

Equation 4.61 shows that the electric work produced by the fuel cell amounts to the decrease of Gibbs' free energy of the reaction.

With this expression it is possible to derive the maximum potential for a cell; therefore from (4.61) and (4.54) it is obtained:

$$E_0 = -\frac{\Delta G}{n \cdot z \cdot F} \tag{4.62}$$

In case hydrogen and oxygen react to form water in standard conditions[22] then it is possible to obtain $E_0 = -1{,}186$ V.

The ideal thermodynamic efficiency (assuming the reversible transformations) is:

$$\eta_t = \frac{\Delta G}{\Delta H} \tag{4.63}$$

In the case in point, always in standard conditions, $\eta_t = 0.9458$ is obtained.[23]

Gibbs' free energy and enthalpy variations depend on temperature and pressure; therefore, as they vary, both the ideal voltage (open circuit) and the ideal thermodynamic efficiency vary accordingly. Without lingering on the details of calculations, it is possible to conclude that the ideal voltage and the thermodynamic efficiency slightly increase as pressure increases (Fig. 4.31), whereas they decrease as temperature increases (Fig. 4.32).

The previous reactions are obtained by considering all processes as reversible. Taking into consideration irreversibility, indicating with $\Delta S^*$ the entropy variation in case of real transformations (irreversible), with $\Delta S_{\text{irr}}$ the variation due to irreversibility and with $\Delta S$ the one of reversible transformations, it follows that:

$$\Delta S^* = \Delta S_{\text{irr}} + \Delta S. \tag{4.64}$$

---

[22] The reaction $\Delta G$ of water formation in standard conditions ($T = 25°\text{C}$ and $p = 1$ bar) amounts to 228,888 J/mo, the number of moles ($n$) involved in the reaction is 2, and the number of electrons transported by each (hydrogen) ion is 1.

[23] The reaction $\Delta H$ to water formation in standard conditions amounts to 242,000 J/mol.

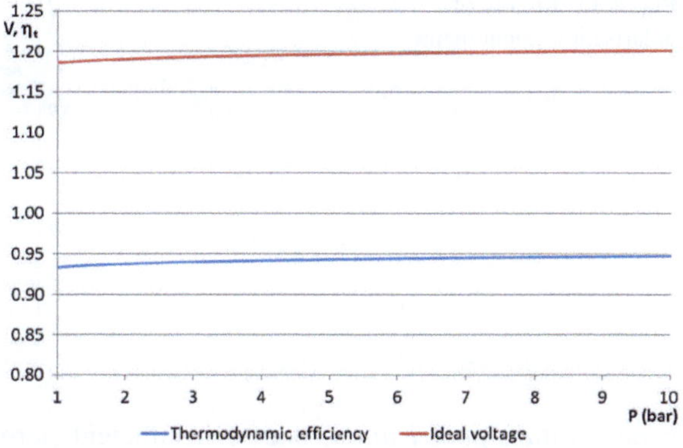


**Fig. 4.32** Variation of the ideal voltage (*V*) and of thermodynamic efficiency according to pressure

Having considered that, according to the II Law of Thermodynamics, $\Delta S_{irr} > 0$ (4.64) shows:

$$\Delta S^{*} > \Delta S \tag{4.65}$$

In the case of real transformations, (4.58) becomes:

$$\Delta G^{*} = \Delta H - T \cdot \Delta S^{*} \tag{4.66}$$

Considering (4.65) and being T > 0, then (4.66) shows:

$$\Delta G^{*} < \Delta G \tag{4.67}$$

Therefore, in the case of irreversible transformations, the electric work is lower. Indicating with *V* the closed circuit voltage[24] (irreversible), it follows that:

$$V < E \tag{4.68}$$

As concerns efficiency, recalling (4.67), it follows:

$$\eta_t^{*} = \frac{\Delta G^{*}}{\Delta H} < \eta_t \tag{4.69}$$

*Efficiency of Fuel Cells*

The main cause of irreversibility of fuel cells can be identified in the heat developed by current due to the Joule effect in its passage through the electrolyte. Furthermore, irreversible phenomena occur when the circuit is closed, causing a decrease in the electric potential. These phenomena occur in three different forms: *ohmic polarisation*, due to the resistance of the electrolytic membrane at the passage of ions and the resistance of electrodes at the passage of electrons;

[24] The passage of current generates irreversibility; therefore it is possible to consider V as the driving force in case of irreversible transformations.

Fig. 4.33 Effects of polarisation phenomena

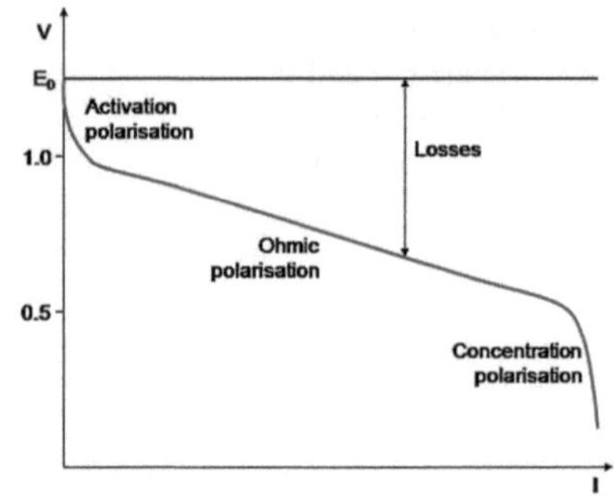


*concentration polarisation*, due to insufficient porosity of electrodes and to the spreading of products; *activation polarisation*, due to the double electric layer on the surface of electrodes, which determines an activation voltage, decreasing the speed of chemical kinetics.

Therefore, although a temperature increase reduces thermodynamic efficiency, the global efficiency increases since it reduces the ohmic polarisation (increasing the ionic conductibility of the electrolyte), and concentration polarisation (improving the transportation and activation phenomena together with, improving the reaction kinetics) and generally improves the tolerance to the impurities present in gas supply. However, too high temperatures accelerate the corrosion phenomena, the problems in terms of sintering and crystallisation of catalysts, and the degradation of the electrolyte.

A pressure rise, on the contrary, increases thermodynamic efficiency and, in producing higher partial pressures of reagents, improves the phenomena of gas transportation and solubility in the electrolyte, contributing to reduce the electrolyte losses, therefore increasing overall efficiency. However, too high pressures entail problems of resistance of the different materials, and require an accurate control of differential pressures (Fig. 4.33).

The efficiency referred to is the one of the single cell; a fuel cell plant is made up of one (or several) stack(s) of cells, an inverter, the auxiliary systems (anodic and cathode supply, refrigeration, etc.). There might be a system of fuel processing as well as a heat recovery system.

As concerns the entire plant, by definition it is possible to express the efficiency of the plant, $\eta$, as the ratio between the useful electric power ($P_{e,u}$) and the product between the mass flow rate $\dot{m}_{\mathrm{comb}}$ of the fuel and its lower heating value:

$$\eta = \frac{P_{e,u}}{\dot{m}_{\mathrm{comb}} \cdot \mathrm{LHV}} \tag{4.70}$$

Indicating with $V_r$ and $I_r$ the real voltage and current to the terminals of the stack(s) (4.70) becames:

$$\eta = \frac{P_{e,u}}{V_r \cdot I_r} \frac{V_r \cdot I_r}{\dot{m}_{\mathrm{comb}} \cdot \mathrm{LHV}} = \eta_{\mathrm{sist}} \cdot \frac{V_r \cdot I_r}{\dot{m}_{\mathrm{comb}} \cdot \mathrm{LHV}}, \tag{4.71}$$

having defined the system efficiency $\eta_{\text{sist}}$ as the ratio between the useful electric power and the one produced by the fuel cell stack(s), namely:

$$\eta_{\text{sist}} = \frac{P_{e,u}}{V_r \cdot I_r} \tag{4.72}$$

By defining the efficiency of cell ε as the ratio between the real and the ideal voltage, (4.71) can be described as follows:

$$\eta = \eta_{\text{sist}} \frac{V_r}{V_{\text{id}}} V_{\text{id}} \frac{I_r}{\dot{m}_{\text{comb}} \cdot \text{LHV}_{\text{comb}}} = \eta_{\text{sist}} \cdot \varepsilon \frac{V_{\text{id}} \cdot I_c}{\dot{m}_{\text{comb}} \cdot \text{LHV}_{\text{comb}}} \tag{4.73}$$

The Faradic current ($I_F$) is the current that would be in the cell according to Faraday's Law in case of absence of irreversibility and other losses, that is to say:

$$I_F = z \cdot \dot{n} \cdot F \tag{4.74}$$

where $\dot{n}$ is the mole rate (moles/s) of hydrogen consumed at the anode.

A *Faradic* (or *current*) *efficiency* is the ratio between the real current and the Faradic one:

$$\eta_F = \frac{I_r}{I_F} = \frac{I_r}{z \cdot \dot{n} \cdot F} \tag{4.75}$$

On the basis of (4.74) and (4.75), (4.73) can be written as follows:

$$\eta = \eta_{\text{sist}} \cdot \varepsilon \frac{I_r}{z \cdot \dot{n} \cdot F} \frac{V_{\text{id}} \cdot z \cdot \dot{n} \cdot F}{\dot{m}_{\text{comb}} \cdot \text{LHV}_{\text{comb}}} = \eta_{\text{sist}} \cdot \varepsilon \cdot \eta_F \frac{V_{\text{id}} \cdot z \cdot \dot{n} \cdot F}{\dot{m}_{\text{comb}} \cdot \text{LHV}_{\text{comb}}} \tag{4.76}$$

Not all the hydrogen that reaches the anode reacts; a fraction goes to the anodic discharge. Having defined $\dot{n}_a$ the mole rate of the anodic supply, the *hydrogen use coefficient* $U_{H_2}$ is the ratio between the hydrogen rate consumed by the anode and the one of the anodic supply:

$$U_{H_2} = \frac{\dot{n}}{\dot{n}_a} \tag{4.77}$$

Having considered (4.77), (4.76), by multiplying and dividing by $\dot{n}_a$ becomes:

$$\eta = \eta_{\text{sist}} \cdot \varepsilon \cdot \eta_F \cdot U_{H_2} \frac{V_{\text{id}} \cdot z \cdot F}{\dot{m}_{\text{comb}} \cdot \text{LHV}_{\text{comb}}} \dot{n}_a \tag{4.78}$$

Recalling (4.54), the molar electric work $L^m_{e,\text{id}}$ results as:

$$L^m_{e,\text{id}} = z \cdot F \cdot V_{\text{id}} \tag{4.79}$$

Therefore, indicating as $M_{H_2}$ the molar mass (kg/mole) of hydrogen and with $\text{LHV}_{H_2}$ its lower heating value (per mass unit), the ideal (thermodynamic) efficiency $\eta_{\text{id}}$ can be described as follows:

$$\eta_{\text{id}} = \frac{z \cdot F \cdot V_{\text{id}}}{M_{H_2} \cdot \text{LHV}_{H_2}} \tag{4.80}$$

Taking into account (4.80), (4.78) becomes:

$$\eta = \eta_{\text{sist}} \cdot \varepsilon \cdot \eta_F \cdot U_{H_2} \cdot \eta_{\text{id}} \cdot \frac{\dot{n}_a \cdot M_{H_2} \cdot \text{LHV}_{H_2}}{\dot{m}_{\text{comb}} \cdot \text{LHV}_{\text{comb}}} \tag{4.81}$$

having considered that the hydrogen mass rate is:

$$\dot{m}_{H_2} = \dot{n}_a \cdot M_{H_2} \cdot \text{LHV}_{H_2} \tag{4.82}$$

*The efficiency of fuel treatment system* is the ratio between the output and the inlet energy flow in the system itself, that is to say:

$$\eta_{\text{stc}} = \frac{\dot{m}_a \cdot \text{LHV}_{H_2}}{\dot{m}_{\text{comb}} \cdot \text{LHV}_{\text{comb}}} \tag{4.83}$$

Thus taking into account (4.82) and (4.83), the efficiency can be expressed as follows:

$$\eta = \eta_{\text{sist}} \cdot \varepsilon \cdot \eta_F \cdot U_{H_2} \cdot \eta_{\text{id}} \cdot \eta_{\text{stc}} \tag{4.84}$$

During the operation of the cell, heat will be produced due to entropy variation ($T\,\Delta\,S$) and to irreversibility. So, heat production per mole of reagent can be expressed as:

$$Q = \left(\frac{T \cdot \Delta S}{n \cdot F} + \Delta V\right) \cdot I = \left(\frac{\Delta H - \Delta G}{n \cdot F} + \Delta V\right) \cdot I \tag{4.85}$$

where $\Delta V$ is the decrease of the electric potential[25] due to waste phenomena and I is current. Such heat can be conveniently used for cogeneration purposes, recovering in this way the energy wasted during the operation of the fuel cells and obtaining global efficiency (electric power + heat) close to one.

Some considerations on the global efficiency of fuel cells must now be made.

As concerns the system efficiency, the useful electric power $P_{e,u}$ can be expressed as the difference between the electric power generated by the stack and the electric power absorbed by auxiliaries. The global efficiency is:

$$\eta_{\text{sist}} = \frac{P_{e,u}}{V_r \cdot I_r} = \frac{V_r \cdot I_r - P_{e,\text{aux}}}{V_r \cdot I_r} = 1 - \frac{P_{e,\text{aux}}}{V_r \cdot I_r} \tag{4.86}$$

The electric power absorbed by auxiliaries is the one relating to pumps (for the circulation of liquid and/or humidification of cells), compressors or blowers (anodic or cathode supply): if the cell works at low pressures (close to atmospheric pressure) the supply is made by blowers, whereas if the cell works at higher pressures, compressors are necessary. The adoption of high operating pressures on the one side increases the ideal efficiency, on the other side decreases the system

---

[25] $\Delta V = E - E \cdot \varepsilon = \frac{\Delta G}{n \cdot F}(1 - \varepsilon)$.

efficiency, since the power absorbed by compressors is far higher (given the same flow rate) than the one absorbed by blowers.

As concerns *the hydrogen utilisation factor*, it is necessary to observe that cells must be supplied with a quantity of hydrogen always exceeding the one that effectively reacts to the anode. This because, in absence of such "excess", and due to the unavoidably inhomogeneous distribution of hydrogen, there would be areas on the anodic surface in which hydrogen concentration would be lower than necessary for anodic reactions, and this would drastically reduce the (real) potential of the cell. In brief, the real potential of the cell increases as the excess of hydrogen increases; therefore, by increasing the surplus of hydrogen, on the one side there is a decrease in the utilisation factor $U_{H_2}$, whereas on the other side the efficiency of the cell $\varepsilon$ increased. The optimal value of $U_{H_2}$ depends on the kind of cells; it ranges between 0.75 and 0.85. Excess hydrogen can be obviously recovered and reused; in this way it is possible to increase the excess of hydrogen (and, therefore, $\varepsilon$), without decreasing $U_{H_2}$. It is necessary to consider the expenditure for the pumping activity destined to the recovery of excess hydrogen, which decreases the system efficiency $\eta_{sist}$.

As concerns the Faraday efficiency, it must be observed that this is always close to one 1, as a first approximation, it is possible to consider $\eta_F = 1$. In case the plant is directly supplied through hydrogen (produced through other means) the efficiency of the fuel treatment system must be considered as amounting to 1.

*Classification of Fuel Cells: the Different Technological Solutions*

Fuel cells are conventionally classified according to the type of electrolyte they contain, which determines their working temperature:

- *Polymer electrolyte membrane fuel cells* (*PEMFC*), using a high proton conductibility polymer membrane as electrolyte and work at temperatures ranging between 70 and 100°C (up to 120–130°C in case of "high temperature" PEMFC); they are mostly developed for small-size generation/cogeneration and vehicles' powertrain (1–250 kW);
- *Alkaline fuel cells* (*AFC*), using an electrolyte made up of potassium hydroxide and operate at temperatures around 100–250°C (some special 250 designs operate at 25–70°C). They reached a good level of technological development, mainly for special use (military and space applications);
- *Phosphoric acid fuel cells* (*PAFC*), operating at temperatures close to 200°C with an electrolyte made up of a concentrated solution of phosphoric acid. They represent the most mature technology for stationary use, for cogeneration applications in the residential and tertiary applications (100–200 kW);
- *Molten carbonate fuel cells* (*MCFC*): the electrolyte is a solution of alkaline carbonates fused at the working temperature of the cell (650°C) and contained in a porous ceramic matrix; they are promising especially for the generation of electric power and cogeneration of some hundreds of kW and some dozens of MW.

**Table 4.5** Classification and characteristics of the different FC

| Types of fc | Electrolyte (exchanged ions) | Temp. (°c) | Catalysts | Fuel | Oxidiser | Electric efficiency (%) |
|---|---|---|---|---|---|---|
| PEFC | Polymeric membrane (H +) | 70–100 | Pt | $H_2$, Reformed gases | $O_2$, aria | 40–60 |
| AFC | KOH (OH-) | 60–120 | Pt/Pd, Ni | Pure $H_2$ | $O_2$ | 60 |
| PAFC | Phosphoric acid (H +) | 160–200 | Pt | $H_2$, Reformed gases | $O_2$, aria | 40–50 |
| MCFC | Li and K(CO3-) carbonates | 600–700 | Ni | $H_2$, Reformed gases | $O_2$, aria | 45–55 |
| SOFC | Zirconium oxides (O-) | 800–1,000 | – | $H_2$, Reformed gases | $O_2$, aria | 45–60 |
| DMFC | Polymer membrane (H +) | 90–130 | Pt, alloy Pt-Ru | Methanol | $O_2$, aria | 40–60 |

- *Solid oxide fuel cells* (*SOFC*), operating at high temperatures (about 900–1000°C) to ensure a sufficient conductibility to the electrolyte, made up of ceramic material (yttrium oxide–doped zirconium oxide); as carbonate fuel cells, they are promising especially for electric power generation and cogeneration of some kW to some dozens of MW;
- *Direct methanol fuel cell* (*DMFC*), operating at temperatures between 70 and 120°C. Like PEFC, they use a polymeric membrane as electrolyte. DMFC are directly fed by methanol at the anode. Methanol is easier than pure hydrogen to store because it is liquid at ambient conditions.

The fields of application of each type of fuel cells is determined by its operating characteristics, and in particular by the working temperature and by the type of fuel used. Table 4.5 sums up the different types of fuel cells with their main characteristics and the present state of technology.

The operating temperature of the cell influences the reaction kinetics, which increases with temperature. When it is particularly high, the heat generated can be profitably used, improving the general efficiency of the production system, but it generates a series of drawbacks: thermal expansion makes the assembling of components difficult, thermal stresses require materials able to support particularly severe working conditions, the lifecycle of the cell is mostly influenced by the operating cycles, and a rather long starting time is necessary.

Another important aspect tied up to high temperature operating is the possibility of producing $H_2$ inside of the cell itself, starting from conventional fuels, carrying out an internal reforming. This possibility turns out into a higher simplicity of the system, since it allows to eliminate the need for an external processor and to directly supply the cell with a conventional fuel.

High-temperature fuel cells (MCFC and SOFC) are therefore suitable for applications linked to the production of electric power for stationary use with cogeneration of electricity-heat and with gas and/or steam plants.

*Polymer electrolyte membrane fuel cell*: in this type of fuel cells, the electrolyte is made up of a proton conducting polymer membrane.

The working temperature is presently limited to nearly 70–90°C by the use of polymer membranes, therefore the reaction kinetics requires the presence of catalysts (mainly platinum). However, new membranes are being developed, able to work at higher temperatures to favour both the reaction kinetics and the possible exploitation of the thermal energy produced by the cell, especially in stationary applications.

The following reactions take place in a polymer electrolyte cell:

$$\begin{array}{ll} \text{Anodic reaction} & H_2 \Rightarrow 2H^+ + 2e^- \\ \text{Cathode reaction} & \frac{1}{2}O_2 + 2H^+ + 2e^- \Rightarrow H_2O \\ \text{Overall reaction} & H_2 + \frac{1}{2}O_2 \Rightarrow H_2O \end{array}$$

The electrolyte is made up of an ion exchange membrane, with a chlorinated sulphuric acid polymer (usually Nafion 117) with a thickness of 50–250 μm. Electrodes are made of a platinum-impregnated porous material[26] with a coating in Teflon (water-repellent) with a thickness of 5–50 μm.

Electrode with a porous structure are obtained by putting the catalyst on a conductive substratum. The latter consists of a layer of coal dust, bound with a polymer, on a coal tissue porous base.

The present low working temperatures also entail the use of catalysts made of noble metals (platinum), which easily poison when hydrogen is not sufficiently pure. Therefore, in the fuel processing section preceding the cell, there must be the stage of impurity elimination, CO in particular. The supply of cathodes, conversely, is generally made with air.

*Direct Methanol Fuel Cells* (DMFC) represent the latest generation of FC and can be considered as deriving from PEFC, since both use a polymer membrane as an electrolyte. However, DMFC distinguish themselves for the fuel that can be used; in fact, they can be directly powered with methanol, since they are able to electrochemically oxidise it, thanks to the use of a new class of catalysts.

The following reactions occur in a DMFC:

$$\begin{array}{ll} \text{Anodic reaction} & CO_3OH + H_2O \Rightarrow CO_2 + 6e^- + 6H^+ \\ \text{Cathode reaction} & \frac{3}{2}O_2 + 6H^+ + 6e^- \Rightarrow 3H_2O \\ \text{Overall reaction} & CO_3OH + \frac{3}{2}O_2 \Rightarrow CO_2 + 2H_2O \end{array}$$

The possibility of using liquid methanol directly in the cell makes DMFC particularly suitable to be used in the portable electronics and mobility sectors, thanks to the remarkable simplification of the system and to the reduction of its size, also to the detriment of a lower total efficiency.

The theoretical efficiency is nearly 35–40% when operating at temperatures between 70 and 100°C; higher efficiencies can be obtained at higher temperatures.

---

[26] The platinum content for this type of cells presently amounts to nearly 0.6 mg/cm$^2$. The electric power supplied by a cell per unit of anodic surface is nearly 0.4 W/cm$^2$, therefore the Platinum content amounts to 1.5 g/kW. The cost of Platinum is about 40 €/g, which means that the cost of the Platinum alone contained in the anode amounts to about 60 €/kW.

The great interest for this type of technology is sided by the need to solve the still open technical problems that limit their immediate application. DMFC, in fact, still have a rapid decay in time, besides a low efficiency, and also show fuel "cross-over" problems (part of the methanol that feeds the anode crosses the electrolytic membrane and reaches the cathode by deactivating the catalyst), and furthermore the quantity of the necessary catalyst is high if compared to the one of PEFC.

*Alkaline Fuel Cell* In this type of cells, the electrolyte is liquid, and made up of an aqueous solution of potassium hydroxide (KOH) with a concentration (in weight) that varies from 30 to 85% according to the operating temperature (from usually 70 to 250°C). Electrodes are made of a carbon porous base on a mesh of Nickel with a hydrophobic coating polytetrafluoroethylene (PTFE) with a low-platinum content (0.3 mg/cm$^2$).

The following reactions occur in an AFC:

| | |
|---|---|
| Anodic reaction | $H_2 + 2OH^- \Rightarrow 2H_2O + 2e^-$ |
| Cathode reaction | $\frac{1}{2}O_2 + H_2O + 2e^- \Rightarrow 2OH^-$ |
| Overall reaction | $H_2 + \frac{1}{2}O_2 \Rightarrow H_2O$ |

The circulation of the electrolyte (liquid) allows to use the electrolyte itself as a cooling substance and to empty the cell during the periods of down-time, hence avoiding the carbonisation of the cell increases its useful life. On the other hand, the corrosiveness of the electrolyte imposes the use of special materials as stainless steels and polymer materials (PVDF—*polyvinylidenefluoride*, PFA—formed by PVDF and HFP—*perfluoropropene*). These cells must be powered with very pure hydrogen and, furthermore, they show a very low tolerance to $CO_2$, therefore it is necessary to envisage a system to remove the $CO_2$ in the air for cathode supply.

*Phosphoric Acid Fuel Cells*: in this type of cell the electrolyte (liquid) is formed by a concentrated solution (85–100%) of phosphoric acid ($H_3PO_4$) in a matrix of silicon carbide. Electrodes are made of a graphite base on which carbon catalysts with PTFE and platinum are applied (0.1 g/cm$^2$ at the anode and 0.5 g/cm$^2$ at the cathode).

The following reactions take place in a PAFC:

| | |
|---|---|
| Anodic reaction | $H_2 \Rightarrow 2H^+ + 2e^-$ |
| Cathode reaction | $\frac{1}{2}O_2 + 2H^+ + 2e^- \Rightarrow H_2O$ |
| Overall reaction | $H_2 + \frac{1}{2}O_2 \Rightarrow H_2O$ |

To avoid the solidification of the electrolyte, the cell must always be heated[27]; thanks to high operating temperatures, they show a good tolerance to impurities (nearly 1% for CO).

[27] Phosphoric acid solidifies at 42 ± 5°C.

*Molten carbonate fuel cells*: in this type of cell the electrolyte is made up of a mix of alkaline carbonates (lithium, potassium, sodium) in a lithium—aluminium porous matrix. Electrodes are in a nickel alloy with 10% of chromium (anode) and porous nickel oxide with 1–2% of lithium (cathode).

The reactions in a MCFC are the following:

| | |
|---|---|
| Anodic reaction | $H_2 + CO_3 \Rightarrow H_2O + CO_2 + 2e^-$ |
| Cathode reaction | $\frac{1}{2}O_2 + CO_2 + 2e^- \Rightarrow CO_3$ |
| Overall reaction | $H_2 + \frac{1}{2}O_2 \Rightarrow H_2O$ |

These cells need to be fed at the cathode, besides with oxygen (or air), also with carbon dioxide.[28] This makes them particularly suitable to be fed with hydrogen locally produced by gas reforming: in this case it is possible to use the $CO_2$ produced by the reforming system for the cathode supply of the cell. Thanks to the high operating temperatures (to keep molten salts the operating temperature of the cell must exceed 600°C), internal reforming is possible; by powering the cell with $CH_4$ and $H_2O$ (steam), the reforming reactions take place first, followed by the already seen anodic reactions. Furthermore, high temperatures allow these cells to present a high tolerability to impurities.

*Solid Oxide Fuel Cells*: in this type of cell the electrolyte is made up of ceramic material (yttrium-zirconium oxide) with a thickness of 30–40 μm. Electrodes are free from platinum; the anode is of cobalt or nickel zirconium oxide and the cathode of strontium-doped lanthanum magnetite.

The $H_2$ reactions in a SOFC are the following:

| | |
|---|---|
| Anode reaction | $H_2 + O \Rightarrow H_2O + 2e^-$ |
| Cathode reaction | $\frac{1}{2}O_2 + 2e^- \Rightarrow O$ |
| Over reaction | $H_2 + \frac{1}{2}O_2 \Rightarrow H_2O$ |

The co reactions in a SOFC are the following:

In these cells, the presence of gas and solid phases alone reduces the corrosion problems, whereas high temperature reduces the polarisation losses and markedly improves the tolerance to impurities. For SCFC, CO is not a poisoning agent but rather a fuel (namely it actively participates in electrochemical reactions); $CH_4$ can be inert (while not participating in reactions, it does not cause any damages) or fuel; $H_2O$ and $CO_2$ are inert, whereas the compounds of Sulphur are harmful in concentrations exceeding 1 ppm.

---

[28] The quantities of $CO_2$ requested are by far higher than those normally contained in the air: two moles of $CO_2$ every mole of $O_2$. Air contains nearly 20% (in moles) of oxygen and 388 ppm (parts per million) of $CO_2$, which accounts for 0.039%. Therefore, on average, in the air the ratio $CO_2/O_2$ amounts to 0.00195, that is to say over 1,000 lower than the one demanded.

*Fields of Application, Costs and Perspectives of Fuel Cells*

After space applications, recently, fuel cells were extensively used in the field of electric power production, both in large power plants and for the small scale production "on site". Power plants were installed in several places, in Japan, in Europe and in the USA, where a passage occurred from the prototype phase to the technical–economic evaluation phase.

In the military field, fuel cells are used in submarines and in ground vehicles, thanks to their low noise level and thermal "trace" that makes their infrared detection difficult, whereas there are minimum applications under study in the surface naval sector.

In general, the most interesting applications, to the purpose of a large-scale development of the technology of fuel cells, include the electric power generation for stationary use (in cogeneration), the sector of vehicles and that of portable electronics.

***Stationary Use***: For the production of electric power, fuel cells show several advantages compared to conventional systems, including:

- Higher flexibility of use since they are more suitable to meet the needs of a variable load, whereas their efficiency is not linked to the size of the plant
- Higher electric efficiency
- Lower environmental impact
- Higher easiness of planning and construction: thanks to their modular structure, the construction times can be strongly reduced and the power of the plant can be increased over time.

Cogeneration (and distributed generation in general) meet environmental (low emissions) and socio-political (management and control at a local level, competitiveness and quality and flexibility of services) needs, that will become increasingly important in the medium-long term. It is therefore possible to envisage a growing space for small-medium size generation technologies, with a high efficiency and a limited environmental impact, as fuel cells. It is predictable in the short-medium term that the various types of cells, according to their different operational characteristics (operating temperature, efficiency, size, etc.), can be used for distributed generation and cogeneration for civil and industrial use, for powers ranging between some kW to some MW.

***Transports***: in the sector of transports, fuel cells show advantages compared to both internal combustion engines (ICE) and an electric powertrain vehicle with accumulators. The use of hydrogen-supplied fuel cells allows to have zero emission vehicles (ZEV) eliminating in this way the pollution problems linked to the use of ICE without the limits of electric powertrain vehicles with accumulators (high weight, low autonomy, long recharge times). As concerns electric vehicles, it is important to underline that both fuel-cell powered vehicles and vehicles powered through accumulators are driven by an electric motor: the two solutions, therefore, shall not be necessarily considered as alternative to one another. On the contrary, the experiences developed in the last few years show that the adoption of "hybrid" solutions (fuel cells for the production of electric power onboard together with both traditional storage systems—accumulators—and new generation ones—supercapacitors) allow to conjugate the high efficiency of electric vehicles with accumulators without the limits due to the limited range and the long recharge times.

Among all the types of cells, the polymer electrolyte membrane (PEM) fuel cells are those that, in consideration of their working temperature, power, weight and size, are suitable for the applications in the sector of transports.

***Portable applications***: in the field of portable electronic equipments, batteries are at the moment the only possible option for products that require a power up to 100 W. The limits of batteries in the applications, for instance, for notebooks, are determined by weights and sizes, by the limited autonomy, the recharge times and their replacement which entails disposal problems. A fuel cell might, conversely, supply energy for times very much higher than present Li-ion batteries. Furthermore, given the same weight, hydrogen-rich fuels contain electrochemical energy that is two order of magnitude higher than whatever type of battery. These considerations led to the development of testing activities for the use of fuel cells for small electronic units that need high-quality power supply.

The main problems to be solved are linked to the complexity of the internal design of the device, in particular as concerns the storage hydrogen subsystems, fuel and oxidant control, and cooling system. Miniaturisation entails therefore a delicate balance of different factors, such as power, size, handiness and costs.

*Costs and Perspectives*

The present development led to obtain high specific powers; for instance PEMs, that, as already said, are particularly suitable for the drive, have a specific power of about 1 kW/l (corresponding to a volume of nearly 50 litres for a medium-power vehicle). Another crucial factor for the development of fuel cells is duration; presently, in case of low-temperature cells, this duration is around 3,000–3,500 h, whereas it reaches nearly 5,000 h in case of high-temperature cells. Costs vary according to the type of cells and their size.

The following targets in terms of costs and duration were set as far as marketing is concerned: 100 €/kW and 5,000 h of operation for low temperature cells for the powertrain. This duration corresponds, in case the system is assembled on a car, to about 100,000 km of mileage. In case of stationary plants, objectives include a cost not exceeding 1,000 €/kW and a duration exceeding 40,000 h.

### *4.2.4 Chemical Energy-to-Mechanical Energy Conversion Plants*

#### 4.2.4.1 Internal Combustion Engines

In these systems, the chemical energy of a fuel is converted into heat through combustion; such heat is converted into mechanic energy through a

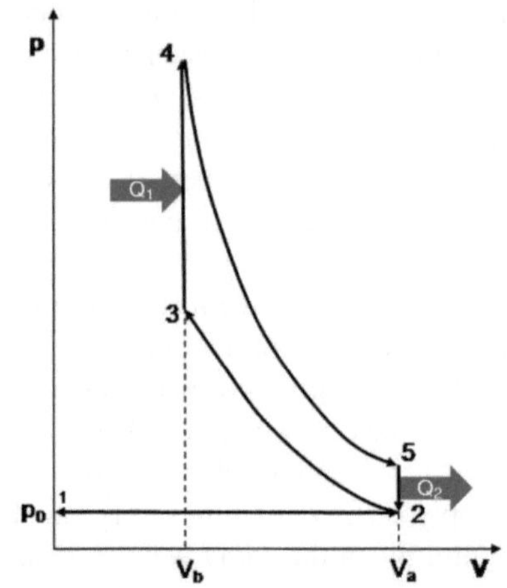


**Fig. 4.34** Ideal Otte cycle on plane *p–V*

thermodynamic cycle. The most widespread application is in the sector of transports; in this case the "useful" energy is the mechanic one. In any case, mechanical energy (on a rotating shaft) can be easily converted into electric energy through a generator coupled to the shaft (ICE power-supply units): in this case the system as a whole performs a conversion from chemical energy to electric power.

ICE can be classified according to the thermodynamic cycle: the main ones are the Otto cycle and the Diesel cycle.

*Otto Cycle*

Figure 4.34 shows the ideal Otto cycle *p–V*. The engines adopting this cycle can be fuelled with petrol, LPG, hydrogen and methane, or a mix of them.

The following phases can be identified in the Otto cycle:

- 1–2: Constant pressure aspiration of the air–petrol mixture;
- 2–3: Rapid compression, ideally an adiabatic one, of the mixture;
- 3–4: Combustion, phenomenon summarised through an isochoric heating;
- 4–5: Rapid expansion, ideally an adiabatic one;
- 5–2: Opening of the discharge valve with isochoric reduction of pressure;
- 2–1: Exhaust at constant pressure.

The ideal efficiency of a thermodynamic cycle is:

$$\eta_{\text{id}} = \frac{L}{Q_1} \tag{4.87}$$

For the Otto cycle there is:

$$L = Q_1 - Q_2 = c_v(T_4 - T_3) - c_v(T_5 - T_2) \tag{4.88}$$

By combining the better two equations and assuming $c_v$ as constant one carry to:

$$\eta_{\text{id}} = 1 - \frac{T_5 - T_2}{T_4 - T_3} \tag{4.89}$$

Applying the equations of the related thermodynamic transformations (adiabatic and isobaric) of the cycle, then

$$\frac{T_3}{T_4} = \frac{T_2}{T_5} \tag{4.90}$$

one comes to:

$$\frac{T_5}{T_4} = \left(\frac{V_b}{V_a}\right)^{\gamma-1} \tag{4.91}$$

Considering (4.90) and (4.91), (4.89) can be written as follows:

$$\eta_{\text{id}} = 1 - \frac{T_5}{T_4} \cdot \left(\frac{1 - T_2/T_5}{1 - T_3/T_4}\right) = 1 - \frac{T_5}{T_4} = 1 - \left(\frac{V_b}{V_a}\right)^{\gamma-1} \tag{4.92}$$

Having defined the compression ratio $r$ as follows:

$$r = \frac{V_a}{V_b} \tag{4.93}$$

the ideal efficiency of the Otto cycle is:

$$\eta_{id} = 1 - \frac{1}{r^{\gamma-1}} \tag{4.94}$$

It is evident how the ideal efficiency increases as the compression ratio increases; in the practical Otto cycles, however, it is not possible to exceed a compression ratio of 8–10 in order to avoid the self-ignition of the mixture that would occur during the compression phase, before the piston reaches the top dead centre, since a fuel–air mix is compressed and in some temperature and pressure conditions, it spontaneously burns out.

With $\gamma = 1{,}4$ and $r = 8$, the ideal efficiency is 56.5%. The real cycle (Fig. 4.35) is very much different from the ideal one, having considered the irreversibility of transformations and most of all the fact that induction and exhaustion do not take place at constant pressure, generating in this way a negative work in the cycle.

The maximum efficiency[29] of the real cycle amounts to about 25%. If the engine operates at a variable speed (automotive powertrain), considering the operation at variable speed the average efficiency drops at nearly 18–20%.

[29] The efficiency of the real cycle (differently from the ideal one) varies according to the number of revolutions of the engine and presents a maximum of nearly 2/3 of the maximum number of revolutions.

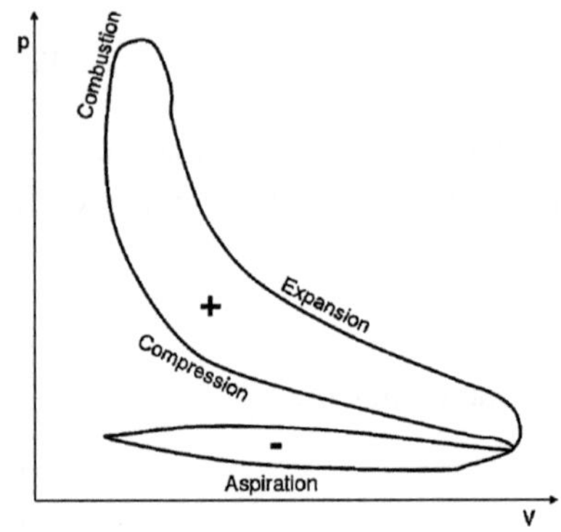


Fig. 4.35 Real Otto cycle on plane $p$–$V$

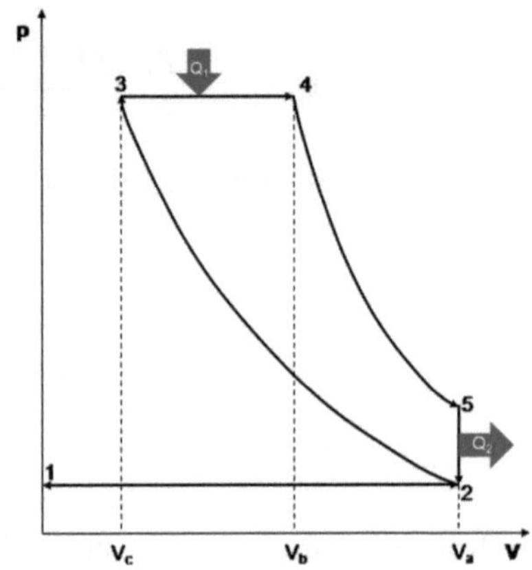


Fig. 4.36 Ideal Diesel con plane $p$–$V$

*Diesel Cycle*

Figure 4.36 shows the Diesel cycle $p$–$V$ diagram. Engines adopting this cycle can be fuelled with gas oil.

In the Diesel cycle the following phases can be identified:

- 1–2: Aspiration at constant pressure of air only;
- 2–3: Rapid compression, ideally an adiabatic one;
- 3–4: Combustion, which takes place by introducing the fuel when air has reached the conditions suitable for air–fuel mixture burns spontaneously, a phenomenon that can be regarded as isobaric heating;
- 4–5: Rapid expansion similar to an adiabatic;
- 5–2: Opening of the discharge valve with isochoric reduction of pressure;
- 2–1: Exhaust at constant pressure.

For the Diesel cycle there is:

$$L = Q_1 - Q_2 = c_p(T_4 - T_3) - c_v(T_5 - T_2) \tag{4.95}$$

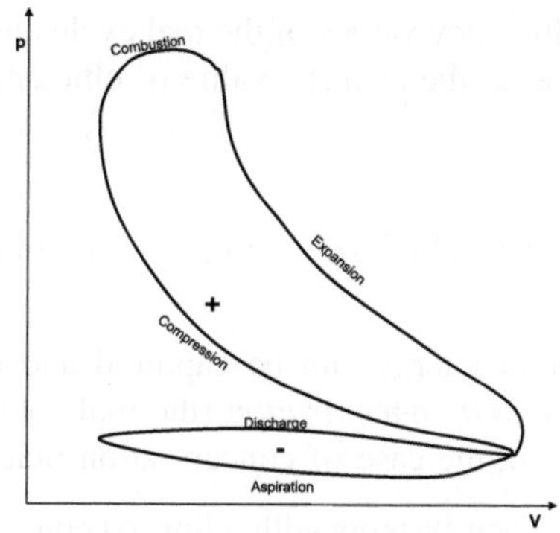


**Fig. 4.37** Real Diesel cycle on plane *p*–*V*

Therefore, considering $c_p$ and $c_v$ as constants, the ideal efficiency is:

$$\eta_{id} = \frac{L}{Q_1} = 1 - \frac{c_v}{c_p}\frac{T_5 - T_2}{T_4 - T_3} = 1 - \frac{1}{k}\frac{T_5 - T_2}{T_4 - T_3} \tag{4.96}$$

Applying the equations of thermodynamic transformations (adiabatic and isobaric) of the cycle, then:

$$\frac{T_4}{V_c} = \frac{T_3}{V_b} \tag{4.97}$$

Considering (4.96), (4.97) becomes:

$$\eta_{id} = 1 - \frac{1}{\gamma} \cdot \frac{1}{(V_a/V_b)^{\gamma-1}} \left( \frac{(V_c/V_b)^{\gamma} - 1}{(V_c/V_b) - 1} \right) \tag{4.98}$$

Defining the following ratios:

$$r = \frac{V_a}{V_b} \tag{4.99}$$

$$e = \frac{V_a}{V_c} \tag{4.100}$$

Replacing (4.99) and (4.100) in (4.98), the efficiency can be written as follows:

$$\eta_{id} = 1 - \frac{1}{\gamma} \cdot \frac{(1/e)^{\gamma} - (1/r)^{\gamma}}{(1/e) - (1/r)} \tag{4.101}$$

In this case as well, efficiency increases as r increases. In the case of the Diesel cycle, however, the real value of $r$ can be higher than in the case of the Otto cycle since, by compressing only air, the problem of self-ignition does not appear; the effective values of $r$ can be 18–20. With $r = 18$ ($e = 9$) a value of the ideal efficiency is obtained as amounting to 63.2%. Also in this case as well, the real cycle (Fig. 4.37) is heavily different from the ideal one for the reasons seen above. is heavily different from the ideal one for the reasons seen above. The maximum efficiency values of the real cycle are around 35% and, during operation at variable speed, the average value of efficiency drops at around 25–28%.

### 4.2.5 Radiant Energy-to-Thermal Energy Conversion Plants

Solar energy can be captured and used in a thermal form, and this can happen *without concentration* (thermal solar at low temperature) or *with concentration.*

In the case of concentration plants, there are two different applications:

1. The systems with a limited concentration, in which case the thermal energy will have a relatively low operational temperature (i.e. up to 250° C) and heat is used for process purposes, whenever a thermal source is needed, i.e. for drying purposes in agriculture or for chemical-industrial processes.
2. The systems with a high concentration, in which case maximum operational temperature is high and applications can include industrial processes or thermodynamic conversion to produce electric power.

#### 4.2.5.1 Low-Temperature Thermal Solar Plants

The main application of this conversion technology is that of capturing solar radiation energy and, by means of it, to heat water (at 40–70°C) for sanitary use destined to civil and residential applications.

There are several types of plants and can be used for different purposes according to the performance of the collector, the place in which they are located and the types of users they have to serve.

A basic solar thermal system is made up of four main elements that are listed below:

- Solar panel (collector)
- Storage reservoir
- Control unit
- System of pipes and circulation pump.

##### Flat-Plate Solar Collectors

A flat-plate solar collector (Fig. 4.38) is made up of a *transparent covering* with one or several glass or plastic sheets placed at the top of the absorbing plate to reduce convective and radiation thermal exchanges between the plate and the atmosphere, and a *black absorbing plate* that absorbs the radiation and transfers the energy collected to a heat-transfer fluid, one layer of *thermal insulation* to reduce as much as possible the losses due to plate conduction, and of one

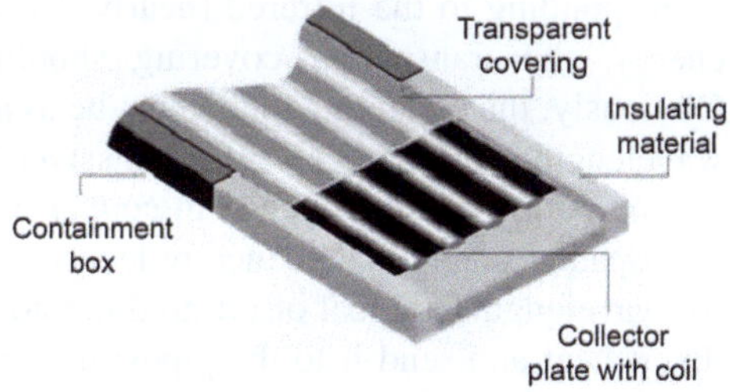


**Fig. 4.38** Elements making up a solar collector

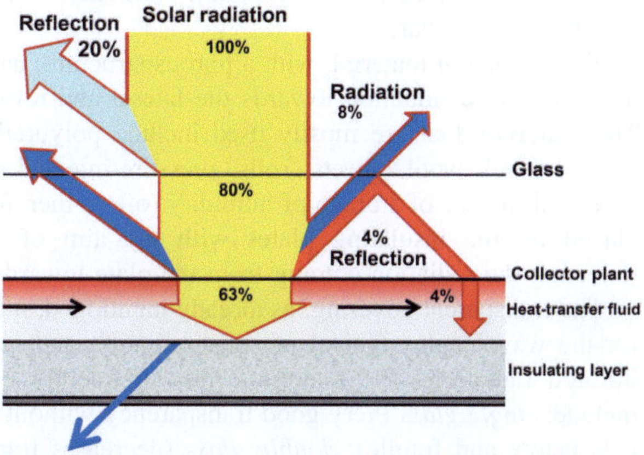


**Fig. 4.39** Thermal flows in a solar collector

parallelepiped-shaped *covering* with a task of containment and protection from dust, humidity, etc.

Figure 4.39 shows the energy flows of a tipical flat-plate solar collector. A part of the incident solar energy on the glass coating is reflected, a fraction is absorbed by the glass itself, another reaches the capturing plate. A part of such energy is reflected, another part is absorbed by the plate. Some amount of the energy reflected by the plate is in turn reflected by the glass coating, hence going back again to the capturing plate, and the phenomenon is repeated originating a greenhouse effect inside of the collector.

On the basis of the description above, it is evident that in order to have a suitable capturing of the incident energy, the covering sheet must be extremely transparent ($\tau \cong 1$) at wavelengths in which there is the maximum possible solar radiation (between 0.4 and 2.5 lm). After being reached by the solar radiation, the absorbing plate absorbs most of it, subsequently emitting energy in the form of thermal radiation, whose wavelength, 1, (in μm) depends on the level of temperature *T* (K) according to Wien's law:

$$\lambda = \frac{2897.8}{\mathrm{T}} \tag{4.102}$$

Since the operating temperature of the collector plate is usually around 100°C (373 K), the energy released by the plate falls within the field of wavelengths corresponding to the infrared (nearly 7.7 μm). Therefore, to maximise the useful energy, the transparent covering should reflect ($\rho \cong 1$) these wavelengths. Obviously, the collecting plate must be as absorbent as possible ($\alpha \cong 1$) for all the wavelengths, and have a low submissiveness for the wavelengths of the infrared.

The collecting plate, made in copper or steel, is superficially treated with dark and opaque paints that reduce reflection and increase absorption (low emissivity for wavelengths, typical of infrared radiation). The main function of the plate is to divert heat and send it to the pipes: it is important to privilege those solutions in which the thermal resistance between said elements is reduced to minimum levels (for instance welded pipes and plate). Normally, canalisations are made to resist pressures of 6–7 bar.

The insulating material, with a porous structure, has to reduce at minimum the losses due to conduction towards the lateral and lower surfaces of the collector. The materials that are mostly used include polyurethane, polyester wool, glass wool or rock wool (sheets, rolls, pressure-injected foams). Thermal insulation gives out in case of storage of humidity (often, therefore, a sheet of aluminium is placed on the insulating plates with the aim of stopping condensation and reflecting the radiation coming from the plate towards the plate itself).

The transparent covering, as already mentioned, must be extremely transparent for the wavelengths typical of solar radiation and, at the same time, opaque to infrared radiation. The materials most frequently used for its implementation include: *single glass* (very good transparency without blocking convection losses, it is heavy and fragile), *double glass* (decreases transparency but increases the thermal insulation capacity), honeycomb polycarbonate (light, inexpensive, resistant, it is characterised by reduced convection losses compared to the single glass, although it is less transparent, tends to be opaque in time and, therefore, its lifecycle is shorter than the collector).

The containment case gives compactness and mechanical strength to the collector, and aims at protecting the internal elements from dirt and atmospheric agents. The materials that are usually used to realise it include stainless steel (generally galvanised and pre-treated), anodised aluminium or, more rarely, fibreglass.

The heat-transfer fluid must have a limited volume (in order to use small-size pipes), a high density and a high specific heat. It shall not be corrosive for the circuit, and must be chemically inert and stable at temperatures around 100°C. It is also necessary that the fluid selected has a low hardness in order to limit the limestone deposition, besides having a low freezing point and a low viscosity (in order to favour an easy flowing of the fluid without introducing excessive load losses in the circuit). In those cases in which the collector is used to supply sanitary hot water, it shall not be toxic (otherwise there would be huge complications of the system for safety reasons). Fluids that corresponding to these characteristics, at limited costs, include *water* (characterised, however, by non negligible problems of hardness and also an easily reachable freezing point in areas with a temperate climate), *a solution of water and ethylene glycol* (toxic and not suitable to systems for the production of sanitary hot water (SHW)) or, in most of cases, *a solution of water and propylene glycol.*

Indicating with $Q$ the incident flow, with $Q_u$ the flow of useful heat, with $Q_p$ the flow of dissipated heat, the balance can be written as follows:

$$Q = Q_u + Q_p \tag{4.103}$$

Indicating with $\tau$ the transmittance of the transparent coating, with $\alpha$ the absorbance of the collecting plate, with I the incident solar radiation and with $A$ the active surface of the collector, the total incident flow is:

$$Q = I \cdot \tau \cdot \alpha \cdot A \tag{4.104}$$

With reference to *specialised* literature, the flow of dispersed heat (sum of thermal losses and of optical losses) can be expressed as:

$$Q_p = U_c(T_p - T_a) \tag{4.105}$$

where $T_a$ is the ambient temperature, $T_p$ is the collecting plate average temperature, $U_c$ is the *thermal dispersion coefficient of the collector* that depends on the convective and radiation thermal exchange coefficients with the environment and can be considered as constant within a limited range of temperatures.

From (4.103), (4.104) and (4.105), it is possible to obtain:

$$Q_u = (\tau\alpha)I - U_c(\mathrm{T}_p - \mathrm{T}_a) \tag{4.106}$$

In (4.106), however, the useful energy is expressed according to the average temperature of the plate, which is unknown. From an operational viewpoint, it is possible to use to this end the Bliss equation that expresses the useful energy according to the temperature of the fluid at the entry point of the collector $T_i$:

$$Q_u = F_r[(\tau\alpha)I - U_c(T_i - T_a)] \tag{4.107}$$

$F_r$ is the *factor of thermal removal of the collector*, always lower than 1. It physically represents the *ratio between the useful energy actually collected and the energy that would be collected if the entire plate had been at the same temperature as the outlet fluid*. This condition might only be assessed in ideal conditions: infinite fluid flow rate and null thermal resistance between fluid and plate. $F_r$ mainly depends on the convection coefficient for fluid and plate, on the thermal conductivity between plate and the canalisation.

Some reference values of the parameters used in the Bliss equation are shown in Table 4.6.

The *efficiency of a solar collector* is defined as the ratio between the useful energy collected in a given period of time and the incident solar energy in the same period:

$$\eta = \frac{Q_u}{I} \tag{4.108}$$

Replacing (4.107) in (4.108), it follows that:

**Table 4.6** Values of the parameters for the calculation of useful energy

| Parameters for the calculation of useful energy | | |
|---|---|---|
| Thermal loss coefficient $U_c$ (W/m$^2$ K) | Non-selective plate and simple glass transparent covering | 7–8 |
| | Non-selective plate and double glass covering | 4–6 |
| | Selective plate and double transparent covering | 3–5 |
| Effective product $\tau\alpha$ | Simple glass covering | 0.9 |
| | Double glass covering | 0.8 |
| Thermal removal factor Fr | Liquid collectors | 0.9 |
| | Air collectors | 0.7 |

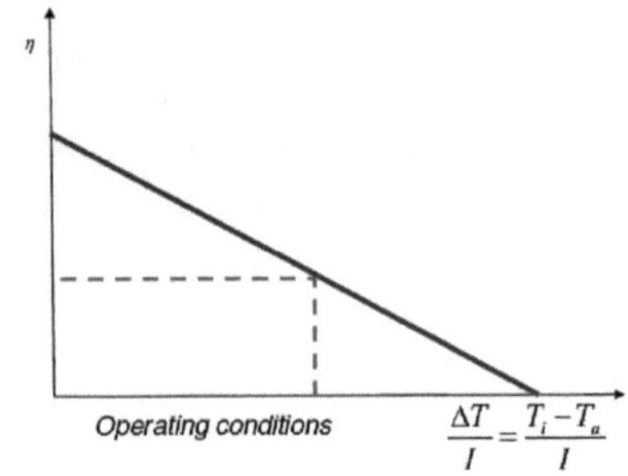


**Fig. 4.40** Efficiency of a solar collector as operating conditions vary

$$\eta = F_r(\tau\alpha) - \frac{F_r U_c (T_i - T_a)}{I} \tag{4.109}$$

Since $F_r(\tau\alpha)$ and $F_rU_c$ can be considered as constant, the efficiency depending on the variable $T_i$-$T_a/I$ is an angular coefficient line $F_rU_c$, as shown in Fig. 4.40.

The results obtained are supported by the experimental analysis, from which it emerges that efficiency has an almost linear trend, sloping down as temperature increases due to the variation of $U_c$. As wind speed increases, efficiency is reduced by some percentage points with more marked variations over the speed of 5 m/s, and as the temperature of collectors increases. The influence of wind is negligible for those collectors endowed with two or more transparent coverings.

The intersection of the efficiency line with the axis of abscissas defines the *stagnation temperature* that, after setting room temperature and the intensity of the incident radiation, represents the temperature at which the plate is taken in conditions of null useful energy.

*Performance of flat-plate collectors influence parameters*

At the usual operating temperatures (40–60°C) for a non-selective flat-plate collector and a glass covering, thermal dispersions are due to 10% only to posterior and lateral conduction losses. The remaining part is equally divided into losses due to convection and radiation between plate and glass. For this reason, it is not possible to obtain remarkable advantages by increasing the thickness of posterior insulation beyond the 5–10 cm that are usually used, whereas there are considerable reductions of losses by taking care of the elimination of all possible thermal bridges, the most common of which is represented by the plate-external covering contact outside of the collector.

Particular importance is also ascribed to the characteristics of the insulating material that shall not degrade over time because of severe thermal stresses. Since the radiation exchange depends on the fourth magnitude of absolute temperature, as temperature increases, radiation dispersions increase as well, which explains the use of selective plates when operating temperatures are higher. This intervention, however, in lowering the equilibrium temperature of glass, stresses the convective losses between plate and glass; this increase, however, is lower than the reduction of dispersions by radiation.

Dispersions by convection can be drastically reduced, creating vacuum between the plate and the glass, with pressures lower than 10 mbar as it happens in vacuum collectors.

To increase $F_r$ it is possible to increase the flow rate which should be included between the minimum value of 0.007 kg/m$^2$·s, necessary to guarantee a sufficient irrigation of all canals, and the maximum value of 0.03 kg/m$^2$·s beside which the advantages of the best thermal exchange can be overcome by an increased pumping activity. As the flow rate decreases, the temperature drop between the input and the output of the collector increases, and therefore also the instantaneous efficiency, which in the long term may cause a penalisation of the overall performance, is reduced. The $F_r$ increase is also obtained by paying attention to the thermal exchange between plate and canalisation, which might also be limited in the case of canalisations welded to the plate. Furthermore, it is important that the plate has a good thermal conductivity, which might be obtained by using metals as copper (which shows the best conductivity), aluminium and steel.

Apparently, performance should be penalised as the thermal capacity of the collector increases, because of the impossibility of exploiting short favourable periods, characteristic of a variable climate.

*Types of flat-plate solar collectors*

Besides the glazed ones mentioned above, there are other types of collectors: *uncovered, integrated-storage* and *vacuum.*

Uncovered collectors, made with plastic pipes (propylene, neoprene, synthetic rubbers, PVC) have the advantage of being cost-effective in some conditions although free from any type of insulation and transparent cover, and provide satisfactory performance only if used for plants working in the summer period, when the great availability from the solar source makes up for their low efficiency.

In integrated storage collectors, a single element replaces the collecting plate, the coil and the external storage; this single element can be made up of siding pipes (diameter of nearly 10 cm), large pipes between two plates facing each other, and a single large container of different shapes. The water necessary for users remains inside of the collector (slow heating with convective motions).

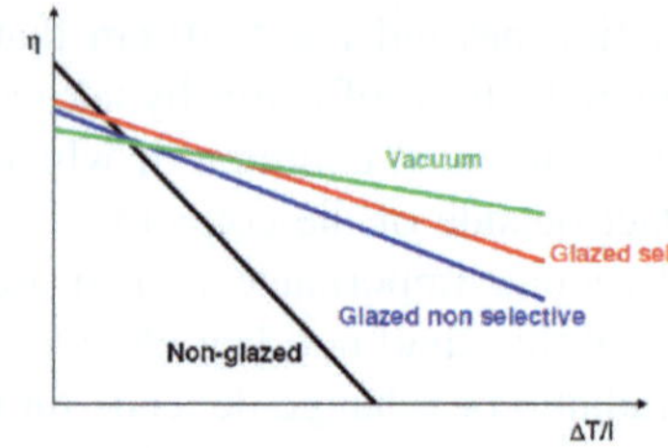


**Fig. 4.41** Efficiency curve of the different types of flat collectors

In order to have an idea on the difference compared to flat collectors, it is enough considering that the water content reaches 80–100 l/m$^2$ against 0.6–2 l/m$^2$ of heat-transfer fluid, typical of the devices with external storage. Since water remains is a "reservoir" insulated only along three sides out of four, this type of collector is characterised by non-negligible leakages and is suitable for particularly mild climates.

Vacuum collectors are those with the highest efficiency; by reducing at minimum levels the presence of air in hollow spaces, it is possible, as a matter of fact, to avoid the losses due to convective motions. Although more expensive compared to the average, these collectors allow good performance also with harsh climates.

From Fig. 4.41, showing the comparison between the efficiency curve of the different types of flat collectors, it is possible to note that the efficiencies of the different types of solar collectors are basically the same for low values of the difference between the temperature of the collector and the ambient temperature. In fact, the different solutions adopted (the covering, eventually selective, the vacuum inside of the collector) aim at reducing leakages. These leakages, for obvious reasons, are in any case limited for the low values of difference between the temperature of the collector and the ambient temperature; as it increases, the leakages for the different types markedly increase. In particular, vacuum collectors keep high efficiencies also at high temperatures.

*Plant solutions*

As concerns the configuration of the plant, *natural circulation* and *forced-circulation plants* are to be considered.

In natural circulation plants, the force of the heat-transfer fluid occurs on the basis of a different density between the hottest water column flowing out of the collector, which tends to increase, and the coldest water column flowing out of the storage reservoir, which tends to increase (Fig. 4.42a). The higher the difference in temperature between the two columns of fluid, the higher the force that allows its circulation. In natural circulation systems, panels must be positioned lower than the storage area. The difference between the two levels as well is proportional to the force generated.

In forced-circulation plants, the fluid is set in motion by a centrifugal pump (Fig. 4.42b); in that case, obviously, position different in the storage can occur.

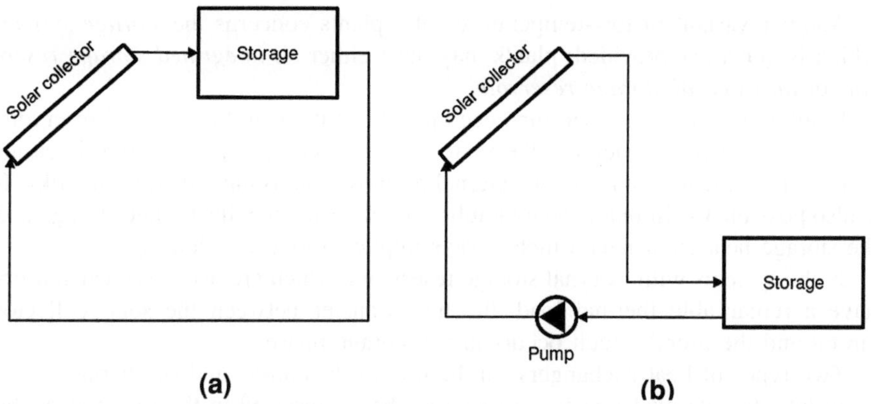


**Fig. 4.42** Natural (**a**) and forced (**b**) circulation plant configurations

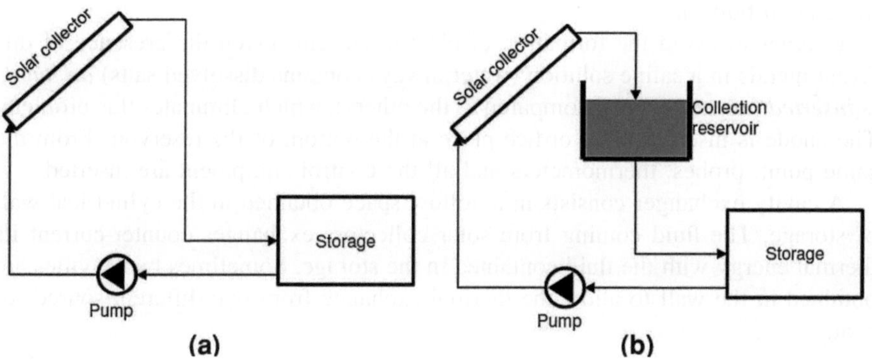


**Fig. 4.43** Plant configuration: closed systems (**a**) and draining systems (**b**)

Another distinction in the plant configuration concerns the state of the circuit when the plant is stopped: there are *closed-loop systems* and *drainback systems* (Fig. 4.43).

In closed-loop systems, solar collectors are always loaded with fluid, even when the circulation pump is stopped. In this case the heat-transfer fluid must be able to face low temperatures without freezing, and impose the use of an antifreeze liquid: this prevents the direct exchange. In closed systems it is always necessary to interpose a heat exchanger between the solar circuit and the users.

In drainback systems, as soon as the circulation pump stops, the solar circuit empties in a drainage reservoir, placed under collectors themselves. Since one better avoid using antifreeze liquids, open systems can be used with a direct exchange without placing any exchanger between the plant and users that, obviously, entails a higher complication of the system and an overall loss of efficiency. Conversely, the disadvantage of direct exchange systems that use running water as heat-transfer fluid is the danger of calcareous deposits on the internal walls of pipes.

Another variant of low-temperature solar plants concerns the *storage system*, which is generally provided: plants may have either an *integrated storage reservoir* or *an external storage reservoir*.

In integrated storage reservoirs, each panel has its own "integrated" reservoir. As a matter of fact, generally, the system is more compact, installation is easier and it is possible to eliminate the external reservoir that is unaesthetic and bulky. It is also possible to eliminate the hydraulic circuit between collector and storage and the storage heat exchanger which always implies a loss of efficiency.

In the systems with external storage reservoirs, which are necessary when users have a remarkable thermal load, the heat transfer between the solar collector circuit and the users' circuit occurs in the storage phase.

Two types of heat exchangers can be used: *tube-bundle* and *cavity* ones.

In tube-bundle exchangers, a series of tubes is inserted in the low part of the storage. Sometimes, two groups are inserted with a view to heating the fluid contained in the storage with two different sources of heat (solar system and integration boiler).

In order to avoid the formation of electric current, given the presence of different metals in a saline solution (water always contains dissolved salts) *an anode is inserted* (negative metal compared to the others), which eliminates this problem. The anode is inserted in an orifice place at the bottom of the reservoir. From the same point, probes, thermometers and all the control equipment are inserted.

A cavity exchanger consists in a hollow space obtained in the cylindrical wall of storage. The fluid coming from solar collectors exchanges counter-current its thermal energy with the fluid contained in the storage. Sometimes two cavities are obtained in the wall to allow the thermal exchange from two different sources of heat.

*Heat storage*

Since solar radiation and consumption of thermal energy do not happen simultaneously, it is necessary to include a heat storage reservoir in the plant. Heat storage can be done on a daily, weekly or seasonal basis. For the sizing of storage (usually daily) it is necessary to refer to the diagrams of daily heat production and needs.

In any case, seasonal storage is also possible. There are two storage systems: the first one (very much used in northern European villages), is obtained by *digging in rocky masses* a cave with a capacity between 100 and 1000 $m^3$. If there are no aquifers nearby, the rock, once filled with hot water, reaches a stable thermal equilibrium. The reservoir is sized to meet the thermal needs from September to the end of February. In March the direct use of water from solar collector begins once again. Very often the heat coming from inter-season storage is integrated with chimney boilers. In the second one (very much used in Middle East) systems called "*Solar Pond*" are made. These are large artificial reservoirs; the bottom of the reservoir is made in black plastic or in any other material provided that it is safe (sometimes also seaweeds). The reservoir is divided into two layers through a polyethylene film or in any case with a transparent plastic: the most

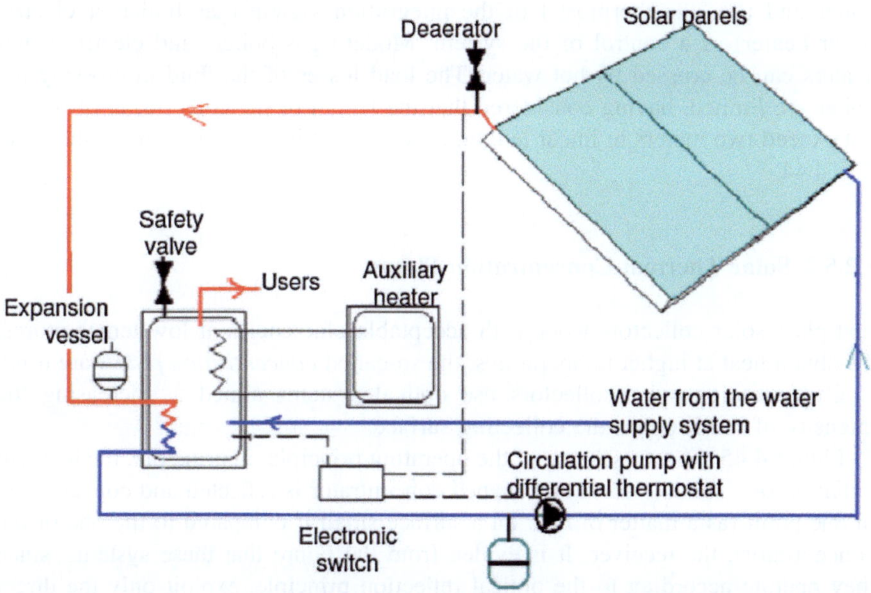


**Fig. 4.44** Diagram of a forced-circulation and closed-circuit integrated system

superficial layer is filled with freshwater, the deepest layer with saltwater. The transparent film acts as a selective material. The temperature of deep saltwater reaches even 90°C.

The weekly storage aims at ensuring the availability of heat in long periods with a low or very low solar radiation. In this case the storage system is a reservoir of such size that guarantees a reserve of thermal energy. Storage can be done through a single reservoir or with several reservoirs, one of which has to be sized for the daily needs of the normal period of use, whereas the secondary ones must be used according to the quantity of storage thermal energy selected. In case of the Mild climate areas, weekly coverage solutions are avoided, since they are more suitable for cold climate regions.

*Integrated plants*

Given the uncertainty of the source, solar collector plants, in order to guarantee the continuous availability of heat to users, must usually be integrated with traditional plants. The heat produced from solar energy will integrate the one produced in traditional boilers: the integration factor is determined according to technical–economic considerations, normally around at least 60%.

The insertion is extremely simple in case of integrated storage in the collector. In fact, having considered that the hydraulic circuit between collectors and storage is not present, any control equipment is necessary (*differential thermostat* and *minimum thermostat*). The solar collector can be installed upstream the integration boiler and uses the thermostat of the integration system (gas boiler or electric water heater) as a control of the system. Modern gas boilers and electric water heaters can be crossed by hot water. The load losses of the fluid in crossing the boiler are limited, having considered that the length of the coils inside of it does not exceed two meters in linear length. A scheme of integrated plants is shown in Fig. 4.44.

#### 4.2.5.2 Solar Thermal Concentration Plants

Flat-plate solar collectors work with acceptable efficiencies at low temperatures. To obtain heat at higher temperatures, the so-called *concentration plants* are used.

Concentration solar collectors use optical systems aimed at increasing the intensity of radiation on the collecting surface.

Figure 4.45 shows a scheme of the operating principle: in principle, the incident radiation of a suitable sized and shaped concentrator is reflected and concentrated in one point (as a matter of fact, on a surface small if compared to the one of the concentrator), the receiver. It is eviden from the figure that these systems, since they operate according to the optical reflection principle, exploit only the direct component of solar radiation, since the diffuse radiation reaches the receiver while being distributed in all directions. Furthermore, for the same reason, they must be endowed with a tracking system, since the angle of incidence of the radiation compared to the normal tangent of the concentrator in a point must remain constant.

When the solar radiation incident on a given surface is concetrated on a smaller one, thermal losses, being them proportional to the exchange surface of the absorbing body, are lower. The ratio between the area of the reflecting surface $A_r$ and the absorbing surface $A_a$ is called *concentration factor*. This value, for the different concentration systems, varies from 1.5 to 10,000; as the concentration factor increases, optical losses become increasingly relevant.

As shown in Fig. 4.46 there are three types of concentrators which are used in concentration solar thermal plants: *linear parabolic concentrators, parabolic dish concentrators, and central receiver concentrators.*

In *linear parabolic concentrators*, the radiation is concentrated along a line that covers the entire length of collectors; along such a line there is a pipe containing the heat-transfer fluid. Collectors follow the sun by rotating around an axis. With these systems, temperatures of around 400°C are reached, with good efficiencies.

In *parabolic dish concentrators*, the solar radiation is concentrated in a point in which there is a device that absorbs the radiation and transfers heat to a gas. The concentrator "follows" the sun rotating around two axes. With these systems, temperatures of around 750°C can be reached.

*Central receiver parabolic concentrators* are made up of a large field of heliostats and a receiver placed on a tower. A heliostat is a flat mirror that moves (to follow the sun) independently from the others by rotating around two axes. With these systems, it is possible to reach temperatures around 1,000°C.

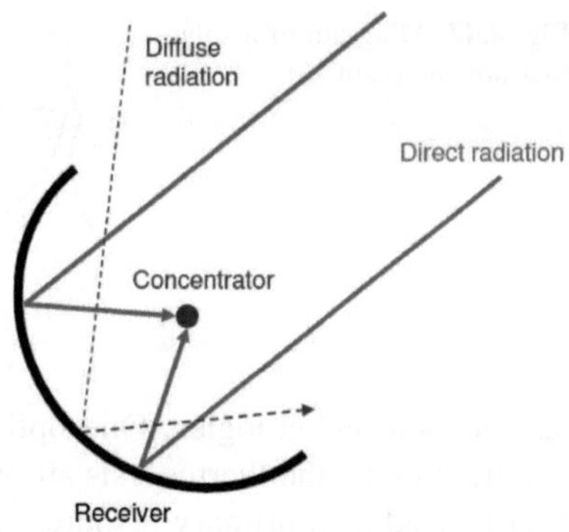


**Fig. 4.45** Scheme of the operation principle of a concentration system

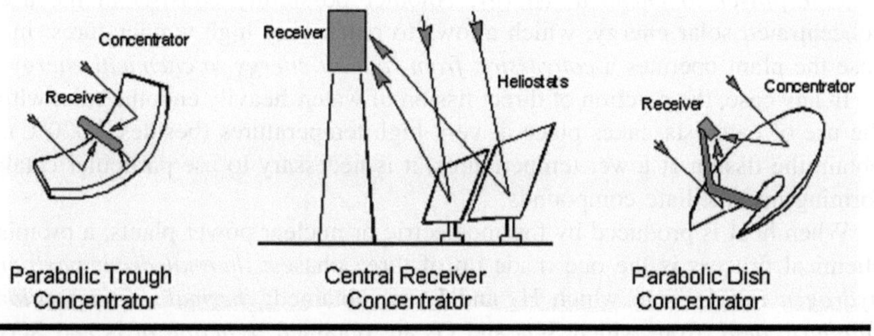


**Fig. 4.46** Different types of concentrators

The heat produced by concentration solar plants has different types of applications. It can be used by the industrial sector as process heat, or to feed a thermoelectric (steam) plant; in this case, the plant as a whole operates a conversion from radiant energy to electric energy. Another possible application (for those operating at high temperatures, that is to say dish concentrators and solar-power towers) consists in providing the necessary heat for the thermodynamic fission of water to produce hydrogen. In that case, the plant as a whole performs a *conversion from radiant energy to chemical energy*.

## *4.2.6 Thermal Energy-to-Chemical Energy Conversion Plants*

### 4.2.6.1 Thermolysis

In general, the thermolysis process consists in the fission of the water molecule into hydrogen and oxygen, using heat for the fission reaction $H_2O + \text{heat} \Rightarrow H_2 + \frac{1}{2} O_2$. Thus it is a system converting thermal energy to chemical energy. Heat can be produced in different ways, and several studies

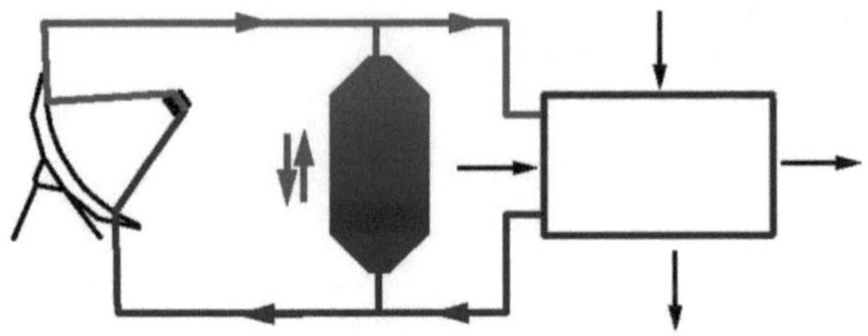

**Fig. 4.47** Diagram of a solar thermolysis plant [3]

assess such technologies. One option is to exploit the heat produced in thermoelectric plants: the thermolysis allow a *transformation* of *chemical energy* (a fossil fuel is used as a primary resource, and another fuel is produced, hydrogen). Also heat produced in nuclear plants can be used: in this we have a *conversion from nuclear power to chemical energy*. Another possibility consists in the use of concentrated solar energy, which allows to reach very high temperatures; in this case the plant operates a *conversion from radiant energy to chemical energy*.

In any case, the reaction of direct fission of water, heavily endothermic, without the use of catalysts, takes place at very high temperatures (besides 2,000°C). To obtain the fission at lower temperatures, it is necessary to use particular catalysts forming intermediate compounds.

When heat is produced by thermoelectric or nuclear power plants, a promising chemical process is the one made up of three phases: *thermal decomposition of hydrogen iodide* from which $H_2$ and $I_2$ are obtained; *thermal decomposition of sulphuric acid*, from which $SO_2$ and $O_2$ are obtained; *reaction of* $I_2$ and $SO_2$, in aqueous solution at ambient temperature, from which HI and $H_2SO_4$ in cycle are generated.

In brief, the process is the following:

$$\begin{array}{ll} \text{Phase } 1(T = 300°\text{C}) & 2\text{HI} \rightarrow \text{H}_2 + \text{I}_2 \\ \text{Phase } 2(T = 870°\text{C}) & \text{H}_2\text{SO}_4 \rightarrow \text{H}_2\text{O} + \text{SO}_2 + \frac{1}{2}\text{O}_2 \\ \text{Phase } 3(T = 25°\text{C}) & 2\text{H}_2\text{O} + \text{SO}_2 + \text{I}_2 \rightarrow \text{H}_2\text{SO}_4 + 2\text{HI} \\ \text{Global reaction} & \text{H}_2\text{O} \rightarrow \text{H}_2 + \frac{1}{2}\text{O}_2 \end{array}$$

The *efficiency of the chemical plant* under consideration (defined by the ratio between the energy content of the hydrogen produced and the heat consumed) amounts to nearly 50%. The overall efficiency of the plant clearly depends on the efficiency of the heat production process.

Another chemical process, used in the plants in which heat is produced through solar concentrators, is based on the use of iron oxide as catalysts. The process is therefore the following:

$$\begin{array}{ll} \text{Phase } 1(T = 2.000°\text{C}) & \text{Fe}_3\text{O}_4 \rightarrow 3\text{FeO} + \frac{1}{2}\text{O}_2 \\ \text{Phase } 2(T = 600°\text{C}) & 3\text{FeO} + \text{H}_2\text{O} \rightarrow \text{Fe}_3\text{O}_4 + \text{H}_2 \\ \text{Global reaction} & \text{H}_2\text{O} \rightarrow \text{H}_2 + \frac{1}{2}\text{O}_2 \end{array}$$

Also in this case, the efficiency of the chemical plant is nearly 50%.

**Fig. 4.48** Typical voltage–current for type I and type II conductors

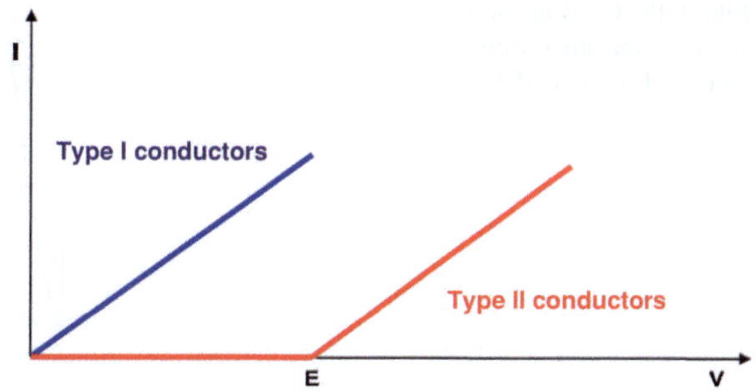


The use of concentration solar plants allows to reach very high temperatures, necessary for the dissociation reactions of the water molecule: in the focus of a parabolic mirror concentrating direct solar reactions, temperatures exceeding 2,000°C are reached.

Figure 4.47 shows the diagram of a solar thermolysis plant. The heat produced by parabolic concentrators is stored in suitable systems since the chemical plant must operate with continuity: in the periods of peak solar radiation, the heat produced is partly used to feed chemical plants, and partly stored. In the hours of absence of radiation, the chemical plant continues to operate, powered by the heat storage system.

Considering an efficiency of the heat collection and accumulation system amounting to 65% and an efficiency of the chemical plant of 53%, a global efficiency relating to hydrogen production from solar energy is obtained and amounts to nearly 35% [3].

## *4.2.7 Electric Power-to-Chemical Energy Conversion Plants*

### 4.2.7.1 Electrolysis

Overview on the Electrolysis Process

Electrolysis is a set of phenomena that take place in an electrolytic solution or in a molten electrolyte at the passage of electric current, usually continuous. In this way, the conversion of electric energy into chemical energy is made.

There are different ways of conducting current, and therefore it is possible to make a distinction between *Types I and II conductors*. In the case of the former, with the passage of continuous current, the Ohm's Law applies, whereas for type II conductors, the validity of this low is not a general one. Figure 4.48 shows the trends of current intensity according to the function of the difference of potential applied to a type I conductor and a type II one.

It is possible to see how in the case of a type II conductor only when the potential difference (PD) exceeds a given value (*E*) a passage of current occurs,

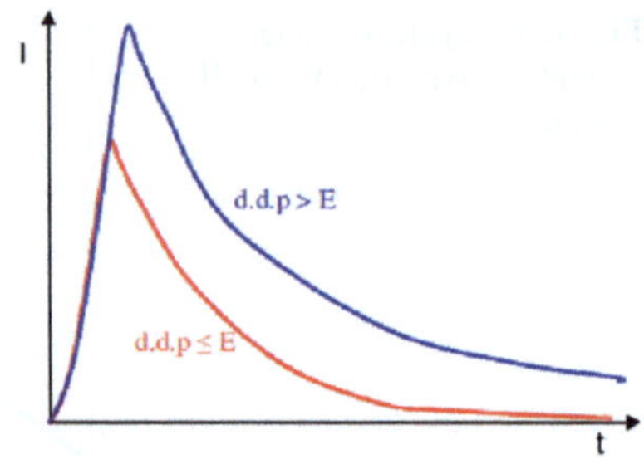


**Fig. 4.49** Current plotted against time for different values of applied PD

which then increases proportionally to the PD applied until reaching a saturation state. Below this potential threshold there is not any passage of current and therefore there is no electrolysis. Electrodes in electrochemical processes are divided in different ways according to their characteristics and the ones of the electrolytic solutions. The latter exclusively represent catalyst elements of the electrode process and a typical example is the platinised platinum electrode.

Figure 4.49 shows the trend over time of current when an electrolysis cell with platinised platinum electrodes in contact with an electrolytic solution is imposed a PD.

If PD is lower than the threshold value *E*, it is possible to observe:

- *A first ascending phase*: in fact, at the beginning there is no PD among electrodes (identical electrodes in the same solution) and no electrode reaction occurs; however, in the moment in which the circuit is closed and the PD set is applied to electrodes, there is a passage of current and electrolysis takes place.
- *A second descending phase*: the molecules that during the passage of current are formed on the electrodes remain absorbed on them and let the cathode and the anode assume the nature of electrolytes relating to the species absorbed. In this way, inside of the electrolysis cell, a stack is generated, whose electromotive force-emf (as opposed to the PD imposed from the outside and also called *counter electromotive force—cemf*), opposes the passage of current. As the ionic species in the solution are exhausted at the electrodes, there is an increase in products' activity and the cemf increases, consequently determining the decrease of the intensity of the electrolysis current.
- *A third constant phase*: as the cemf becomes equal to the PD imposed from the outside, the passage of current basically stops, and remains constant at a value very close to zero.

With a progressively increasing PD, although always lower than E, the qualitative trend of the phenomenon does not change, but any increase of the PD imposed leads to an increase of the activities of electrolysis products on electrodes and therefore of cemf.

In this way, a value of pd imposed is reached, in which the activities of products and reagents becomes such that the pd is higher than cemf and the real electrolysis takes place, that is to say in the third constant previously described there is a passage of current.

The phenomenon described, based on which the electrodes of an electrolytic cell change their chemical nature on the surface during electrolysis, goes under the name of *chemical polarisation of electrodes*.

### *Overvoltage*

On the basis of the above, it follows that:

- If a PD lower than $E$ (threshold) is applied to an electrolytic cell,there is no passage of current and therefore there is no electrolysis. This happens since there are conditions of equilibrium (between cemf and applied PD) and consequently the reaction speeds on electrodes are globally null.
- If the PD applied is higher than $E$, there are no more equilibrium conditions, the passing current has an intensity $I$ and electrolysis takes place.

In order to have electrolysis, then, a PD higher than $E$ must be applied, that is to say an overvoltage indicated as *intrinsic overvoltage* ($V_i$), since it is implied in the phenomenon of electrolysis itself.

Another unavoidable increase of electrolysis PD, necessary to keep constant the value $I$ of current, is due to the discharge of ions on electrodes, causing a decrease of concentration in the solution areas in contact with them, with subsequent migration towards those areas of other ions from the bulk of the solution (where concentration is higher). The slower the ion supply (for instance due to the high viscosity and/or low concentration of the solution), the more marked the concentration decrease in the electrode areas, the higher the cemf. This increase demands, in order to allow $I$ to be constant, an equal increase of the electrolysis PD. In this way, there is another type of overvoltage, of concentration ($V_c$), different from the one described above.

Finally, there is the increase of electrolysis PD due to the ohmic drop in the cell, whose value ($V$) is given by the product of the electric resistance of the cell ($\Omega$) by the intensity of the electrolysis current ($A$); it varies with the distance between electrodes and with the resistance of the solution. This phenomenon represents the so-called *ohmicovervoltage* ($V_h$).

Therefore, having called $\Delta E_{\text{term}}$ the thermodynamic voltage, the PD needed for electrolysis is given by

$$V = \Delta E_{\text{term}} + V_i + V_c + V_h \tag{4.110}$$

In practice, in order to improve efficiency, the maximum reduction of the overvoltage has to be implemented. The intrinsic overvoltage is reduced by using catalysts (on the surface of the electrode), the concentration one by keeping the solution rough, and the ohmic drop by making electrodes as close as possible. Each reduction of $V$, in fact, represents an energy saving, an element which is very much important in the economy of industrial processes.

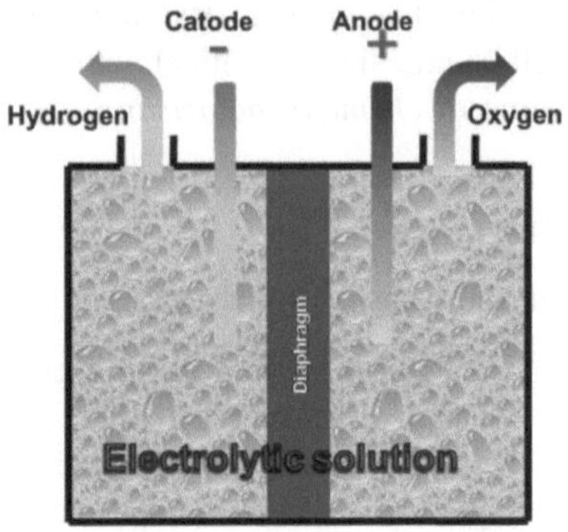


**Fig. 4.50** Diagram of an electrolytic cell

The phenomenon of electrolysis is governed by the Faraday's Law:

- The masses of chemical species formed (or consumed) to the electrodes during electrolysis are proportional to the quantity of electricity that has crossed the cell;
- the masses of chemical species formed (or consumed) for the passage of the same quantity of electricity are, in whatever electrolysis, proportional to the equivalent masses of the species involved in electrolysis itself.
  - The passage of 1 F in an electrolytic solution corresponds to the transformation (oxidation, reduction) of 1 equivalent of the chemical species that is electrolysed. Three different cases are possible:

*Electrolysis of strongly acid solutions*

A solution of $H_2SO_4$ with $[aH_3O^+] = 1$ is taken into consideration. In this case it is possible to admit that $H_3O^+$ ions, which are present in high concentrations at the cathode, are reduced and the $H_2O$ molecules are oxidised at the anode, given the low concentration of $OH^-$ ions. The electrode processes, therefore, are:

| | |
|---|---|
| Cathode (reduction of $H_3O^+$) | $4H_3O^+ + 4e \rightarrow 2H_2 + 4H_2O$ |
| Anode (oxidation of $H_2O$) | $6H_2O \rightarrow O_2 + 4H_3O^+ + 4e$ |
| Global reaction | $2H_2O \rightarrow 2H_2 + O_2$ |

*Electrolysis of strongly alkaline solutions*

A solution of KOH with $[a_{OH}] = 1$ is taken into consideration. In this case, the following electrode processes take place:

| | |
|---|---|
| Cathode (reduction of $H_2O$) | $4H_2O + 4e \rightarrow 2H_2 + 4OH$ |
| Anode (oxidation of $OH^-$) | $4OH^- \rightarrow O_2 + 2H_2O + 4e$ |
| Global reaction | $2H_2O \rightarrow 2H_2 + O_2$ |

*Electrolysis of neutral solutions* A solution of $K_2SO_4$ is considered, since the $K_2SO_4$ does not hydrolyse, the concentration of ions $H_3O^+$ and of ions $OH^-$ is small and the processes to the electrodes are:

| | |
|---|---|
| Cathode (reduction of $H_2O$) | $4H_2O + 4e \rightarrow 2H_2 + 4OH^-$ |
| Anode (oxidation of $H_2O$) | $6H_2O \rightarrow O_2 + 4H_3O^+ + 4e$ |
| Global reaction | $2H_2O \rightarrow 2H_2 + O_2$ |

In the three cases described, the global reaction is the same, and therefore the values of $\Delta E = 1.23$ V coincide. Figure 4.50 shows the scheme of an electrolytic cell.

*Electrolysis Efficiency*

The *efficiency of the electrolysis process* can be defined as the ratio between the electric power theoretically necessary to the process according to the Faraday's Law, and the one that is actually used. Indicating with F the Faraday constant, with $z$ the number of electrons transported by each ion, with $\gamma$ the speed (in mole/s) at which the reagent is consumed, the current, $I_F$, necessary to the process according to the Faraday's Law is:

$$I_F = z \cdot F \cdot \gamma \tag{4.111}$$

The necessary voltage according to the Faraday Law is $\Delta E_{\text{term}}$, therefore efficiency, indicating with $V$ and $I$ the effective voltage and current, is:

$$\eta = \frac{\Delta E_{\text{term}} \cdot z \cdot F \cdot \gamma}{V \cdot I} \tag{4.112}$$

Introducing the *voltage efficiency*$\eta_V$, defined as ratio between thermodynamic and real voltage:

$$\eta_V = \frac{\Delta E_{\text{term}}}{V} \tag{4.113}$$

And the *current efficiency*$\eta_F$, defined as the ratio between the Faraday current and the effective one:

$$\eta_F = \frac{z \cdot F \cdot \gamma}{I} \tag{4.114}$$

Replacing (4.113) and (4.114) in (4.112), the efficiency of the electrolysis process can be expressed as a product of the two efficiencies defined above:

$$\eta = \eta_V \cdot \eta_F \tag{4.115}$$

As already said, the thermodynamic potential of water electrolysis amounts to 1.23 V; however, considering the inevitable overvoltage, at least 1.4 V are necessary to obtain an acceptable reaction speed. In the current density conditions typical of commercial electrolysers (amounting to nearly 1 A/cm$^2$), there is a cell voltage between 1.7 and 2.0 V; current efficiency is close to the unit, thus to these values efficiencies of the electrolysis process of 60–70% correspond. As concerns

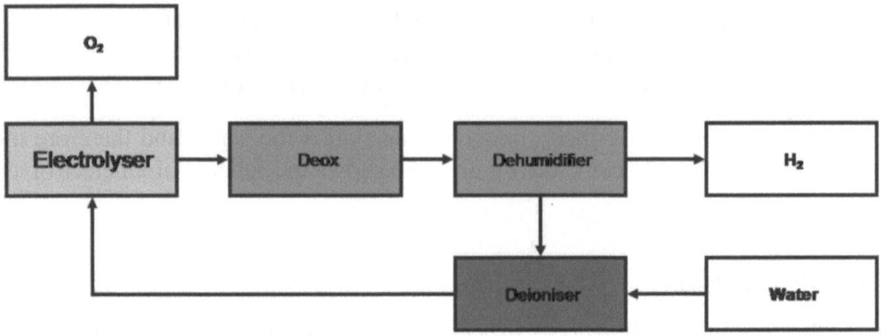


**Fig. 4.51** Scheme of an electrolysis system

the efficiency of an electrolyser, auxiliary consumptions (mainly the pumping of hydrogen and oxygen) must be considered which cause, given the same hydrogen produced, an increase of electric power consumption by nearly 20%.

*Electrolysers*

Figure 4.51 shows the chart of an electrolysis system; the electrolyser, fed with deionised water, produces hydrogen and oxygen. The hydrogen produced undergoes first of all a de-oxidation process and subsequently dehumidification to eliminate the (distilled) water present, which is then reused to fuel the electrolyser; at the outlet of the dehumidifier there is hydrogen with a very high degree of purity (>99.9%).

Water electrolysis always takes place in solution and not in pure water, since its conductibility would be very low and the current crossing the cell would be negligible; very often a basic solution of KOH (alkaline electrolysers) is used. As the concentration of the solution increases, two contrasting phenomena can be observed: on the one side, the number of charge conveyors increases (which improves conductivity); on the other side the interactions between ions which obstacle movement increase, hence leading to a decrease of the conduction movement. The maximum electric current in the cell is obtained with a concentration of the electrolytic solution of nearly 30%.

Electrolysers can be classified as follows:

- *Conventional alkaline electrolysers*: the concentration of the electrolytic solution (normally KOH) ranges between 20 and 30%, operating at temperatures ranging from 70 to 100°C and pressures between 1 and 30 bar;
- *Advanced alkaline electrolysers*: characterised by a lower distance between electrodes (which reduces the ohmic resistance of the electrolyte), by new techniques for the catalysation of active surfaces (which decrease overvoltage) and higher operating temperature;

**Table 4.7** Characteristics of conventional and advanced conventional alkaline

| Parameter | Conventional | Advanced |
|---|---|---|
| Rated current density Jt (A/cm$^2$) | 0.2 | 0.5 |
| Superficial resistance K (W/cm$^2$) | 1 | 0.4 |
| Ohmic overvoltage (V) | 0.2 | 0.2 |
| Electrolysis overvoltage (V) | 0.5 | 0.4 |
| Cell rated tension (V) | 2–1.8 | 1.78 |
| Tension efficiency | 0.747–0.83 | 0.839 |
| Faradic efficiency | 0.985 | >0.99 |
| Total efficiency (Ref. LHV) | 0.62–0.70 | 0.71 |
| Efficiency (ref.HHV) | 0.74–0.82 | 0.83 |
| Specific energy consumption (kWh/Nm$^3$) | 4.83–4.29 | 4.22 |

- *Solid polymer electrolysers*: in which the PEM unifies the separator and electrolyte functions; they show potentially very high efficiencies (94%), but presently the actual efficiency is lower than the in alkaline ones and entail high costs;
- *Steam electrolyser*: the electrolyte is ceramic (ion conductor); the efficiency is potentially very high, but not yet available in the market.

Table 4.7 shows the main characteristics of conventional and advanced alkaline electrolysers.

## 4.2.8 Radiant Energy-to-Chemical Energy Conversion Plants

### 4.2.8.1 Photoelectrolysis

This process is based on the coupling of photovoltaic technology and electrolysis: in practice, these are electrolytic cells with electrodes in photovoltaic semiconductor material. The theoretical efficiency accounts for 45%, and experimentally it was possible to achieve efficiencies of 20% with amorphous silicon cells.

Solutions are under study which envisage the use of multilayer thin films that will reach efficiencies close to the theoretical one.

## 4.2.9 Electric Energy-to-Thermal Energy Conversion Plants

Through thermodynamic reverse cycles it is possible, by absorbing energy, to transport heat from a lower to a higher temperature, that is to say to decrease the heat of a room temperature source by refreezing it up to low temperatures. Usually, the energy that is absorbed by the systems that make these thermodynamic operations is electric. Among the plants that operate as recalled herein, there are *refrigerators* and *heat pumps*. In the formers, the main purpose is to refrigerate a

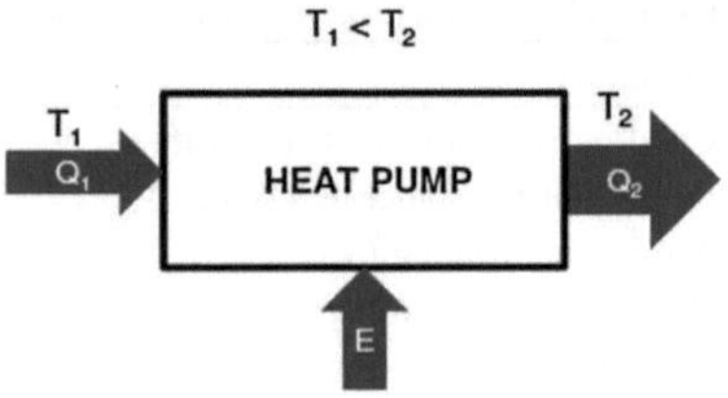


**Fig. 4.52** Operation scheme of a heat pump

body, that is to subtrac heat from a body and reduce its temperature (from a few degrees above 0°C even up to cryogenic values of −250°C and less). In the latter two different functions are possible: can be provide both heat and cool, for winter heating purposes as well as for summer conditioning.

#### 4.2.9.1 Compression Heat Pump

Overview

The *heat pump* is a machine able to transfer heat from a lower temperature room to another higher temperature area (Fig. 4.52). In general, the purpose is to heat a room at higher temperatures; sometimes the objective is to refrigerate a body at the lower temperature. Reversible heat pumps operate in both conditions, simply by changing the interval of operating temperatures.

The heat pump is made up of a closed circuit crossed by a fluid (refrigerant) with suitable characteristics that, according to its temperature and pressure conditions, takes a liquid or gaseous state, usually steam.

In its standard configuration, the closed circuit is made up as follows (Fig. 4.53):

- A compressor, C
- A condenser
- An expansion valve, V
- An evaporator

The condenser and the evaporator are two heat exchangers, namely two pipes in which a refrigerant fluid flows exchanging energy with a service fluid (either water or air). The refrigerant releases heat in the condenser and removes it in the evaporator. The components of the circuit can be grouped into a single block or divided into two parts ("split" systems) linked by the pipes in which the refrigerant fluid flows. In its operation, the refrigerant fluid, within the circuit, undergoes the following transformations (Fig. 4.54):

- *Compression* (1–2): the gaseous and low pressure refrigerant fluid, coming from the evaporator, is taken to high pressure during compression, and heats itself by absorbing a given quantity of heat.

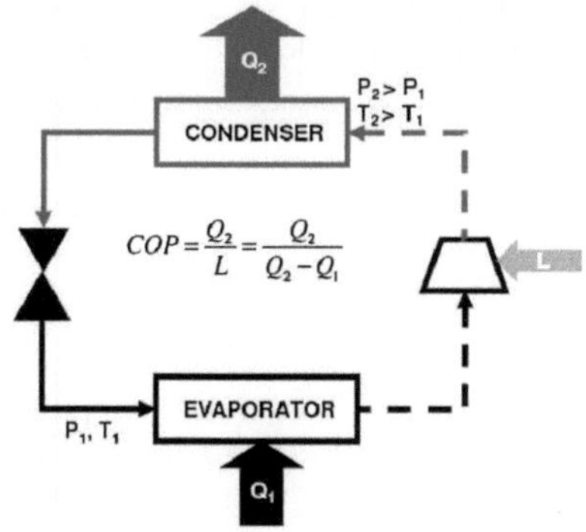


**Fig. 4.53** Cycle of a compression heat pump

- *Condensation* (2–3): the refrigerant fluid, coming from the compressor, passes from the gaseous to the liquid state, releasing heat outside.
- *Expansion* (3–4): passing through the expansion valve, the refrigerant fluid partially transforms into steam and cools down.
- *Evaporation* (4–1): the refrigerant fluid absorbs heat from outside and evaporates completely.

The overall set of these transformations is the *heat pump cycle*: providing energy to the refrigerant fluid with the compressor, it absorbs heat from the surrounding medium in the evaporator and, through the condenser, releases it to the body to be heated.

In the compression heat pumps, that are the far most widespread ones, the energy absorbed by the cycle ($E$ in Fig. 4.52) is the mechanic energy necessary to activate the compressor; as it often happens, from an operative viewpoint, the energy supplied to the machine is electric power and the compressor is set in motion by an electric motor, therefore the plant, as a whole, operates a conversion from electric energy into thermal energy.

*Thermodynamic Cycle*

While discussing about the thermodynamic cycle, reference will be made to compression heat pumps.

The phases of the cycle of a heat pump are shown in Fig. 4.53. The refrigerant fluid, at the liquid state and at a temperature $\mathbf{T}_1$ and pressure $p_1$, absorbs heat from the outside, evaporating; as it is known, the passage from one phase to another takes place at constant temperature and pressure. The steam is then taken to pressure $p_2 > p_1$ through an adiabatic compression; therefore, the temperature at the exit of the compressor (which absorbs the mechanical work $L$) will be $T_2 > T_1$. In these new conditions, the refrigerant fluid, at the gaseous state, releases heat to the outside by condensing at a temperature $T_2 > T_1$. At the exit of the condenser, the fluid is expanded adiabatically up to conditions $p_1$ and $T_1$ and the cycle starts again. Since the refrigerant fluid is at the liquid state, the (useful) work obtainable

**Fig. 4.54** The ideal thermodynamic cycle

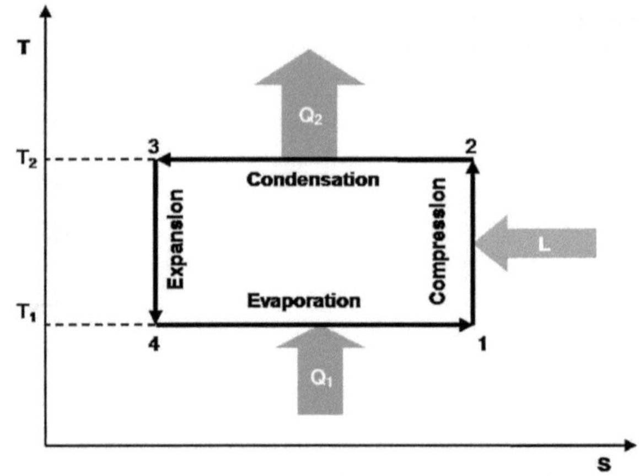


in its expansion is absolutely negligible, therefore this expansion takes place in an expansion valve, that is to say the recovery of the expansion energy is given up.

Since expansion and compression are adiabatic transformations, and the passages of phase (evaporation and condensation) are isothermobaric, the reference thermodynamic cycle is the Carnot cycle, shown in Fig. 4.54, considering the fluid as a perfect gas and the ideal transformations the reversible ones.

From Fig. 4.53, it is possible to determine the energy balance of the cycle as follows:

$$Q_2 = Q_1 + L \Rightarrow L = Q_2 - Q_1 \tag{4.116}$$

If $Q_2$ is the *useful effect* (heating) the *coefficient of performance* (COP) is the ratio between the useful effect and the energy supplied from the outside. Therefore, having considered (4.116), it follows that:

$$\mathrm{COP} = \frac{Q_2}{L} = \frac{Q_2}{Q_2 - Q_1} \tag{4.117}$$

With regard to Fig. 4.54 we have:

$$Q_2 = T_2(S_3 - S_2) = T_2 \cdot \Delta S \tag{4.118}$$

$$Q_1 = T_1(S_3 - S_2) = T_1 \cdot \Delta S \tag{4.119}$$

By replacing (4.118) and (4.119) in (4.117), it follows

$$\mathrm{COP} = \frac{T_2}{T_2 - T_1} \tag{4.120}$$

The refore, the COP, depends on the limit temperatures of the cycle and decreases as the difference between the two temperatures increases.[30] Considering,

[30] For this reason, heat pumps are more suitable for mild climate.

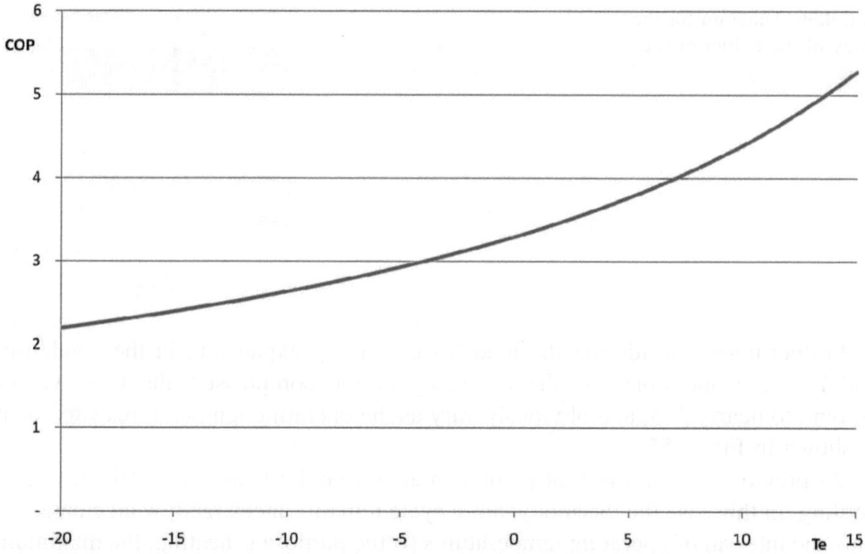


**Fig. 4.55** COP of a heat pump as the external temperature varies

for instance, $T_1 = 5°C = 278$ K and $T_2 = 20°C = 293$ K a COP $= 18.3$ is obtained.

For the calculation of the real value of the COP, it is necessary to consider the actual operating conditions.

First of all, as concerns the previous example, the external temperature amounts to 5°C, and the temperature of the fluid at the evaporator $T_2$ must be lower, to allow the thermal exchange between the external source and the fluid itself. Similarly, if the internal temperature is 20°C, the temperature of the fluid at condenser $T_1$ must be higher. Considering, for thermal exchanges, a $\Delta T$ of 10°C, then $T_1 = -5°C = 268$ K and $T_2 = 30°C = 303$ K. With these values, a COP $= 7.7$ is obtained.

Furthermore, it is necessary to consider the real transformations of the thermodynamic cycle: the real compression work $L_r$ is higher than the ideal work $L_i$, being:

$$L_r = \frac{L_i}{\eta_c} > Q_2 - Q_1 \tag{4.121}$$

where $\eta_c$ is the *compression thermodynamic efficiency*. Furthermore, the electric energy consumed for compression ($L$ in Fig. 4.53) is higher than the real work since it is necessary to consider the *efficiency of the electric motor* $\eta_{ME}$:

$$L = \frac{L_r}{\eta_{ME}} = \frac{L_i}{\eta_c \cdot \eta_{ME}} = \frac{Q_2 - Q_1}{\eta_c \cdot \eta_{ME}} \tag{4.122}$$

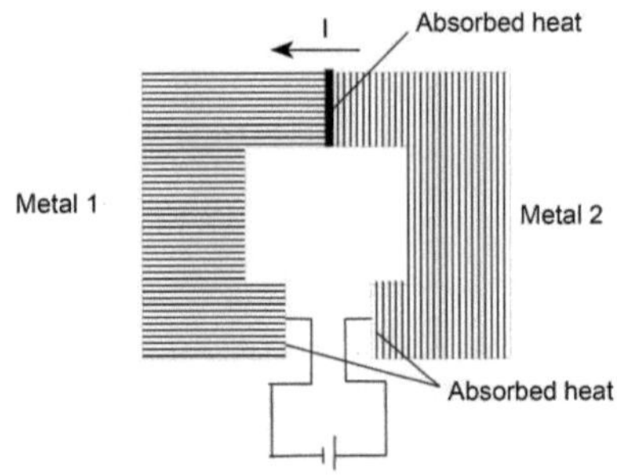


**Fig. 4.56** Diagram for the study of the Peltier effect

Furthermore, considering the load losses (during expansion, in the condenser and in the evaporator) and the efficiency of the compressor, the COP values amount to nearly 3–5, and obviously vary as the operating temperatures vary, as it is shown in Fig. 4.55.

As previously said, the heat pump can also have the feature of refrigerating or cooling. In this case the thermodynamic cycle remains unchanged; what changes is only the interval of operating temperatures (if the purpose is heating, the maximum temperature will be, for instance, 45 ÷ 50°C; for refrigeration, it will be close to room temperature, having considered that the purpose in this case is to keep temperature $T_1$ lower than the room one).

From the plant and the thermodynamic viewpoint, the same considerations apply: in this case the useful effect is the heat absorbed during condensation ($Q_2$). The COP value of cooling is normally lower than the one obtained in heating.

In conditions of partial load, if the compressor, set in motion by an electric motor, works at a constant number of revolutions, then the compression efficiency (and, therefore, the COP) falls suddenly. The use of inverters, that change the frequency of the supply current of the electric motor, allows to vary the number of revolutions of the compressor, so obtaining a good efficiency also in the operation with partial loads.

#### 4.2.9.2 Peltier Effect Heat Pumps

This type of heat pumps operates on the basis of the *Peltier effect*. To briefly recall the characteristics of this effect, one can refer to Fig. 4.56, in which a thermoelectric circuit is connected to a battery that allows direct current *I* to flow in. As it is known, two different metals have free electrons at different levels of Fermi,[31] characterised by separate values of kinetic energy and concentration; as a consequence of this, at the contact of the two metals the *e.m.f.* allows each electron that crosses the interface to deliver a given quantity of energy (positive in one

[31] At absolute zero the *Fermi level* or *Fermy energy* is, at absolute zero, the highest energy level occupied by the "free" electrons of the conduction band. It is typical of the material considered and weakly depends on temperature. At higher temperatures, some free electrons can be at higher levels, according to the statistical distribution function provided by the Fermi–Dirac theory.

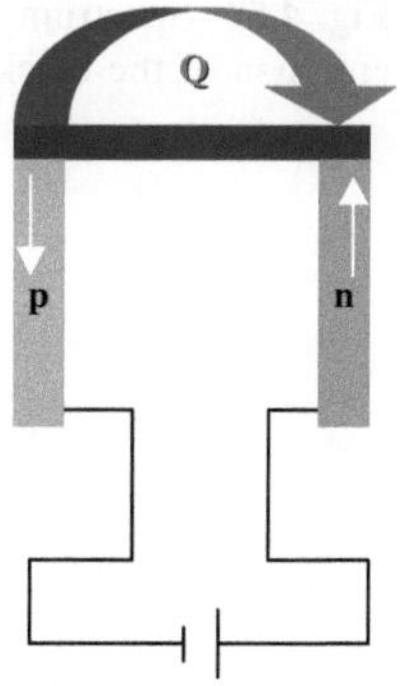


**Fig. 4.57** Diagram of a Peltier effect heat pump

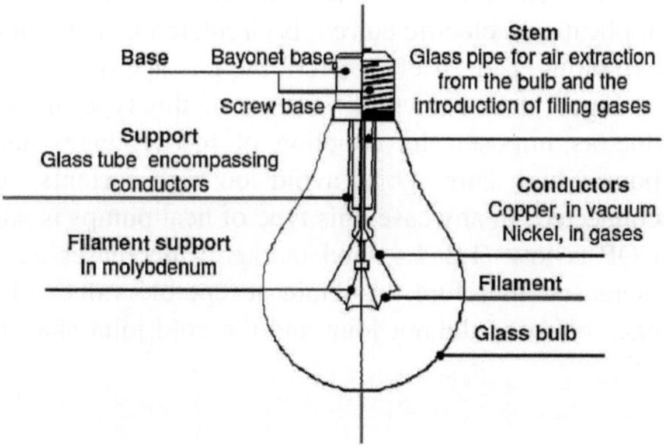


**Fig. 4.58** GLS incandescent lamp [4]

direction, negative in the other). This energy exchange translates into a heat absorption in one of the junctions and into release in the other.

The electric power $W_p$ developed and absorbed in each junction is given by the ratio:

$$W_P = \alpha_{P_{1,2}}(T) \cdot I \tag{4.123}$$

where $\alpha_{P,1,2}$ (*T*) is the *Peltier coefficient*. It has the same physical mean of a *e.m.f* and, depends on the materials used and varies according to temperature. If dependency upon temperature is put aside[32], then the quantity of heat absorbed in a time unit at a junction is exactly the same as the one developed at the other.

It is useful to consider the Peltier coefficient as follows:

$$\alpha_{P_{1,2}} = \alpha_{P_1} - \alpha_{P_2} \tag{4.124}$$

By convention, the Peltier coefficient is positive if heat is absorbed when the current flows from the material with a Fermi level higher than the one with the lower level 0.

[32] This admission does not introduce a significant error.

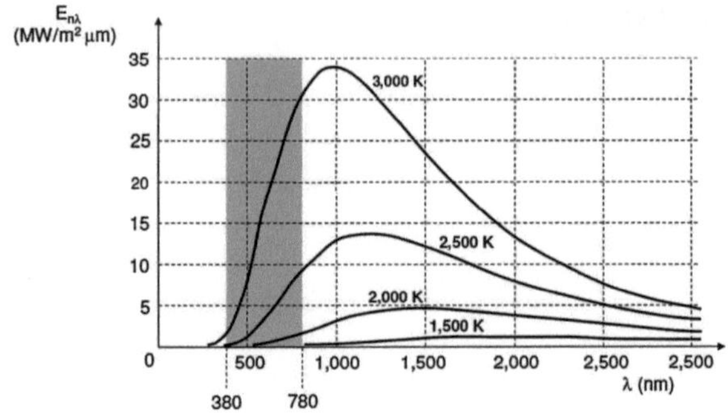


**Fig. 4.59** Spectrum of emission of the black body

The Peltier effect can be exploited for the carrying out of heat pumps; in these applications, electric current is circulated in a circuit as the one shown in Fig. 4.56: exploiting as a "useful effect" the passage of heat from one junction to the other.

Figure 4.57 shows the scheme of this type of heat pump: the use of semiconductors imposes the adoption of low voltages and, therefore, given the same power, high currents; to avoid too high currents, several cells in series must be connected. In any case, this type of heat pumps is only used for small powers. The COP is low (1.2–1.4) and markedly decreases as the difference of temperature increases: therefore, to obtain acceptable values of COP, the temperature difference between the hot joint and the cold joint shall not exceed 20°C.

## *4.2.10 Electric Power-to-Radiant (Luminous) Energy Conversion Plants*

### 4.2.10.1 Incandescent Lamps

GSL Incandescent Lamps

Historically, incandescent lamps are the previous ones and nowadays are still the most widespread; their success over time is due to the good compromise they make between mean life, luminous efficiency and price.

The most common model of incandescent lamps is represented by *general lighting service* (GLS) lamps (Fig. 4.58). The first patent of these lamps dates back to Edison, in 1841, but it was only in 1880 that they were actually commercialised.

They are made with a base and a glass bulb inside of which there is a filament that, when the lamp is connected to a power circuit, is crossed by an electric current that heats it until reaching incandescence, leading it to emit radiant energy including different wavelengths. The filament emission is regulated by the Planck's Law, since in general it is a material similar to a high-emission black or grey body (Fig. 4.59). Therefore, it emits on the entire spectrum, for wavelengths between 0 and $\infty$, and the emission decreases until becoming null for $\lambda = 0$ and $\lambda = \infty$.

According to the Wien's Law, the maximum emission is obtained when:

$$\lambda_{\max} = \frac{2.9 \times 10^6}{T} \tag{4.125}$$

As temperature increases, this maximum moves from the infrared ($T = 1.500$ K) to the field of visible light ($T = 3.000$ K). In particular, in order to have it at $\lambda = 555$ nm, to which corresponds the maximum level of visibility, it is necessary that the filament reaches a temperature of 5,225 K, too high for whatever known material: therefore, in practice, the maximum emission mainly occurs in the field of infrared radiation. In the first models, until 1913, the filament was linear and made up of coal; it worked at nearly 1,800°C, since it was not possible to reach higher temperatures in order to avoid the *sublimation of carbon*. This phenomenon must be generally avoided for two reasons: because it produces a too rapid destruction of the filament that, in loosing matter, weakens and breaks, and because of the subsequent deposit of the vapours released on the internal surface of the glass bulb darkening it and therefore decreasing the transparency of the glass and, subsequently, the luminous flow emitted outside.

At 1,800°C the maximum emission is at $\lambda = 1,400$ nm, whereas only a small portion of the radiation energy falls in the field of visible radiations. Furthermore, the lamps work with very low pressures in the bulb: in fact, the vacuum was created in it to delay the oxidation of the filament and prevent it from burning; however, the low pressure had, together with high temperature, the unwanted effect of *accelerating carbon sublimation*. This problem has represented for a long time a limitation to temperature increase, since it was kept very much below the melting temperature, until tungsten was adopted, once all the techniques necessary for its processing in rather thin wires of suitable electrical and mechanical characteristics were fine-tuned.

This material, due to its very high melting point (3.770 K), allows to reach very high temperatures (it should not exceed 2,500°C in vacuum and 2,800°C in inert gas), hence guaranteeing high-energy emissions with wavelengths that fall in the visible light.

Furthermore, it shows outstanding features in terms of ductility and electric and mechanical resistance, which allow in the lamps that are presently available in the market, differently from the first types of products, to coil it up in a double-helix shape to increase its mechanical resistance and allow a minimum heat dispersion.

In case of powers exceeding 25 W, inert gases are generally introduced in the bulb, usually argon and nitrogen mixes (but also, in case of special applications, kripton or hydrogen) that reduce the sublimation of the filament, although in this way a certain heat transmission is activated by convention between the filament, the gas and the glass, which lowers the temperature of the filament and subsequently reduces its efficiency. In general the best filling gas is kripton since, besides having a sufficient chemical inertia, it also shows a low value of thermal conductivity.

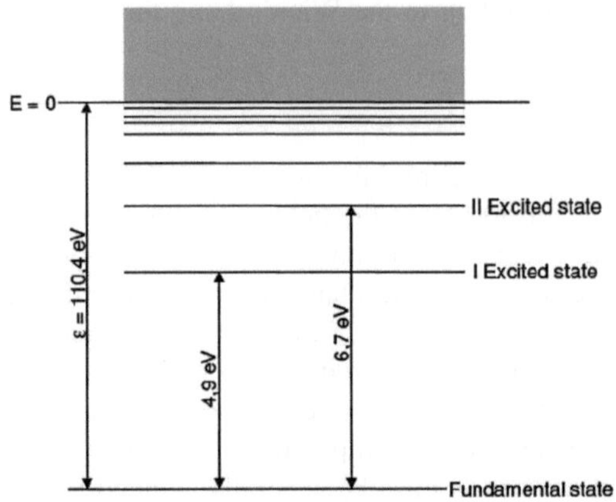


**Fig. 4.60** Diagram of the energy levels in mercury atom

Incandescent lamps are however characterised by a heat emission that clearly prevails compared to the one of luminous radiations, and for this reason they have a limited efficiency, which varies between 10 and 20 lm/W, with higher values for greater powers. Their mean life is nearly 1000 h, with a final luminous flow accounting for 87% of the initial one. The colour temperature coincides with the filament temperature which is variable from 2,700 to 3000 K. To increase its value, the glass of the bulb is generally coloured in light blue: in this way it is possible to obtain a less reddish light that is more suitable for some applications. However, in doing so, the luminous flow emitted is reduced—as it is partly absorbed by the coloured glass—as well as efficiency.

*Incandescent Halogen Lamps*

A further development of this type of light source is represented by *halogen lamps*, in which a small quantity of halogen gas (generally iodine, bromine or chlorine) is inserted in the bulb to effectively counter the evaporation of tungsten.

The presence of these substances, in addition to inert gases, allows the triggering of a regenerative cycle of tungsten, given by the following reactions:

$$\begin{aligned} I_2 &\rightarrow 2I \\ W + 2I \rightarrow WI_2 &\quad 2.000\,K \\ WI_2 \rightarrow W + 2I &\quad 2.800\,K \end{aligned}$$

In the area of the lamp with the lowest temperature, the vapours of tungsten freed from the filament are chemically combined with the iodine present, forming gaseous tungsten iodine ($WI_2$) or other similar compounds. This compound, while migrating to the area of the hot filament, decomposes in its elements, forming metallic tungsten that newly deposits on the filament. In this way, most of the sublimated tungsten deposits once again, markedly slowing down both the erosion process of the filament and the deposition on the inside of the bulb.

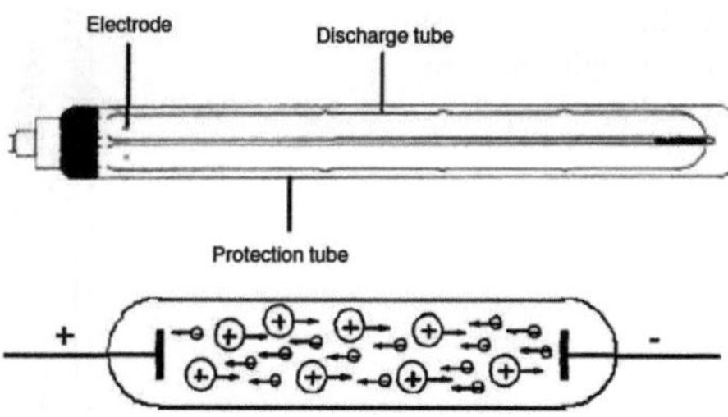


**Fig. 4.61** U-shaped discharge tube filled with low-pressure gas [4]

Furthermore, in these conditions the filament can be crossed by a more intense current, allowing to reach higher temperatures ($\sim$3200 K) and subsequently higher flows (this entails, however, that the bulb consists of quartz glass, in consideration of its capacity to resist to the high temperatures to be reached); therefore, there is an increase both in the luminous efficiency, that is on average 25 lm/W, but which can also reach 30 lm/W, and the colour temperature (white light). *The colour-rendering index* amounts to 100, the mean life is 2,000 h, and the final luminous flow is 94% of the initial one. For their particular characteristics, these lamps find application both in interior lighting, and in exterior lighting in case of large complexes, for photograph shooting, in the headlights of vehicles, etc.

#### 4.2.10.2 Gas-Discharge Lamps

In discharge lamps, light production is due to the emission of luminous radiations by a preventively excited gas[33]: when an electron goes back to its orbit, luminous energy is emitted in the amount of the difference of energy between the two states. Since electronic orbits are defined, energy gaps are defined as well and typical of the gas filling the bulb; to a given energy corresponds therefore a well-defined wavelength and therefore a precise colour (Fig. 4.60). Consequently, these sources emit a light made up of the typical radiations of gas, and the spectrum is discontinuous (with lines).

Differently from incandescent lamps, in which the light emitted can be considered as a by-product of the transformation of electric power into thermal energy, in the discharge ones the light is obtained through the *direct transformation of electric power into luminous energy*. For this reason they are characterised by a light efficiency far higher than the filament ones, ranging between 35 and 200 lm/W. The mean life as well is higher than incandescent lamps and varied between 5,000 and 12,000 h according to the type of lamp. Conversely, generally these lamps do not meet in the same way the colour rendering[34] requirements.

[33] Excitation consists in the removal of electrons from the stable orbit, made by other electrons accelerated by an electric field.

[34] Due to the fact that these sources emit a light whose spectrum is discontinuous.

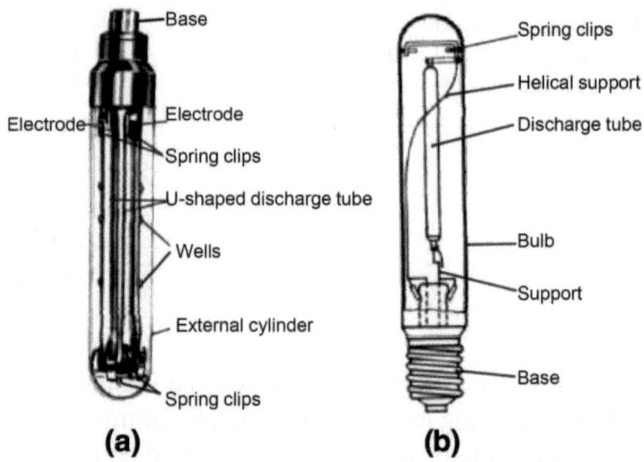


**Fig. 4.62** Low pressure (**a**) and high pressure (**b**) sodium lamp [4]

They are essentially made up of a container, generally in glass or quartz with two electrodes, in which a gaseous body is fit (metallic vapours, sodium or mercury in particular, since the radiations they emit are found in the visible spectrum, xenon or rare gases) at a suitable pressure (Fig. 4.61).

Sodium in particular, being solid at the ordinary temperature, must pass to the gaseous state to be able to emit: therefore it is plunged in a neon atmosphere which allows the triggering of the discharge and the necessary temperature increase. According to the value of pressure of the discharge pipe, this type of lamp can, in turn, be divided into two sub-groups: *high* and *low pressure*. In low-pressure sodium steam lamps (0.5 Pa) the discharge takes place in a glass U-shaped pipe, protected by an external glass tubular bulb, whose internal surface is covered with indium oxides to ensure a good thermal insulation, having considered that the lamp works better at high temperatures (270°C) (Fig. 4.62).

At low pressures, sodium emits a practically monochromatic yellow light (a doublet of yellow-orange spectral lines, very close to the maximum light sensitivity, at 589 and 589.6 nm), and cannot be used for internal lighting and for all those applications in which the distinction of colours is important: for this reason, their use is destined to those cases in which colour rendering is not important; in particular, in Italy this lamp is mostly used for road lighting. Light efficiency in low-pressure lamps reaches 200 lm/W, whereas their mean life ranges between 8,000 and 10,000 h. Colour rendering is obviously completely insufficient (Ra = 0): in particular the low-pressure sodium lamps are disadvantaged from this viewpoint. In high-pressure sodium lamps (40 kPa) the discharge still takes place at high temperature, with the emission on a line spectrum or continuous spectrum in the visible light, therefore the colour is not monochromatic yellow any longer, but rather white-gold (Ra = 60) with a colour temperature varying from 2,200 to 4,200 K (Fig. 4.62).

The luminous efficiency is slightly lower than the one obtained at low pressure, in any case exceeding 100 lm/W. Its lifetime goes from 5,000 to 12,000 h, and the flow at the end of the useful life is reduced on average to 90% of the initial one.

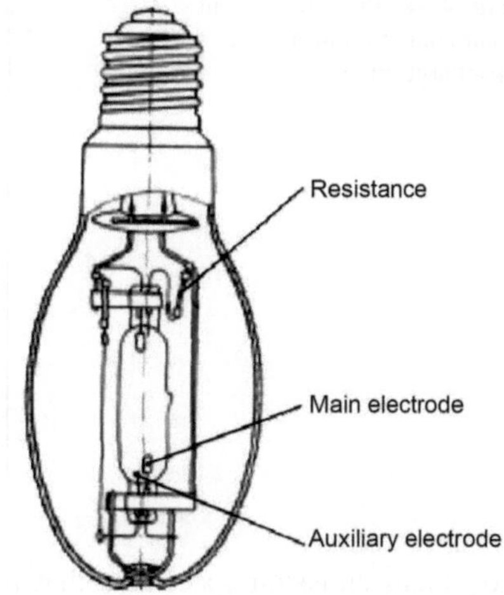


**Fig. 4.63** Very high pressure mercury steam lamp [4]

On the contrary, at a low pressure mercury usually has an emission spectrum characterised by a doublet of lines placed in the ultraviolet area (254 and 185 nm), whereas at high pressure there are other lines in the field of the visible light (blue, green); with the further pressure increase, the lines widen into bands and the emission spectrum becomes a continuous one, although lacking in the wavelengths corresponding to red. Therefore low pressure bands are not used for lighting purposes, but rather for particular technological operations that require the use of UV radiation. Their widespread use is possible, in fact, only after correcting the quantity of light emitted, as it happens in fluorescent lamps.[35]

Efficiency in high-pressure lamps is relatively low and ranges from a minimum of 35 lm/W to a maximum of 50 lm/W. Also the colour-rendering index (Ra = 60) and the colour temperature (2.150 K) are relatively low, and this explains the present low use of these lamps (in the case of street lighting, they have been almost entirely replaced by the sodium-vapour ones). In some types of lamps, sodium halide (hydride), thallium and indium are added to correct the characteristics of the emission spectrum and to improve the light efficiency, which can reach 95 lm/W. These lamps are able to cover the same market of halogen lamps and fluorescent lamps, with the advantage compared to the former of a higher efficiency and to the latter of the availability of higher flows.

In order to improve colour rendering (Ra = 65–80) recently small halide lamps were produced with the addition of holmium, dysprosium, etc., comparable to halogen lamps but with the advantage of a 2/3 energy saving. These are in practice discharge lamps miniaturised with a luminance of 15 Mcd/m$^2$, whose colour temperatures range between 3,000 and 6,000 K, and with a mean life exceeding even 6,000 h. Finally other discharge lamps use a xenon atmosphere, which allows

[35] See Sect. 1.2.10.2.1.

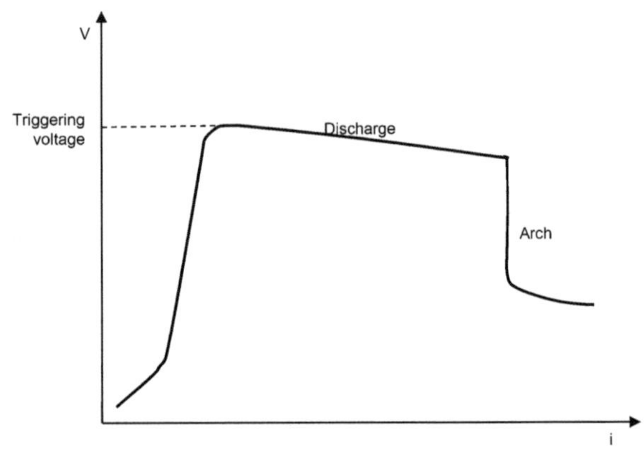


**Fig. 4.64** Voltage–current characteristic curve in a discharge tube

to obtain emission spectrums that reproduce, with a good approximation in the visible field, the one of solar radiation. It follows a faithful rendering of colours, and therefore these lamps are well used for the lighting of indoor and outdoor areas.

A brief mention is finally deserved by very high pressure mercury vapours (3.5–10 MPa), with very high luminance (100,000 stilb) and are mostly used in scientific laboratories (Fig. 4.63).

The gas discharge phenomenon can be described as follows. In a mass of rarefied gas there are always free electrons due to photoelectric effects or to a cosmic radiation. At ordinary temperatures, the energy owned by molecules (3kT/2) is of the same magnitude as the energy exchanges among molecules or between electrons and molecules caused by their collisions.

If however temperature is analysed or if particles are accelerated with a suitable electric field, then there are the conditions for the ionisation of gas and the subsequent phenomena that are at the basis of the operation of a discharge lamp. In particular, when the two electrodes are connected to a source of direct voltage the characteristic curve $V(I)$ is diagrammed, then it follows that (Fig. 4.64): for very low values of gas, which usually has a very low thermal conductivity, it acts as a perfect insulator; in these conditions there is only a weak passage of current due to the free electrons that are present and near the first section of the curve there is a trend of growing $i$ with $V$.

For the highest voltage values, electrons will acquire kinetic energy according to the electric field existing between electrodes, moving towards the anode and colliding in their pathway against gas atoms. A collision occurring at low speeds is elastic and causes the deviation of the collision electron, which releases a small part of its kinetic energy to the gas atom that is heated.

On the contrary, as speed increases, the quantity of energy exchanges caused by the collision between particles increases; consequently the likelihood of determining the ionisation of a molecule. A collision taking place at high speeds causes the change of orbit of one of the external electrons to a higher energy level, exciting the atom. After a very short time, ranging between $10^{-9}$ and $10^{-8}$ s, the electron goes back to its initial level and in passing from a higher energy status $E_2$ to a lower one $E_1$ the atom emits a quantum of light (photon) amounting to the difference:

$$E_2 - E_1 = h \cdot v \tag{4.126}$$

with h as Planck's constant (6.63 × $10^{34}$ Js) and $v$ as frequency of the emitted radiation (in $s^{-1}$). Since the energy released by the electron with the collision amounts to eV, the wavelength of the photon emitted will be given by:

$$\lambda = \frac{c}{v} = \frac{c \cdot h}{e \cdot V} \tag{4.127}$$

When finally the collision takes place at a speed so high that a peripheral electron is taken away from the atomic system, the atom is charged positively and therefore it ionises. The freed electrons behave exactly like a colliding electron, whereas the ion can recombine with another electron producing light, or release heat through collision against the walls. If an appropriate continuous voltage is applied to the discharge tube, there will be a continuous formation of ions moving towards the cathode and a production of other electrons that will very quickly move towards the anode. In the characteristic curve of the tube, this phenomenon turns into a rapid increase of current as voltage increases. Under the bombardment of ions, the cathode heats and emits in turn electrons due to the thermionic effect: therefore the ionisation increases and a number of electrons is produced that is sufficient to allow the discharge to become independent from the original free electrons (in this case it is generally said that it is self-sustained). The gas has become in this way a conductor, the voltage drop on the lamp decreases and the current increases. The minimum voltage able to trigger this process, that corresponds to the voltage at which the lighting of the lamp takes place, is called in fact *triggering voltage*.

In the typical curve there is then a section of negative resistance (corresponding to the operation of discharge lamps) in which the potential drop between electrodes decrease as current increases. The voltage at the terminals sufficient to keep the discharge is lower than the triggering voltage. Another section corresponding to the disruptive or arch discharge finally follows that is only exploited for arch lamps, used in special applications.

The trend of the curve (triggering voltage, arch voltage, etc.) depends on the form of the tube and on the nature of gas. In particular, the triggering voltage depends on the product of pressure by the distance between electrodes (Paschen's Law), therefore the higher their values, the higher it will be: this can be explained by making the following considerations.

At high pressures the *mean free time* (the time between the two subsequent collisions) is rather short, therefore they do not succeed in gaining a sufficient kinetic energy to produce the ion ionisation in collisions. Only by increasing the pd at the terminals it is possible to suitably accelerate the particles in such a way as to allow collisions to have the ionisation of other atoms: this translates, therefore, into an increase of the value of the triggering voltage.

At low pressures, when the gas is very much rarefied, the mean free time is much larger, therefore the possibility of having a collision before the accelerated particle reaches the opposite electrode is rather low. Also in this case, only by increasing voltage it is possible to reduce the mean free time producing a higher agitation of particles and this, once again, translates into an increase of the triggering voltage.

Furthermore, for a given gas and at a given pressure, the triggering pressure increases as the diameter of the tube decreases. This can be explained by observing that a small diameter facilitates the possibility of ion and electron recombination on lateral surfaces before causing other ionisations. The value of the triggering voltage can however be reduced by using both *rare gases*, and *appropriate electrodes*. The former show an excitation voltage very close to their ionisation voltage, which facilitates the definition of the discharge through the gas (furthermore these gases, as all the metallic vapours, have a monatomic molecule, and this allows not to disperse energy in the collisions with electrons to excite oscillations and rotations of the atoms of the molecule). As to the appropriate electrodes, it is possible for instance to coat the cathode, which is generally made in tungsten, with layers of alkaline-terrous oxides, such as BaO, SrO and CaO, with a low exit potential, which facilitate ionisation phenomena). It is also possible to adopt auxiliary electrodes, that is to say the heating of electrodes.

In order to allow the phenomenon of the discharge not to be exhausted, it is necessary that the gas remains ionised, with shifts towards the cathode of positive ions and of electrodes towards the anode. In this way a progressively higher current is however determined, which must be stabilised by using some *current limiters*, called *reactors*, inductive ballast resistors that are used to limit power drops.

### Fluorescent Lamps

To exploit the emission of ultraviolet radiations by some gases and vapours (mainly mercury) used in the discharge tubes (that, exception made for specific applications, is unusable) in this type of lamps the phenomenon of fluorescence is exploited, a property that some substances have in terms of absorbing small wavelength electromagnetic radiations and releasing energy in the form of higher wavelength. This phenomenon allows the recovery of the energy emitted outside of the visible, also allowing the correction of the colour of the light emitted, with a very high colour rendering. It is based on the property owned by some substances called *phosphor* of exciting when they are invested by a radiation, emitting in turn energy radiations with a lower energy than the excitement one, and therefore of a higher wavelength. The emissions stop within a very short period of time ($10^{-8}$ s) from the interruption of the exciter radiation. With these substances, that are made

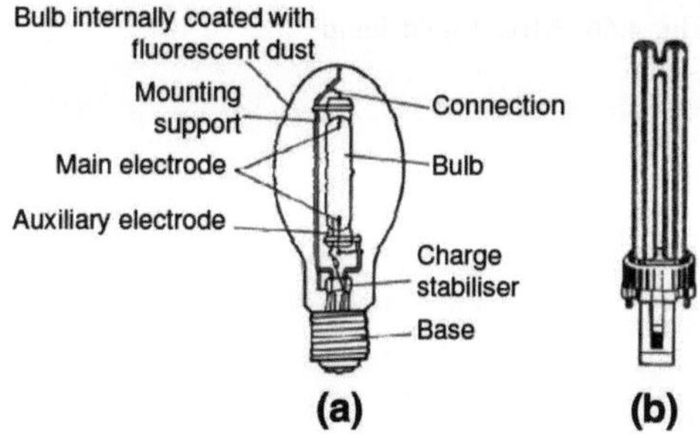


**Fig. 4.65** Mercury vapour (**a**) and compact (**b**) bulb fluorescent lamp [4]

up of oxisulphide, aluminates, phosphate tungstate and silicate of Calcium, Magnesium and Zinc, associated to heavy metals as copper or antimony, the internal surface of the glass cover is therefore involved (Fig. 4.65). Furthermore, very often in fluorescent salts the particular traces of substances called *activators* are dispersed.

A similar phenomenon often consistent with fluorescence is furthermore *luminescence* or *phosphorescence*, that is conversely extended in time. The colour of the resulting light varies with the nature of phosphors and of activators, and goes from hot white ($\approx$3.000 K), that is close to the colour of the light emitted by incandescence lamps, up to the white-light blue of the daily light (temperature of colour $\approx$ 6.500 K). Tubular fluorescent lamps are on the contrary generally filled with argon, with the addition of a small quantity of mercury, in order to facilitate the triggering of the discharge. A common characteristic of this type of lamp is the low luminance of the emitting surface (4,000–7,000 cd/m$^2$): this feature sometimes makes it possible for the use of these sources without any direct shielding to avoid glare, as it is conversely necessary with other types of lamps.

In consideration of the capacity of these lamps to recover energy, the luminous efficiency is generally very high (85–95 lm/W). Their mean life is nearly 10,000 h, whereas the final luminous flow accounts for 85% of the initial one. The colour-rendering index is included between 85 and 95.

Additionally, in the last few years small-sized fluorescent lamps were produced, which are destined to the same market as incandescent ones, of which they basically offer the same chromatic quality of light (Ra = 85), allowing however considerable energy savings (Fig. 4.66).

Their cost, still higher compared to incandescent lamps, is in fact offset by a higher efficiency (50–80 lm/W) and a longer mean life (6,000 h).

*Mixed-Light Lamp*

These are mercury vapour bulb fluorescent lamps, where inside of the bulb, a filament of tungsten is added and connected in series to the discharge tube (Fig. 4.66). The filament, in emitting through incandescence with its continuous

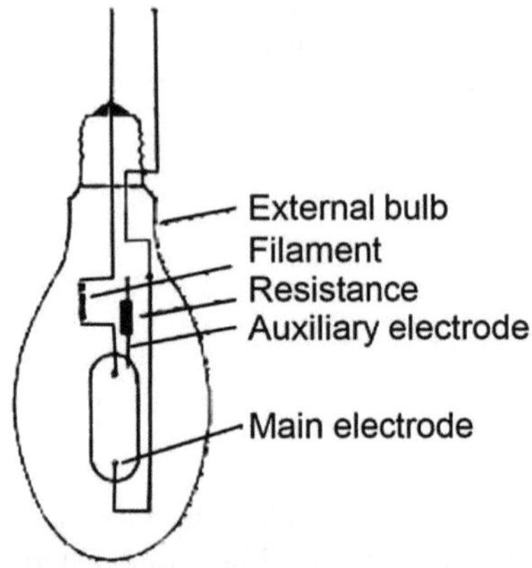


**Fig. 4.66** Mixed-light lamp [4]

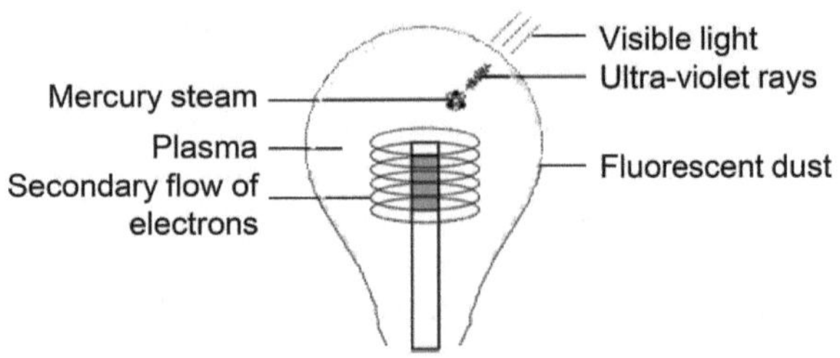


**Fig. 4.67** Operation diagram of an induction lamp [4]

spectrum rich in large wavelength radiations, offers a useful contribution to the production of the luminous flow and improves colour rendering. A further advantage is given by the fact that these lamps do not require any reactor. The values of luminous efficiency are intermediate between incandescent lamps and fluorescent lamps (20–30 lm/W), the colour temperature is around 3,000–4,000 K, the colour-rendering index goes from 40 to 75 and the mean life from 5,000 to 7,500 h.

*Induction Lamps*

These are new generation lamps, whose functioning is based on the fluorescent lamps, but in which electrodes are eliminated since they represent the weak point in terms of duration. The power requested is in this case provided by a high-frequency electronic generator (2.65 MHz) that, once coupled with a coaxial cable to an antenna, creates an alternate magnetic field that in turn induces a secondary electric field (Fig. 4.67). Therefore, an induced current is generated which circulates in the mix of mercury vapour and rare gas contained in the bulb, triggering the ionisation phase previously described.

The efficiency is around 65 lm/W, the colour temperature around 3,000–4,000 K, the colour-rendering index is higher than 80 and the estimated mean life is 60,000 h.

**Table 4.8** Energy transformation processes

| Conversion | Plant | Conversions and transformations |
|---|---|---|
| Chemical | Reformer | Conversion of fuel's chemical energy to be reformed into thermal energy<br>Conversion of thermal energy into chemical energy of the fuel produced |
| | Partial oxidation | Conversion of fuel's chemical energy to be reformed into thermal energy<br>Conversion of thermal energy into chemical energy of the fuel produced |
| | Gasification | Conversion of fuel's chemical energy to be reformed into thermal energy<br>Conversion of thermal energy into chemical energy of the fuel produced |
| | Pyrolysis | Conversion of fuel's chemical energy to be reformed into thermal energy<br>Conversion of thermal energy into chemical energy of the fuel produced |
| | Fermentation and digestion | Direct transformation |
| | Photoproduction | Direct transformation |
| Heat | Absorption heat pumps | Direct transformation |

# 4.3 Energy Transformation Plants

Also in the case of *transformation plants* aimed at producing energy vectors, the same concepts and criteria exposed for conversion plants apply. In general, in a plant, different energy transformation and conversion processes take place that allow to transform the resource used into the desired form of energy. For instance, in the steam reforming process there is also a conversion of chemical energy into thermal energy (part of the fuel to be reformed to produce hydrogen is burnt to produce the necessary heat to trigger reactions); the heat produced is in turn converted into chemical energy. Therefore, globally, *a steam reforming plant operates* a *transformation of chemical energy*, producing a fuel with given characteristics starting from a different fuel with different characteristics.

Table 4.8 shows the transformation plants tackled in this paragraph.

## *4.3.1 Chemical Energy Transformation Plants*

### 4.3.1.1 Steam Reforming

This process, which since the mid-60 ranks first in the industrial preparation of hydrogen (presently, the most cost-efficient, the most common and with the highest efficiency), is based on the reaction of a hydrocarbon (mainly methane) with water vapour according to the reaction $CH_4 + H_2O \leftrightarrow CO + 3H_2$. This is an endothermic reaction and the relevant balance, at low temperatures is heavily shifted to the left side. Therefore, this reaction is carried out at high temperatures (nearly 800°C[36]), in the presence of catalysts (nickel salts supported on aluminium oxide), and at a pressure of nearly 30 bar.

The $CO + 3H_2$ mix produced, also containing low percentages of $CO_2$,[37] is indicated with the name of water gas, is then subtracted from the equilibrium, enriched with a suitable quantity of water steam and passed through catalysts (with iron and cobalt) at temperatures of nearly 400°C (shift reaction). In this way there is a reaction called *water gas conversion reaction* $CO + H_2O \rightarrow CO_2 + H_2$.

In brief this process, with a global efficiency of around 75%, will be:

| | |
|---|---|
| First phase | $CH_4 + H_2O \rightarrow CO + 3H_2$ |
| Second phase | $CO + H_2O \rightarrow CO_2 + H_2$ |
| Global reaction | $CH_4 + 2H_2O \rightarrow CO_2 + 4H_2$ |

In the ideal case, as it can be derived from the global reaction, there is the emission of 1 mole of $CO_2$ for every 4 mole of hydrogen produced. Considering molar weights, there are 5.5 g of $CO_2$ for every g of hydrogen produced. As concerns the methane produced, there is the emission of a mole of $CO_2$ for every mole of $CH_4$ consumed, that is to say 2.75 g of $CO_2$ every g of $CH_4$ consumed. As already mentioned, as a matter of fact, a part of methane is burnt to produce the heat necessary to reactions. The methane combustion reaction is $CH_4 + 2O_2 \rightarrow CO_2 + 2H_2O$, from which it is possible to infer than in the combustion there is an emission of 2.75 g of $CO_2$ every g of $CH_4$ burnt. Therefore, every gram of methane consumed in the reforming process (burnt for heat production or reformed to produce hydrogen) there is the emission of 2.75 g of $CO_2$.

The actual consumption can be expressed according to the process efficiency. Indicating with $m_{H_2}$ the mass of hydrogen produced and with $LHV_{H_2}$ its lower heating value, with $m_{CH_4}$ the mass of methane consumed and with $LHV_{CH_4}$ its lower heating value, by definition, the efficiency of the reforming process $\eta_R$ is:

$$\eta_R = \frac{m_{H_2} \cdot LHV_{H2}}{m_{CH_4} \cdot LHV_{CH4}} \tag{4.128}$$

Thus the mass of methane consumed is:

$$m_{CH4} = \frac{m_{H_2} \cdot LHV_{H_2}}{\eta_R \cdot LHV_{CH_4}} \tag{4.129}$$

Indicating with $EM_{sCO_2}$ the $CO_2$ emissions per g of $CH_4$ consumed, the $CO_2EM_{CO_2}$ emissions are:

[36] The necessary heat is produced by the combustion of a given quantity of methane.

[37] Also at high temperatures and pressures, the reaction is never complete.

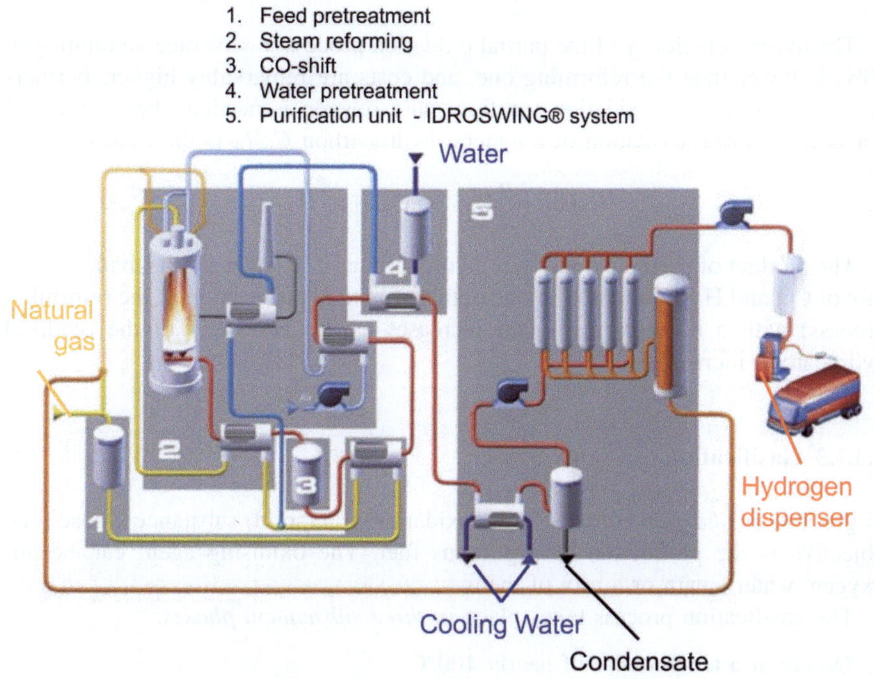


**Fig. 4.68** Scheme of a reforming plant

$$EM_{CO_2} = m_{H_2} \frac{LHV_{H_2}}{\eta_R \cdot LHV_{CH_4}} \cdot EM_{sCO2} \tag{4.130}$$

Assuming an efficiency of the reforming plant scheme accounting for 75%, from (4.130) it is possible to obtain:

$$EM_{CO_2} = m_{H_2} \cdot 9.13 \tag{4.131}$$

Therefore, there are emissions of 9.13 g of $CO_2$ for every gram of hydrogen produced.

Figure 4.68 schematically shows a reforming plant. These are generally, large plants, with a production capacity around 800 $MNm^3$/year of hydrogen.

#### 4.3.1.2 Partial Non-catalytic Oxidation of Hydrocarbons

Partial oxidation consists in the reaction of hydrocarbons with oxygen. The temperature at which the oxidation reaction occurs depends on the hydrocarbon and on the presence of catalysts: light hydrocarbons react, with the catalyst, at nearly 600°C, whereas heavy hydrocarbons (for instance *naphtha*), in the absence of catalysts, need temperatures around 1,400°C.

The overall efficiency of the partial oxidation process, on average amounting to 50%, is lower than the reforming one, and costs are remarkably higher. Furthermore, for the partial oxidation reaction, pure oxygen is needed: the exothermal reaction of partial oxidation of a general hydrocarbon $C_nH_m$ is the following:

$$\mathrm{C}_n\mathrm{H}_m + \frac{n}{2}\mathrm{O}_2 \rightarrow n\mathrm{CO} + \frac{m}{2}\mathrm{H}_2$$

The product of partial oxidation of hydrocarbons is a synthesis gas made up of a mix of CO and $H_2$ (similar to the one obtained in the first reaction of the reforming process) with a content in $H_2$ that increases as the H/C ratio of the oxidised hydrocarbon increases.

#### 4.3.1.3 Gasification

In general, *gasification* consists in the oxidation of a (solid) substance whose final objective is the production of a gaseous fuel. The oxidising agent can be air, oxygen, water steam or a mix of them.

The gasification process takes place in *three subsequent phases*:

1. *Drying* at a temperature of nearly 100°C.
2. *Pyrolysis* at temperatures between 200 and 700°C.
3. *Oxidation–reduction processes* at temperatures around 900°C.

The reaction processes are in variable percentages, according to the starting substance, the oxidation agent and the gasification technology used, hydrogen, carbon monoxide, carbon dioxide, steam, methane and hydrocarbons exceeding the status of vapours or oils and ashes. The general gasification reaction of a $CH_\alpha O_\beta$ substance is the following:

$$\begin{aligned}\mathrm{CH}_\alpha\mathrm{O}_\beta + \mathrm{wH_2O} + \mathrm{yO_2} + \mathrm{zN_2} \Leftrightarrow{} & \mathrm{x_1C} + \mathrm{x_2H_2} + \mathrm{x_3CO} + \mathrm{x_4H_2O} + \mathrm{x_5CO_2} \\ & + \mathrm{x_6CH_4} + \mathrm{x_7N_2} + \mathrm{x_8O_2} + \ldots\end{aligned}$$

Coal Gasification

In *coal gasification*, carbon reacts with water steam according to the reaction $C + H_2O \rightarrow CO + H_2$. This is an endothermic reaction and the relevant balance, at low temperatures, is heavily shifted to the left. Therefore, this reaction must be carried out at high temperatures (800–1,000°C), which are kept through the combustion of a part of coal with dosed inlets of air or oxygen or heat from a combustion chamber. The reaction described represents the first phase of the process. Subsequently, the mix $CO + H_2$ produced (synthesis gas called *water gas*), also containing small percentages of $CO_2$, is subtracted from the equilibrium, enriched with a suitable quantity of water steam and passed through catalysts (with iron and cobalt) at temperatures of nearly 400°C, similar to the case of the reforming process. In this way a reaction occurs, called *water gas conversion reaction*: $CO + H_2O \rightarrow CO_2 + H_2$.

In brief this process, with a global efficiency of around 50%, is:

$$\begin{aligned} &\text{First phase} \quad C + H_2O \rightarrow CO + H_2 \\ &\text{Second phase} \quad CO + H_2O \rightarrow CO_2 + H_2 \\ &\text{Overall reaction} \quad C + 2H_2O \rightarrow CO_2 + 2H_2 \end{aligned}$$

Proceeding similarly to the reforming process, it is possible to calculate $CO_2$ emissions for the production of hydrogen from coal gasification: assuming an efficiency of the gasification plant of 50% $CO_2$ emissions amounting to 30 g per g of hydrogen produced are obtained.

### Gasification of Biomass

One of the most advanced systems for the gasification of biomass is represented by two fluid beds in which combustion and gasification occur separately to produce a gas not diluted with nitrogen. In this case (*fluid bed indirect heating gasificator*) an oxygen generator is not necessary and there are no problems in the materials for heat transportation to high temperatures. This type of plant allows a high conversion of the carbon present in the biomass, which can be used also in a non-dried state.

The output synthesis gas from the gasificator must be treated before being used in combustion engines, turbine or high temperature fuel cells. There are two main gas conditioning methods that are substantially used: *cold* and *hot*.

The *cold methods* consist in the mechanical removal of particles through porous membranes (bag filters) or sand (sand filters) or water jets (scrubber); although sophisticated in their realisation, they are conceptually very simple and do not need in-depth analyses.

In *hot methods* the first cleaning stage is always represented by a cyclone, which removes the entire char[38] and part of the TAR (*Topping Atmospheric Residue*)[39] and of the particulate. The subsequent stages remove the fractions of TAR and particulate (ceramic filters) not removed by cyclones. Where possible these methods, are be preferred, since they mostly retrieve the thermal and chemical content of the gas produced, do not have to cool the gas and do not produce polluted wastewater.

The *efficiency* of these plants, defined as the ratio between the energy content of the synthesis gas produced and the one of the biomass used, is on average 70%.

---

[38] Carbonaceous solid product mainly made up of carbon, containing residues at a high molecular weight, as furan-derivatives and phenol compounds.

[39] "Liquid oily" fraction containing water and compounds with a low molecular weight as aldehydes, acids, ketones, alcohols, heavy hydrocarbons condensed at temperatures lower than 20–100°C.

#### 4.3.1.4 Pyrolysis (or *Dry Distillation*)

*Pyrolysis* is a process of thermochemical decomposition of organic matter based on the administration of heat at temperatures between 400 and 800°C, in a total absence of an oxidising agent, or with a reduced quantity of oxygen. In practice, thanks to this process, the organic macromolecules break into chains with a lower molecular weight. The products of pyrolysis are gaseous, liquid and solid, in proportions that depend on the methods of pyrolysis (fast, slow or conventional) and on the reaction parameters:

- from *conventional pyrolysis* that occurs at temperatures lower than 600°C and with moderate reaction times, it is possible to obtain approximately *three fractions (solid, liquid and gas)* in equal proportions;
- in *fast pyrolysis* the conversion reactions occur fast and at temperatures between 500 and 650°C, and with short contact times in order to reduce the new formation in intermediate compounds, hence favouring the production of the *liquid fraction* up to 70–80% in weight of the organic matter supplied;
- the *flash pyrolysis* takes place at temperatures exceeding 700°C and with contact times lower than 1s, in such a way as to favour the production of a *liquid fraction* (around 80% in weight of supply), but with a more reduced composition variation.

#### 4.3.1.5 Fermentation

*Fermentation* is an anaerobic process, similar to the known process of anaerobic digestion, in which the glucose of biomass is decomposed, through bacteria, into hydrogen, carbon dioxide and acids according to the following reaction:

$$C_6H_{12}O_6 + 2H_2O \Leftrightarrow 2CH_2COOH + 2CO_2 + 4H_2.$$

The theoretical efficiency is 4 mole of hydrogen per mole of glucose (0.5 $Nm^3$ of hydrogen per hk equivalent glucose). The final co-products of fermentation as acetic acid can be further fermented or submitted to photosynthesis processes, increasing the maximum theoretical efficiency to 12 mole of hydrogen per mole of equivalent glucose. The bacteria necessary to fermentation can be cultured in pure or mixed cultures. Pure cultures of *caldicellulosirupter saccharolyticus* and *thermotoga elfii* can produce hydrogen at temperatures of 65°C. Mixed cultures of *clostridia* use non-sterile environments in continuous reactors at nearly 30°C on substrates of carbohydrates including cellulose.

The research is oriented towards obtaining mixed cultures in non-sterile environments with a high hydrogen efficiency and able to inhibit the consumption of the hydrogen produced.

#### 4.3.1.6 Photoproduction

*Hydrogen photoproduction* can be obtained by using some photosynthetic organisms including green algae. It is performed by conditioning algae through illumination cycles alternated to darkening cycles. Algae cause bio-photolysis by splitting water into $H_2$ and $O_2$. In some conditions, the *hydrogenase* enzyme contained in green algae acts as a catalyst causing the reduction of protons to form hydrogen. *Hydrogenase* is synthesised after many hours of incubation in anaerobic (without oxygen) conditions and without light. *Hydrogenase* contains iron and is oxygen-sensitive, so that when after a short period of light exposure in which $H_2$ and $O_2$ are produced, the production of hydrogen decreases. Green algae as *chlamydomonas reinhardtii*, blue algae as *cyanobacteria* and photosynthetic bacteria as *rhodobacter sphaeroides* have been extensively tested. Blue algae and bacteria produce hydrogen by using the *nitrogenase* enzyme, normally involved in the formation of $NH_3$. *Nitrogenase* is less efficient than *hydrogenase* (due to a higher energy demand and a slower turnover). The reactor must be designed to allow the collection of the hydrogen produced, be penetrable to light, water and culture.

## *4.3.2 Thermal Energy Transformation Plants*

#### 4.3.2.1 Absorption Heat Pumps

Operational Principle

Absorption heat pumps operate on the basis of the capacity of some substances to absorb large quantities of steam of others, hence forming a mix or forming weak chemical bonds.

The principle at the basis of the operation of absorption machines is based on the behaviour of liquid mixtures in equilibrium with their own vapour.

To analyse the phenomena determining the operation of these machines, two containers shall be considered: A (absorber) and *E* (evaporator), the former (*A*) containing a mix of substances $\alpha$ and $\beta$ (i.e.: liquid in equilibrium with its vapour) and the latter (*E*) substance $\alpha$ alone (also in this case in liquid phase in equilibrium with vapour). The two containers communicate through an initially closed tap, as shown in Fig. 4.69. In these conditions it is possible to express pressure in *A* ($p_A$) and pressure in *E* ($p_E$) according to the concentrations ($X_\alpha$ and $X_\beta$) and the partial pressures of the vapour phase ($p_\alpha$ and $p_\beta$)[40] of the two solutions through the Raoult's Law. In container A there is a solution of $\alpha$ and $\beta$, therefore:

[40] In the presence of the two phases of liquid and steam, the pressures of the steam phase ($p_\alpha$ and $p_\beta$) only depend on the temperatures exciting in containers ($T_A$ and $T_E$).

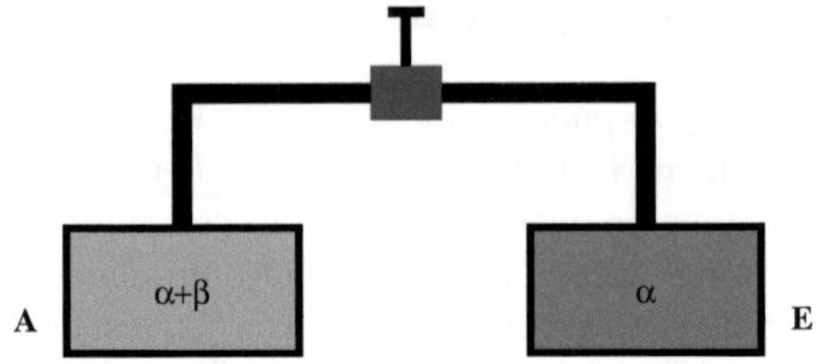


**Fig. 4.69** Diagram of the operation of absorption heat pumps

$$P_A = X_\alpha P_\alpha + X_\beta P_\beta \tag{4.132}$$

In container $E$ there is pure $\alpha$, therefore:

$$P_E = P_\alpha \tag{4.133}$$

If substance $\alpha$ is more volatile than $\beta$, that is if $P_\alpha \gg P_\beta$ (4.132) can be as follows:

$$P_A = X_\alpha P_\alpha \tag{4.134}$$

The Clapeyron's Law links temperature and pressures in a biphasic system through the following ratio:

$$\frac{dP}{dT} = \frac{\Delta H}{T\Delta V} \tag{4.135}$$

where $\Delta V = V_v – V_1$ is the variation of volume in the passage of phase from vapour ($V_v$) to liquid ($V_1$) and ΔH is the latent heat of evaporation.

Putting aside the volume of the liquid compared to the one of vapour ($\Delta V \cong Vv$), considering vapour as a perfect gas and ΔH constant, it follows that:

$$\frac{dP}{dT} = \frac{\Delta H}{RT^2} P \Rightarrow \frac{dP}{P} = \frac{\Delta H}{R} \frac{dT}{T^2} \tag{4.136}$$

Integrating (4.136) and indicating with K the constant of integration, it follows:

$$\ln P = -\frac{\Delta H}{R} \frac{1}{T} + K \tag{4.137}$$

Therefore the pressure of the steam phase of substance $\alpha$ according to temperature is:

$$P_\alpha = e^{-\frac{\Delta H}{R}\frac{1}{T} + K} \tag{4.138}$$

From (4.133) it follows:

$$P_E = e^{-\frac{\Delta H_\alpha}{R}\frac{1}{T_E}} + K_1 \tag{4.139}$$

From (4.134) it is obtained:

$$P_A = X_\alpha \cdot e^{-\frac{\Delta H_\alpha}{R}\frac{1}{T_A}} + K_1 \tag{4.140}$$

By putting the two containers in communication, obviously the pressures in A and E will be the same; from (4.139) and (4.140) therefore it follows:

$$e^{-\frac{\Delta H_\alpha}{R}\frac{1}{T_E}} = X_\alpha \cdot e^{-\frac{\Delta H_\alpha}{R}\frac{1}{T_A}} \tag{4.141}$$

Moving on to logarithms:

$$-\frac{\Delta H_\alpha}{R}\frac{1}{T_E} = \ln X_\alpha - \frac{\Delta H_\alpha}{R}\frac{1}{T_A} \tag{4.142}$$

From (4.142) it is obtained:

$$\frac{1}{T_A} = \frac{R}{\Delta H_\alpha}\ln X_\alpha + \frac{1}{T_E} \tag{4.143}$$

Being $\Delta H_\alpha > 0$ and $0 < X_\alpha < 1$, therefore $\ln X_\alpha < 0$, it is possible to obtain:

$$\frac{1}{T_A} > \frac{1}{T_E} \Rightarrow T_A < T_E \tag{4.144}$$

The variation of concentration (in E there is a pure $\alpha$ whereas in A there is a solution of $\alpha$ and $\beta$) at the same pressure conditions, therefore the temperature in A is lower than temperature in E. From (4.144) it is evident that (R being a constant) the difference between temperature in A and in E increases as both $X_\alpha$ (that is to say the concentration of $\alpha$ must be low) and $\Delta H_\alpha$ decrease (which means that the latent heat of evaporation of $\alpha$ must be as low as possible[41]).

*Description of the Absorption Heat Pump Plant*

As regards Fig. 4.70, providing heat ($Q_E$) in $E$ there is the production of steam of $\alpha$, which migrates to A. By subtracting heat ($Q_A$) from A the steam previously formed goes back to the liquid state and, thanks to the affinity between $\alpha$ and $\beta$ is absorbed by the solution contained in A. This system can be therefore used for the operation of a heat pump taking heat ($Q_E$) to a certain temperature and releases it ($Q_A$) at a lower temperature (4.144). However, in the process described, the variation of concentrations in $E$, as well as the exhaustion of $\alpha$ in $E$, allow that, at a certain point, the process itself stops. In order to make this system cyclical, it is necessary to bring the two containers back to their initial conditions. To this purpose, other two exchangers are used (Fig. 4.71): a steam generator ($G$), that absorbs heat $Q_G$ at temperature $T_G$ and a condenser that releases heat $Q_C$ at temperature $T_C$.

Summing up, as regards Fig. 4.71, the components of an absorption heat pump are the following:

[41] That is in line with the hypothesis according to which $\alpha$ is far more volatile than $\beta$.

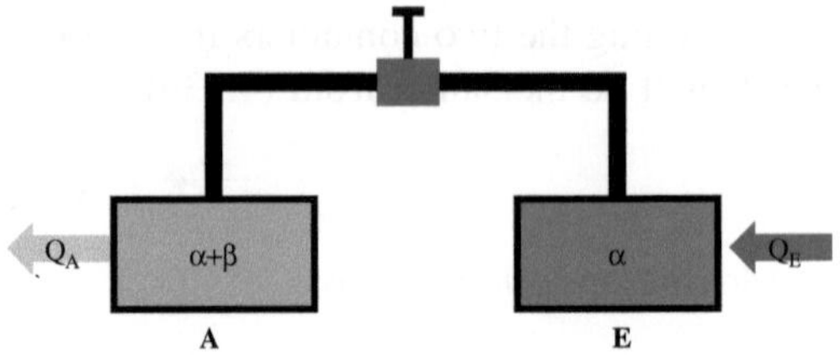


**Fig. 4.70** Operational diagram of the absorber–evaporator system

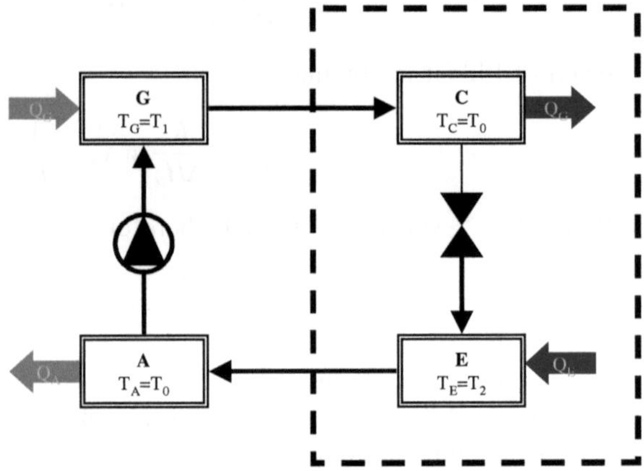


**Fig. 4.71** Operation scheme of an absorption heat pump

- *Steam generator* (*G*): here the fluid absorbs heat $Q_G$ at temperature $T_G$. This entails an evaporation of the mix with its subsequent dilution[42];
- *Condenser* (*C*): here the fluid releases heat $Q_C$ at temperature $T_C$ by condensing it at high pressure;
- *Evaporator* (*E*): between it and the condenser there is a pressure difference (through a valve) ($p_E < p_C$) that allows the refrigerant to evaporate by absorbing heat $Q_E$ at temperature $T_E$.;
- *Absorber* (*A*): here the refrigerant is reabsorbed by the solution, completing the cycle. The absorption causes the condensation of the refrigerant, therefore it is necessary to subtract heat $Q_A$ (at temperature $T_A$).

As mentioned above, the cycle of an absorption heat pump is based on the use of a binary mix of fluids, for instance a solution of water and lithium bromide ($H_2O$–BrLi), or ammonia and water ($NH_3$–$H_2O$). The first of the two substances of the mix behaves as a *refrigerant fluid*; the second as *solvent*, in which the refrigerant is dissolved in a more or less high concentration according to the point of the plant.

To keep the necessary difference of pressure between condenser and evaporator[43] (and, therefore, between steam absorber and generator) it is necessary to

[42] Being $\alpha$ more volatile than $\beta$, the steam will be richer than $\alpha$ compared to the liquid phase.

[43] Normally the pressure at the condenser is nearly 10 times as much as higher than the one at the evaporator.

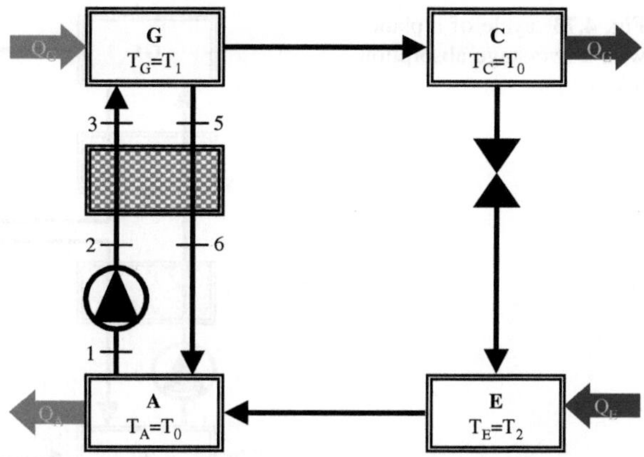


**Fig. 4.72** Scheme of a plant with an absorption heat pump

insert a laminating valve between condenser and evaporator and a pump between steam absorber and generator.[44]

Figure 4.72 shows a more detailed diagram of the operation cycle of an absorption heat pump. The cycle uses a solution of water and lithium bromide, which is represented here as an example with the values indicating the typical values of the cycle (concentrations, temperatures and pressures).

At the exit of the absorber (section 1) the solution is rich in refrigerant ($X_\alpha = 59\%$, $X_\beta = 51\%$) and is at a pressure of 800 Pa and at a temperature of 41.5°C. Unless load losses occur, the pressure in the evaporator is the same available in the absorber, a pressure at which water evaporates at around 5 ÷ 6°C.

The solution (at the liquid state) taken by the absorber is compressed at a constant temperature through the pump and in section 2 the pressure is nearly 8,000 Pa: from here the solution enters the heat exchanger (where it exchanges heat with the solution outgoing from the steam generator) heating up to a temperature of nearly 77°C, at a constant pressure and concentration. In the steam generator (where heat $Q_G$ is absorbed) there are two phases: in the first one there is a heating (up to nearly 85°C) at constant pressure and concentration; in the second one, being the solution in equilibrium with its steam, at an increase of temperature corresponds a separation (through evaporation) of the refrigerant from the solution. Pressure remains constant also in this second phase, at the end of which there is a pure refrigerant at the vapour state and a solution at the liquid state rich in solvent ($X_\alpha = 36\%$, $X_\beta = 64\%$) at the temperature of nearly 94°C. The liquid solution is introduced in the exchanger where it heats the solution coming from the absorber; from here it is sent to the absorber itself, providing it with a solution, rich in solvent, able to absorb the refrigerant. The refrigerant is sent to the condenser, where it condenses by releasing heat $Q_C$ at a high temperature (94°C). From here it

[44] Also in this type of pumps, therefore, there is a compression: however, compared to the case of compression heat pumps where a gas is compressed through a compressor, in the absorption ones a liquid is compressed through a pump, and therefore the compression work is much smaller.

**Fig. 4.73** cycle of a plant with a two-stage absorption heat pump

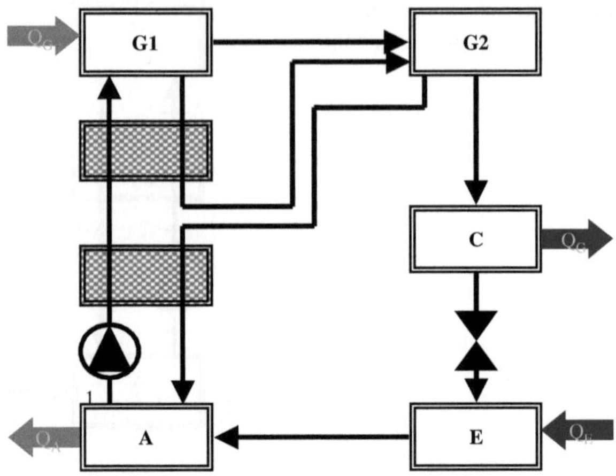


crosses the laminating valve, decreasing its pressure (from 8,000 to 800 Pa) and it is injected in the evaporator, where it evaporates by absorbing heat $Q_E$ at the temperature of nearly 5°C. At this point the refrigerant (at the vapour state) is taken from the evaporator and introduced in the absorber by closing the cycle.

The absorption plant can take a more complex configuration: in fact in order to improve the efficiency of the cycle two-stage concentrators are used (Fig. 4.73). In this case the steam generator is divided into two sections, one at *high* and one at *low* temperature.

*Performance of Absorption Heat Pumps*

The performance of an absorption heat pump are generally indicated with a defined Coefficent Of Performance (COP) code; in this case, as a ratio between the useful (thermal power) heat flow ($Q_u$) and the thermal power introduced in the steam generator $Q_G$

$$\mathrm{COP} = \frac{Q_u}{Q_G} \tag{4.145}$$

As regards Fig. 4.72, in the heating operation there is:

$$\mathrm{COP} = \frac{Q_A}{Q_G} \tag{4.146}$$

Whereas in the cooling phase:

$$\mathrm{COP} = \frac{Q_E}{Q_G} \tag{4.147}$$

The COP values in the heating operation are around 1.5, whereas in the cooling phase they are markedly lower. For simple machines, they are typically lower: the value of COP varies in the range 0.6–0.75, with higher values for lithium bromide machines compared to the ammonia ones: In the case of double effect machines, the COP can reach values exceeding the unit, generally in the range of 1.1–1.3. Furthermore, performance vary if the machine works in a partial load. In this connection, it is possible to observe that an absorption machine shows a lower consumption compared to a compression machine and in particular a partial load cooling COP can outperform the corresponding full load value. This is due to the absence, in absorption machines, of auxiliaries with parts in motion, whose performance always decrease with partial loads, and with the increase in the exchange efficiency in system exchangers when the system operates with ranges lower than the design ones.

## References

1. Caputo C (1979) Le turbomacchine vol III. Editoriale ESA. ISBN:88.405.3216.1. Also: Logan E (1983) Turbomachinery: basic theory and applications, CRC press, ISBN 9780824791384, and Baskharome EA (2006) Principles oh turbomachinery in air breathing engines, Cambridge University Press, ISBN 9780521858106
2. Pello' PM (1990) Impianti di accumulo mediante pompaggio per la generazione elettrica: il caso italiano in L'energia elettrica no° 9 p 369
3. Maccari A, Vignolini M (2001) Progetto di massima di un impianto pilota per la produzione di 2000 $m^3$/giorno di idrogeno solare basato sul processo ut-3
4. Nucara A, Pietrafesa M (2001) Elementi di illuminotecnica, Universita' degli Studi "Mediterranea" di Reggio Calabria; Facolta' di Ingegneria; Dipartimento di Informatica Matematica Elettronica e Trasporti. Also: Dilouie C, Advanced lighting controls: energy saving, productivity technology and applications, The Fairmont Press, ISBN 0-88173-510-8

# Chapter 5
# Distributed Generation and Cogeneration

## 5.1 Distributed Generation

The distributed generation of electric power, rather than a real technology, represents a paradigm of energy production, being complementary and/or alternative to the centralised and/or localised production in large-scale plants. It is based on the use of small-size generators that operate close to end users and with a power ranging from a few kW to some MW. To date, the spreading and the incidence related to distributed generation on global electric power production are marginal, although increasing. On the other hand, a considerable and articulated series of technical–economic evaluations exist, describing the present state of the art and setting out generally growing development scenarios for the next decade.

The subject of electric power distributed generation was introduced in the late '90s with some considerations on technologies, strategic aspect, advantage and problems [1]. In this section, these elements will be resumed, and integrated with further considerations stimulated by recent events.

### *5.1.1 General Features of Conversion Plants*

In brief, conversion plants can be classified according to the following features:

- Energy source used
- Environmental impact
- Size and flexibility.

Obviously, the source used determines the technology as well as the conversion and processing procedures adopted to obtain the form of energy desired. In the case of non-renewable sources (fossil and nuclear) the resource used usually in plants is not the primary source, but rather a resource obtained from it (secondary

F. Orecchini and V. Naso, *Energy Systems in the Era of Energy Vectors*,
Green Energy and Technology, DOI: 10.1007/978-0-85729-244-5_5,

source) through suitable transformation processes. In the case of renewable sources, on the contrary, the plant normally directly uses the primary source, exception made for biomass: in this case, suitable processes are needed to make this source usable in most efficient technologies. The type of process varies according to the type of biomass (and therefore the plant technology): in the case of biomass usable in boilers or gasifiers (biomass suitable for combustion) the process simply consists of preparing the biomass according to a size suitable to feed the plant. On the contrary, in the case of biomass usable in biochemical processes (fermentation, digestion and esterification), the preparation consists of a real transformation process in which, from the biomass, it is possible to obtain a gaseous fuel (biogas) usable in traditional boilers.

In general, energy conversion plants interact with the environment consuming resources, producing waste, occupying the soil they are located in, etc. It is therefore possible to state that these plants cause a certain environmental impact: according to whether this impact produces effects at a global and local level, reference is made to global or local impact.

When talking about the size of the different technologies used in energy conversion plants, it is suitable to make some considerations. It is evident that, at least in theory, a plant—whatever technology is used—can be of any size: as a matter of fact, for some technologies, the main characteristics of the plant itself (cost and performance) are strictly related to its size. This is true, for instance, for steam thermoelectric plants: as it is known, these are highly performing plants (exceeding 40% efficency) and allow to produce electric power at low costs. These features, however, occur in large-size plants (normally exceeding 1,000 MW) thanks to the scale effect: the same type of small plant (for instance, of some tens of kW) would not have such a high performance and the cost of the energy produced would markedly increase. These considerations lead to state that steam plants are large sized ones. Conversely, for other technologies (for instance photovoltaic), the absence of the scale effect and the problems that the adoption of large sizes would entail (think of the necessary surfaces) "discourage" the setting up of large plants. For these reasons, photovoltaic plants are normally classified as small-size plants.

The main characteristic that defines the currently needed operational features of a plant is flexibility, that is to say the capacity of the plant to regulate the power it supplies. As far as this characteristic is concerned, it is possible to distinguish three main categories of plants:

- plants that directly use *aleatory sources* (sun and wind), and in this case normally there is no possibility to adjust the power supplied (without systems for the systems for the accumulation of the energy produced);
- plants in which, although an adjustment is technically feasible, it entails a remarkable worsening of performance; it is the case of large steam thermoelectric plants that, in order to keep a good efficiency, must work as much as possible at normal conditions and with continuity, reducing switching off and on;

- plants that allow to adjust—within wide ranges—the power supplied according to the load, keeping their efficiency nearly constant; it is the case of turbogas plants and, more evidently, of fuel cells.

### 5.1.2 Scale Economies

The centralisation of electricity production is traditionally based on consolidated evaluations of scale economy, deriving from the low management and investment charges per kWh produced in large plants. The need and/or the convenience to delocalise plants imposed to resort to burdensome infrastructures for the transportation, processing and distribution of the energy produced (networks), which however accounts for an increasing percentage of total costs furthermore, considerable energy losses. Further difficulties are found in the case of new plants, not only for the considerable investment charges, but also for the possible oppositions by local communities vis-à-vis the new transportation plants and facilities.

On the other hand, the small size of distributed generation plants does not entail the need for high-medium voltage distribution networks and, at the same time, allows to use low-voltage for local connections. Furthermore, their progressive and non-traumatic introduction is favoured, both in those cases in which a production structure is missing, and when replacing the existing networks is considered as suitable or convenient.

### 5.1.3 Energy Sources and Distributed Generation Technologies

*Distributed generation plants* (including *stand alone ones*) must meet specific requirements, including:

- *plant-engineering modularity* which means reduced scale economies and installation costs almost independent from the size; this allows to scale power according to users' needs;
- *reduced management and maintenance needs*: this characteristic is due to the an installation closer to the users' place is hardly suitable for plant solutions requiring frequent maintenance with specific professional skills;
- *reduced emissions* installations at users, including in towns, impose particularly strict limits to emissions.

As stand-alone units, macro-generators in distributed generation allow the use of a potentially unlimited range of energy sources and, in any case, much wider than central systems, which mainly operate with traditional fossil fuels. In particular, the small scale entails flexibility in the choice of energy sources, depending on local availability, geographic characteristics, as well as on the technological and economic conditions of users. In this framework, the provisions of the United

**Table 5.1** Characteristics of electric power distributed generation technologies

| Technology | Diesel ICE | Gas ice | Gas turbines | Micro-turbines | FC | PV |
|---|---|---|---|---|---|---|
| Scale (kW) | 20–10,000 | 50–5,000 | >1,000 | 30–200 | 50–1,000 | >1 |
| Efficiency (%) | 36–43 | 28–42 | 21–40 | 25–30 | 35–54 | n.a. |
| $CO_2$ emissions (kg/MWh) | 650 | 500–620 | 580–680 | 720 | 430–490 | 0 |
| NOx emissions (kg/MWh) | 10 | 0.2–1.0 | 0.3–0.5 | 0.1 | 0.005–0.01 | 0 |

Nations conferences and summits on sustainable development (starting from Rio de Janeiro 1992), relating to both the energy needs of the underdeveloped areas of the planet, and to the concerns on the anthropogenic effects of development, can be suitably conjugated with the operational characteristics of distributed generation at all application levels.

As a matter of fact, the low spreading of distributed generation, and the subsequent and continued marginality of the relevant technologies, limit the field of usable sources. Among fossil sources, natural gas, oil and reforming hydrogen have the lead. Among renewable sources, thermal and photovoltaic sources prevail. The reference to hydrogen implies its use both in a long-term perspective, as a main energy vector in fuel cells, and in a short/medium-term perspective as a component of fuel mixtures (i.e., hydrogen-methane mixtures) in heat engines and turbines.

The production technologies available include reciprocating engines (diesel and gas), microturbines, fuel cells and photovoltaic conversion. Wind production is usually considered as a variant of central systems, although at a small scale. Table 5.1 shows the semi-quantitative analytic data on the different production technologies.

### *5.1.4 From Concentrated Production to Distributed Production*

As already mentioned above, the present organisation of the energy system at a national level is mainly based on the production of energy vectors concentrated in very few large plants, located far from users, and on the transportation and distribution of the vectors produced to end users. This organisation does not only concern production plants, but also the other infrastructures (the distribution grid) that represent the energy system as a whole. A passage from concentrated to distributed production, therefore, cannot only concern energy production technologies, but also all the components of the energy system, that is to say the infrastructures for energy wheeling, vectors themselves (since a distributed generation system shall provide for the use and therefore the supply of hydrogen together with electricity and heat) up to the technologies for end use.

The present electricity distribution grid as well as the energy network as a whole, is essentially conceived and organised to transport energy towards one direction only (from production centres to users); namely it is unidirectional,

**Table 5.2** Today's grids and smart grids-ETP

| |
|---|
| **TODAY's GRIDS** |
| large centralised generation |
| geographical distribution of generation resources |
| power flow in one direction from the power stations |
| dispatching of power and network control → centralised facilities (controlling several regions from one place) |
| There is little or no consumer participation and no end-to-end communications |
| **SMARTGRIDS** |
| they accommodate bi-directional power flows |
| they allow: |
| • distributed generation management |
| • renewable energy resources management |
| • optimisation of DSM (demand site management) |
| • optimisation of Storage management |
| Coordination of local energy management and full integration of DG and RES with large-scale central power generation |

passive and able to absorb power only from higher voltage grids. Therefore, it is limitedly suitable for a bidirectional system in which distributed production has a particular relevance from the quantitative viewpoint. Therefore, the introduction of distributed generation in this type of network, entails problems that can be overcome only until limited quantities are involved. Conversely, a massive penetration would entail a high degradation of the quality of the service as well as problems in terms of operation and protection.

The momentum given by low-carbon generation technologies, associated to a higher efficiency (therefore a lower energy consumption), will allow consumers a higher interaction with the networks.

Making consumers active players in the process of energy supply, and promoting the spread of renewable sources and distributed generation: these are the main objectives of Smart Grids.

In Europe, in fact, the SmartGrids European Technology Platform (ETP) was set up in order to face the problems linked to an electric grid system no more unidirectional. The aim of the SmartGrids ETP for electricity networks was to formulate and promote a vision for the development of Europe's electricity networks looking towards the future [2].

Compared to traditional electric grids, characterised by large plants with a technology consolidated over decades, centralised control and optimal management at a regional level, Smart Grids exploit at best the advantages offered by the digital era (Table 5.2), but for this purpose they have to be:

- *flexible*, fulfilling customers' needs while responding to the changes and challenges ahead;
- *accessible*, granting connection access to all network users, particularly for RES and high efficiency local generation with zero or low-carbon emissions;

**Fig. 5.1** Future network vision [2–5]

- *reliable*, assuring and improving security and quality of supply, consistent with the digital era demands;
- *cost-effective*, providing the best value through innovation, efficient energy management and "level playing field" competition and regulation.

In such a project, a smart grid scheme (Fig. 5.1) must envisage that:

- the system operation will be shared between central and distributed generators.
- the control of distributed generators might be aggregated to form micro-grids or "virtual" power plants to facilitate their integration both in the physical system and in the market.

In a smart grid perspective, and always bearing in mind the things said about energy systems, sources and vectors, it is necessary to increasingly adopt the larger concept of Smart Energy Networks, as "renewable" as possible.

It is readily evident that the passage towards distributed production neither does necessarily envisage the complete elimination of large plants, nor the possibility of siding them with small production plants directly at users'; it is rather possible to envisage intermediate levels of energy production and distribution, as well as energy vectors at the city or district level.

Figure 5.2 shows a possible organisation of the energy system based on distributed generation.

In an organisation as a **Smart Energy Network**, thanks to a "smart" use of a different mix of energy vectors (priority given to zero emission usability), like electricity, heat and hydrogen, the coverage of the energy needs of single users will no longer be entirely met by the network but rather partly covered by the integrated plants that produce energy directly for users (micro energy networks in Distributed Generation) and partly by the grids. Therefore, the energy taken from the network by a single user will partly come from the national grid (powered by large concentrated production plants), partly from medium-sized local plants, and partly from other users at the city and/or district level (Fig. 5.3) shows that during the production phase it had no consumption needs. This type of system therefore, will have a series of energy production and storage microstations locally.

Furthermore, from an energy viewpoint, the use of distributed generation allows to employ cogeneration, which is very much efficient; this is clearly impossible in

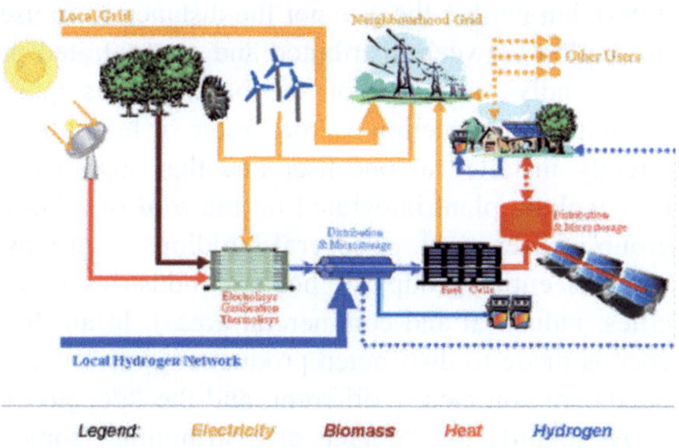


**Fig. 5.2** Organisation chart as a smart energy network

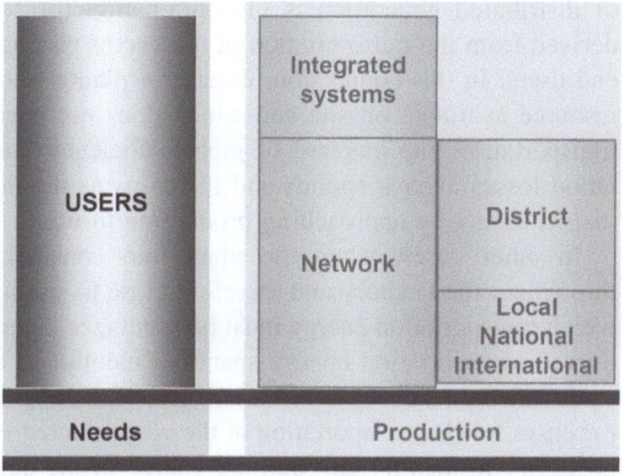


**Fig. 5.3** Possible distribution of energy consumption by end users

the case of concentrated generation, since heat transportation for long distances is not feasible from the energy viewpoint.

A Smart Energy Network has to be flexible, accessible, reliable and cost-efficient. It should also guarantee; renewability of energy resources; efficiency in energy conversion, distribution and use; lowering of the environmental impact; increasing of energy access; tailor making of energy systems on local social-economic-environmental conditions.

## *5.1.5 Type One and Two Distributed Generation*

Let's now some considerations on distributed production. First of all, it is evident that there is no sharp division between concentrated and distributed production: concentrated and distributed production refers to the size and the localisation of the plant (large conversion plants far from users, or small-sized plants close to users), but neither the size nor the distance from users indicate any values of sharp distinction between distributed and concentrated production.

Secondly—although on the basis of this qualitative distinction alone—it is evident that production can be "more or less" distributed: it ranges from the plant directly installed at one user and that supplies that given user (for instance a photovoltaic plant integrated on the roof of a house), to the plant that supplies a group of users (one or several buildings or groups of users), up to the plant that supplies entire groups of houses, industries or commercial activities (districts, cities, industrial and commercial areas). In all the cases mentioned above, reference is made to distributed production, but it is evident that the characteristics of plants are completely different, and the "degree of distribution" varies as well.

As regards the "degree of distribution" some considerations are needed; as already mentioned, from the global energy efficiency point of view, an advantage of distributed generation is the elimination (or strong reduction) of the losses derived from the transportation of the vector produced from the production site to end users. In this connection, in case of plants whose resource used is a primary resource available on site and which does not need any energy spending for its transportation, the increase of global efficiency due to the decrease of transportation losses always applies and increases as the degree of distribution increases that is, to say by approaching production to users.

In other cases, when the plant into consideration, although "distributed" throughout the territory and therefore close to end users, uses an energy vector for which transportation energy must be spent (see chapter on vectors), the decrease in overall transportation energy spending mentioned above shall be carefully evaluated and on a case-by-case basis since, as the degree of distribution increases, the expenses for the transportation of the vector increase. In these cases, therefore, it is necessary to identify an "optimal value" of the degree of distribution.

On the basis of the above, from the energy analysis perspective, not always distributed production is better than the concentrated one. Distributed production of energy, in fact, refers to an energy conversion process close to end users which can start from an energy vector produced in a remote place and distributed up to the final conversion point, or a process that directly starts from a primary energy resource available at the user.

It is therefore necessary to make a differentiation between the two situations already in their definition, since they are completely different from one another. To do so, it is necessary to introduce a specific characterisation besides the indication of distributed production. Therefore we define *distributed production of type I* as the production through plants that use primary resource locally available as a resource. Conversely, we define *distributed production of type II* the production through plants that use as a resource an energy vector made available to the end user and for whose transportation energy was spent.

In the case of distributed production of type I, global efficiency increases with the degree of distribution, whereas in the case of distributed production of type II the degree of optimal distribution must be evaluated.

On the basis of the definition, examples of distributed production of first type include solar plants (photovoltaic, thermal at low, medium and high temperature), and wind plants. It is evident that, although for these plants global efficiency increases as the degree of distribution increases, for the real localisation of plants it is necessary to consider other aspects, such as the characteristics of the plant and the availability of the source. In some cases, as in the case of photovoltaic plants and solar collectors for heat production at low temperature, the characteristics of the plants themselves (modularity, possibilities of integrated solutions in buildings), as well as the characteristics of availability of the source[1] make them particularly suitable for production at the users. In other cases, the availability of the source (wind[2]) and the characteristics of plants[3] impose some constraints to the localisation of plants themselves. In any case it is possible to state that the distributed production plants of type I must be localised as close as possible to users, irrespective of the energy vector produced and compatibly with other types of constraints.

As concerns the plants for type II distributed production, the optimal "degree of distribution" firstly depends on the vector used and on the one produced.

In the case of heat production, because of the relevant losses for the transportation of heat itself, plants must necessarily be sufficiently close to users. Within some limits, in any case, production can be more or less distributed[4]: in this case, it is necessary to identify the different factors that determine the global energy efficiency. In particular:

- energy spending for the distribution of the vector used for heat production purposes;
- energy spending for heat distribution;
- eventual efficiency variations of the conversion plant for the different solutions analysed.

For electric power production, in the case of type II plants, the low energy spending for its transportation[5] suggests a centralised production or, more generally, a distribution of plants that limits at the transportation of the vector used at a minimum level. For instance, if natural gas is used for the production of electric power, the expenditure for the transportation of the vector used is higher than the one for the transportation of the vector produced: in fact, if we consider an

[1] Incident solar radiation depends on the latitude and on the cloudiness of a given area. In practice, the solar radiation available can be considered as constant within rather wide macro-areas.

[2] In the case of wind energy, the availability of the source varies according to micro-areas; this makes necessary a detailed assessment of the resource.

[3] Concentration solar plants, for instance thermodynamic solar plants, are necessarily large and demand large areas with particular infrastructures.

[4] Such solution includes the possibility of producing heat for domestic heating at the level of single house, single flats or groups of flats or at a district level.

[5] See Sect. 3.4.

efficiency of the conversion plant as amounting to 40%, in order to produce 1 kWh of electricity we need 2.5 kWh of gas. On the basis of the considerations made in the Sects. 3.4.3.1 and 3.4.4.1, in case of gastransportation the energy spending, given the same energy transported is nearly as much as doubled compared to electric power[6]. This means that distributing methane for the production of electric power entails an energy spending five times higher than the one needed for the distribution of the electricity produced starting from methane.

Let us now turn to the production of hydrogen from the reforming of natural gas. In this case, the efficiency of the processing plant is high, and amounts to nearly 75%; therefore, to produce 1 kWh of hydrogen, 1.33 kWh of gas is consumed. The energy spent for the transportation of hydrogen is, given the same energy transported, nearly three times[7] as much as higher than the one for the transportation of natural gas. Therefore, the energy spending for the distribution of the hydrogen produced is over two times higher than the one for the distribution of the gas used. In this case, then, the global energy efficiency increases as the degree of distribution increases.

A particular case is the one in which biomass is used; due to the low energy density (in weight and volume) its transportation is relatively energy consuming. The specific characteristics vary according to the type of biomass; in any case, generally speaking, to limit the energy spending for transportation into acceptable values (10 $\div$20% of the energy transported), the transportation distance shall not exceed 100–150 km. Therefore, in this case, it is convenient from the energy viewpoint to limit the transportation of the resource used; this however does not mean centralising production, but rather to distribute it near the resource and not near users. Obviously, more in detail, for an optimal "distribution" it is necessary to consider the energy vector produced by the plant; electricity, heat and synthetic fuels (solid, liquid or gaseous) for whose transportation energy spending varies considerably.

## 5.2 Combined Production of Electric Power and Heat: Cogeneration

The word *cogeneration* indicates *the set of operations targeted to the combined* production (*cascade*) *of electric power and thermal energy, both considered as a useful effect*. This solution consists in the recovery of heat otherwise released in the environment by electric power production plants, with subsequent increase of the overall efficiency compared to the separate production of electric power and heat.

---

[6] Also considering the necessary voltage transformations for the transportation of electric power for long distances.

[7] See Sects. 3.4.3.1 and 3.4.5.2.

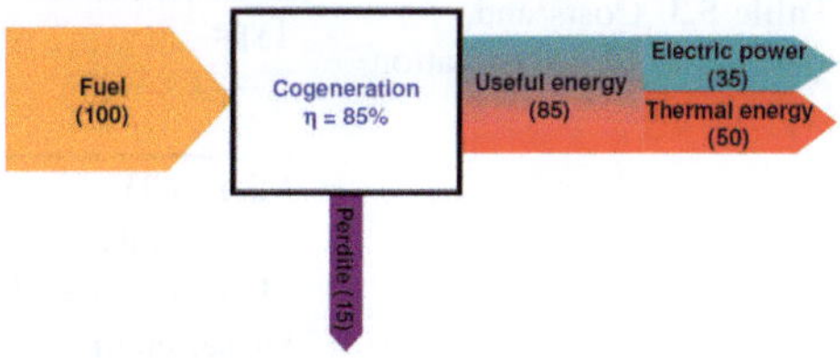


**Fig. 5.4** Energy balance of a cogeneration system (kWh)

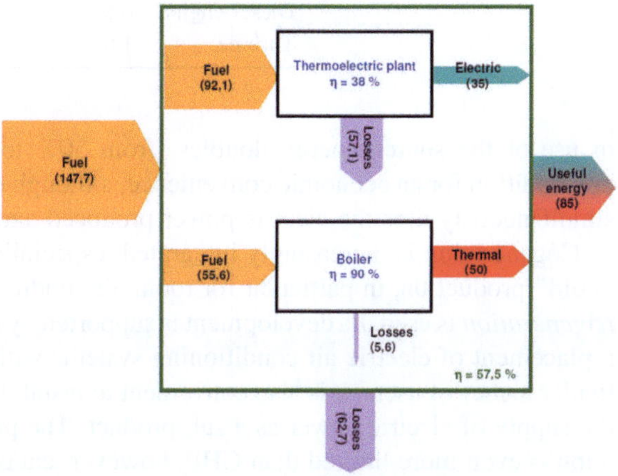


**Fig. 5.5** Energy balance for the separate production of electric power and heat (kWh)

Figure 5.4 shows an example of energy balance in a cogeneration system. This chart is referred to a cogeneration system with a turbogas: the ratio between electric power produced and thermal power strongly depends on the technology used, whereas the efficiency does not heavily vary from one technology to the other; therefore, the example shown can be considered as a general validity. Therefore, if we consider a fuel consumption of 100 kWh, it is possible to obtain a total useful energy of 85 kWh (35 kWh electric energy and 50 kWh thermal energy).

Comparatively, Fig. 5.5 shows the energy balance of the separate production of electric power and heat, given the same useful effect and considering an efficiency of electric power production amounting to 38% (almost the average value of the Italian electric system) and an efficiency of 90% of boilers for heat production. In this case, to obtain 35 kWh of electric power and 50 kWh of thermal energy, fuel consumption amount to 147.7 kWh.

The conversion efficiency of distributed generation microgenerators into electric power is generally lower than in large plants. However, their proximity to end users (which in fact reduces energy trasmission losses and complexity), offers the possibility for considerable energy recovery at low temperatures, otherwise dispersed, mainly for the heating of rooms and of household water. The simultaneous generation and use of electric energy and heat (*Combined Heat Power*—CHP), or *cogeneration*, does not alter thermoelectric efficiency, while substantially increases the portion totally used of source energy. In general, as a consequence of a 20% increase of investment costs for CHP devices, depending on the size, the efficiency

**Table 5.3** Costs and efficiencies of cogeneration technologies

| Type | Size (MW) | Initial cost (€/kW) | Electric efficiency (%) | Efficiency of use (%) |
|---|---|---|---|---|
| Micro-CHP (Stirling) | <0.015 | 2,700 | 15–25 | 85–95 |
| Micro-turbine | 0.1 | 1,970 | 29 | 59 |
| Diesel engine | 0.1 | 1,380 | 28 | 75 |
| Fuel cells | 0.2 | 3,800 | 36 | 73 |
| Diesel engine | 0.8 | 980 | 31 | 65 |
| Turbine | 1.0 | 1,600 | 22 | 72 |

of use of the source energy doubles (from 30% to over 60%) (Table 5.3). The precondition for an economic convenience, although not always met, is that the user simultaneously uses the electric power produced and the heat recovered.

Cogeneration is increasingly integrated, especially in commercial sectors, with "cold" production, in particular for room air conditioning, in which case the word *trigeneration* is used. Its development is supported by the huge savings following the replacement of electric air conditioning systems with gas-absorption ones. In particular niches of users, it is also convenient to install hot/cold water generators with the supply of electric power as a sub-product. The present penetration of trigeneration is even more limited than CHP; however, cautious estimates indicate for the future a considerable market growth, especially in the most advanced EU Countries.

### *5.2.1 Cogeneration Technologies*

Cogeneration systems, depending on the technology used and on the thermodynamic cycle, can be summed up as follows

- steam turbines;
- gas turbines;
- combined-cycle;
- fuel cells;
- internal combustion engines (ICE);
- gas esternal combustion engines (Stirling).

Table 5.4 shows the values relating to the size and the *characteristic electric index*, which is the ratio between the electric power and the overall power (electric + thermal) of these plants.

Indicating with $\eta_c$ the cogeneration efficiency, with $\eta_e$ the electric efficiency, $P_e$ the electric power, $P_t$ the thermal power, with $\dot{m}_c$ the mass flow of the fuel and with LHV its Lower Heatic Value, it follows that:

$$\eta_c = \frac{P_e + P_t}{\dot{m}_c \cdot \mathrm{LHV}} \tag{5.1}$$

**Table 5.4** Characteristics of the main cogeneration technologies

| Type of plant | Power (kWe) | Electric index |
|---|---|---|
| ICE | 10 – 100 | 0.35 – 0.45 |
| Stirling | 1 – 25 | 0.20 – 0.65 |
| Gas turbine | 400 – 100,000 | 0.20 – 0.40 |
| Gas microturbines | 10 – 80 | 0.20 – 0.40 |
| Fuel cells | 10 – 1,000 | 0.55 – 0.65 |
| Steam turbine | 1,000 – 150,000 | 0.5 – 1 |
| Combined cycles | >10,000 | 0.45 – 0.60 |

thus:

$$\eta_c \cdot \dot{m}_c \cdot \text{LHV} = P_e + P_t \tag{5.2}$$

By definition, electric efficiency is

$$P_e = \eta_e \cdot \dot{m}_c \cdot \text{LHV} \tag{5.3}$$

Indicating with $I_e$ the electric index, from its definition, we have:

$$I_e = \frac{P_e}{P_e + P_t} \tag{5.4}$$

Replacing (5.2) and (5.3) into (5.4), one comes to:

$$I_e = \frac{\eta_e \cdot \dot{m}_c \cdot \text{LHV}}{\eta_g \cdot \dot{m}_c \cdot \text{LHV}} = \frac{\eta_e}{\eta_g} \tag{5.5}$$

#### 5.2.1.1 Cogeneration with ICE

In these plants, electric efficiency generally amounts to nearly 30%, whereas the overall cogeneration one accounts for about 80%.

As shown in Fig. 5.6 the heat available can be recovered, more or less in the same quantities, at the discharge: at temperatures of about 350–400°C, and from the cooling circuit at lower temperatures (85–90°C).

#### 5.2.1.2 Cogeneration with Fuel Cells

These plants are characterised by high electric efficiency (from 50% to over 60%, according to the type of cells used).

The heat flow (thermal power) generated by a cell[8] is:

[8] See Section "The Thermodynamics of Fuel Cells" of Chap. 4.

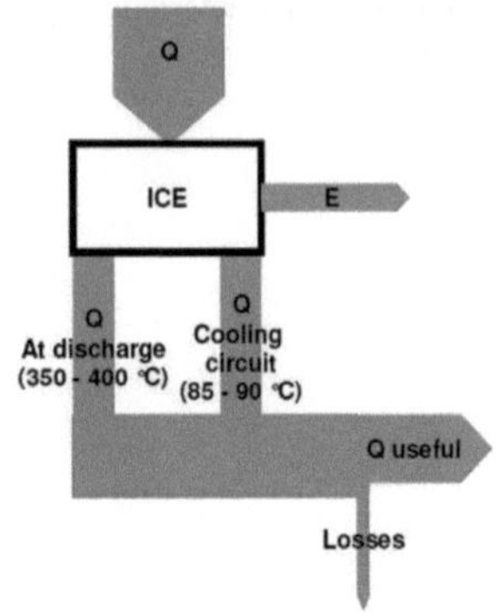


**Fig. 5.6** Energy balance with ICE

$$Q = \left(\frac{T \cdot \Delta S}{n \cdot F} + \Delta V\right) \cdot I = \left(\frac{\Delta H - \Delta G}{n \cdot F} + \Delta V\right) \cdot I \tag{5.6}$$

The voltage drop $\Delta V$ is:

$$\Delta V = E - E \cdot \varepsilon = \frac{\Delta G}{n \cdot F}(1 - \varepsilon) \tag{5.7}$$

Replacing (5.7) in (5.6) it is possible to obtain

$$Q = [\Delta H - \Delta G + \Delta G(1 - \varepsilon)]\frac{I}{n \cdot F} \tag{5.8}$$

from which,

$$Q = [\Delta H - \Delta G \cdot \varepsilon]\frac{I}{n \cdot F} \tag{5.9}$$

Assuming, the Faradic efficiency of the cell as amounting to 1, (5.9) becomes:

$$Q = \Delta H - \Delta G \cdot \varepsilon \tag{5.10}$$

The ideal electric work is equal to $\Delta G$. The real one, having assumed the Faradic efficiency and the fuel processing system as amounting to one, is:

$$L_{e,r} = \Delta G \cdot \varepsilon \cdot U_{\mathrm{H2}} \cdot \eta_{\mathrm{sist}} \tag{5.11}$$

From (5.10), the total useful energy (electric + thermal) is:

$$Q + L_{\mathrm{e,r}} = \Delta H - \Delta G \cdot \varepsilon + L_{\mathrm{e,r}} \tag{5.12}$$

Replacing (5.11) in (5.12) it is possible to obtain:

$$Q + L_{\mathrm{e,r}} = \Delta H - \Delta G \cdot \varepsilon + \Delta G \cdot \varepsilon \cdot U_{\mathrm{H2}} \cdot \eta_{\mathrm{sist}} \tag{5.13}$$

By definition of cogeneration efficiency, we have:

$$\eta_c = \frac{Q + L_{e,r}}{\Delta H} = 1 - \eta_i \cdot \varepsilon + \eta_i \cdot \varepsilon \cdot U_{H2} \cdot \eta_{sist} \tag{5.14}$$

(5.14), indicating with $\eta_e$ the electric efficiency of the fuel cell (5.14), can be written as follows:

$$\eta_c = \frac{Q + L_{e,r}}{\Delta H} = 1 - \frac{\eta_e}{U_{H2} \cdot \eta_{sist}} + \eta_e = 1 - \eta_e \left( \frac{1}{U_{H2} \cdot \eta_{sist}} - 1 \right) \tag{5.15}$$

(5.15) gives the cogeneration efficiency according to the electric efficiency of the fuel cell: $\eta_e$ exceeds 85%. The temperature at which heat is available deeply changes with the type of cell used: it ranges from 70°C of PEMFC up to over 1,000°C of SOFC. Since electric efficiency is around 50–55%, the electric index takes values of 0.55–0.65.

#### 5.2.1.3 Cogeneration with Steam Plants

This system is very much used for industrial cogeneration, since it is possible to make the steam available for thermal use at pressure values that are more frequently requested by processing, that is to say between 2 and 10 bar. It is also used in the civil sector for district heating.

The characteristic parameter of this type of plants is the so-called *degree of recovery*, which expresses the ratio between the mechanical energy produced by the steam flow rate recovered for thermal use and the mechanical energy overall produced.

Two different plant solutions are possible for the production of electric power and heat with steam cycle:

- *backpressure plants*;
- *condensation plants* with controlled bled-steam.

The most relevant aspects of the plant include:

- high investment costs (they sharply vary according to the kind of plant, such as either backpressure or controlled bled-steam);
- possibility of using different types of fuel (external combustion);
- good reliability with reduced maintenance costs;
- low electric index value;
- electric index depending on the temperature at which heat is requested.

Backpressure Plants

The scheme of this type of plant is shown in Fig. 5.7.

According to thermal users' needs, turbine expansion is stopped at suitable pressure and temperature values, and the remaining condensation heat of the steam

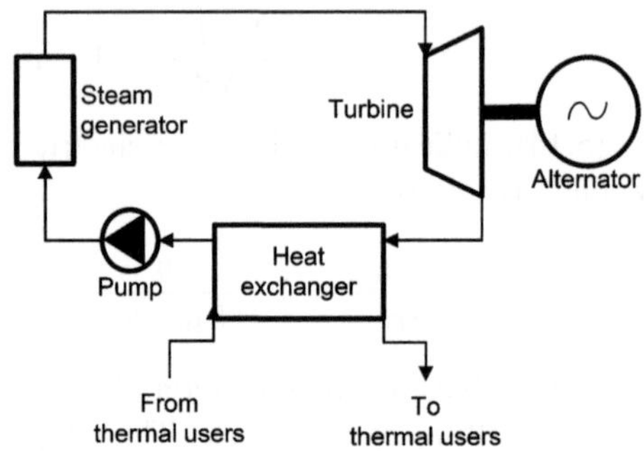


**Fig. 5.7** Scheme of a cogeneration plant with a backpressure steam cycle

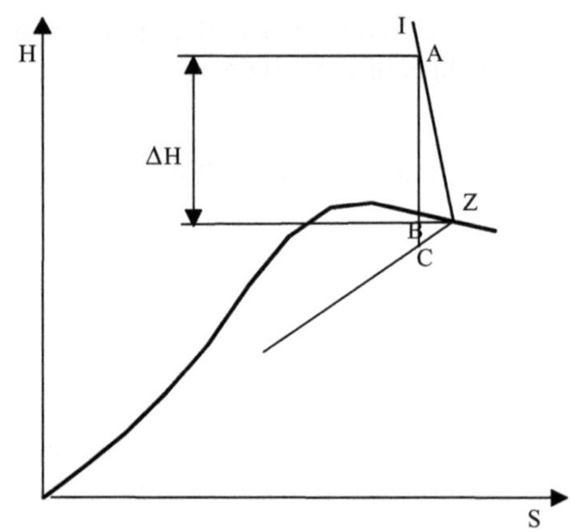


**Fig. 5.8** Expansion enthalpic variation in the turbine of Fig. 5.7

is used for heat production purposes. This solution is the easiest and cheapest one from the plant viewpoint (the condenser is replaced by the heat exchanger), although showing a scarce operational flexibility; electric and thermal power are produced in a pre-fixed ratio, and an increase of the former necessarily implies an increase of the latter.

As concerns heat demand, the steam status and its range (that is to say the thermal power requested) are defined. As regards Fig. 5.8, *Z* being the point representing the status of steam requested by the industrial process (or, more generally, by users), after tracing the isothermobaric process passing from point *Z* and estimating a value for the efficiency of turbine $\eta_t$ it is possible to trace the expansion line, being:

$$\eta_t = \frac{\overline{AB}}{\overline{AC}} = \frac{\overline{AB}}{\overline{AB} + \overline{BC}} = \frac{1}{1 + \frac{\overline{BC}}{\overline{AB}}} \tag{5.16}$$

In fact from (5.16), it is possible to identify point *A* through the following equation:

$$\overline{AB} = \overline{BC}\frac{\eta_t}{1 - \eta_t} \tag{5.17}$$

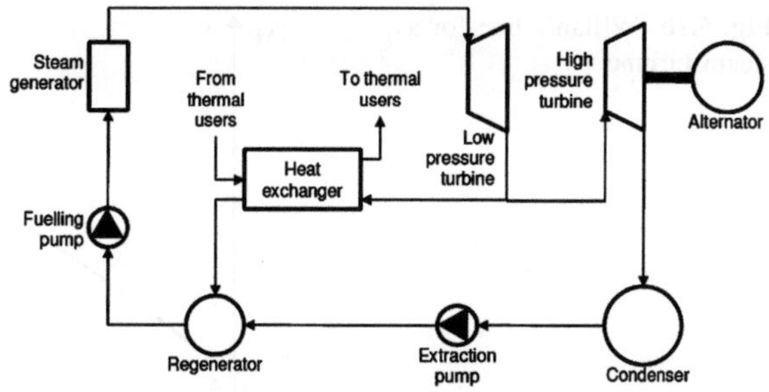


**Fig. 5.9** Scheme of a cogeneration plant with a controlled spilling steam cycle

The point $I$ at the beginning of the expansion must be identified on the line for points $A$ and $Z$ according to the ratio between thermal power and electric power to be supplied: the mechanical power obtainable, having defined $M$ as the steam range and $\eta_m$ as the mechanical efficiency, is

$$P = M \cdot \Delta H \cdot \eta_m \Rightarrow \Delta H = \frac{P}{M \cdot \eta_m} \tag{5.18}$$

Such a procedure makes it is possible to identify the point $I$ at the beginning of the expansion, while the isobaric process passing for point $I$ indicates the pressure that the fluid must have outside of the steam generator. Please note that, after setting the heat power requested, as the obtainable electric power increases, the pressure at the beginning of the expansion also increases (even much faster), that is to say the operating power of the steam generator, with obvious consequences on the plant and management costs. This practically limits the electric index: IE $\leq 0.2$. In any case, after fixing point $I$, as the temperature (and thence the pressure) at which expansion stops increases, heat production increases and electric power, i.e., electric efficiency, decreases.

Since electric power and heat are obtained from the same range of steam, it is possible to vary the total power supplied only by changing this range, and leaving IE unchanged. The *degree of recovery* in this type of plants, for the same reason, always amounts to one.

### Condensation Plants with Controlled Bled-Steam

By combining a condensation plant with a backpressure plant, it is possible to obtain a good operational flexibility with regard to the electricity/heat ratio produced. The scheme of such a plant is shown in Fig. 5.9. Expansion process occurs into two machines. In the high-pressure turbine the expansion of the entire steam flow occurs; at the end of this expansion, which correspond to steam conditions appropriate for users' needs, a part of the steam flow is destined to thermal utilisation (heat exchanger), while the remaining part continues its expansion in the low pressure turbine, reaching the condenser. The two flows rejoin in regenerator R.

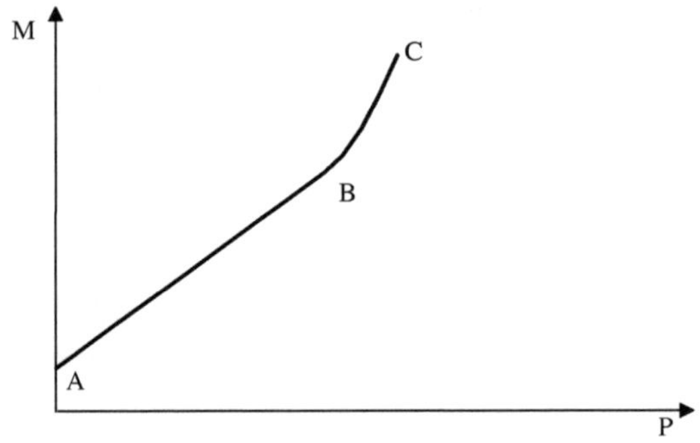


**Fig. 5.10** Willan's line for a steam turbine

Acting on the flow of the bled-steam, giving the same flow at the input of or acting on the flow of such a turbine giving the same bled-steam, the value of the electric index IE changes.

The maximum and minimum flows of the bled-steam (and, therefore IE values) are defined by the steam flow (both maximum and minimum) of the condenser, as well as by the maximum flow of the steam generator. The minimum steam flow to the condenser sets a lower threshold to the electric index at around 0.2. The maximum flow to steam condenser and generator impose a minimum value to the flow of the bled-steam, then a maximum value to the electric index. While developing the project it is evidently possible to envisage maximum flows to condenser and steam generator so as to allow the working of the plant without any bleeding: in this case, the electric index clearly amounts to one.

For the management of the plant, besides the project, it is necessary to know the correlation between the different variables: *mechanic power*, *thermal power*, *steam flow rate* in the turbine $T_1$ and *flow of bled-steam*. This correlation is given by the working diagram of the plant, obtained by tracing the two turbines of the Willan's lines for the different flows obtained as the spilled flow or the input flow of high pressure turbine change.

The Willan's line for a turbine is the diagram of flow $M$ absorbed according to the power $P$ supplied (Fig. 5.10): it is nearly linear from point $A$ of no-load running (to which corresponds a given steam flow $M_0$ necessary to keep the turbine rotating without supplying any power) up to point $B$ (project conditions at which there is the highest performance of the turbine), beyond which the slope increases up to point $C$ corresponding to maximum overload. For industrial applications, the linear Willan's line is adopted in the entire field of operation of the turbine.

As the spilled steam flow changes, the *degree of recovery* of these plants takes values ranging between 0 and 1.

#### 5.2.1.4 Cogeneration with Gas Plants

Systems based on gas turbine as a prime mover have the advantage of making independent, in a wide field of variation, production of electric power from thermal one. The latter takes place downstream in the machine at the expense of the heat of fuel gases that would be in any case released into the atmosphere, at temperatures of 400–500°C; the air excess of exhaust gases allows, in case heat is requested at higher temperatures, to operate an afterburning that increases the level of gas temperature.

The main characteristics of this type of plants, whose scheme in shown in Fig. 5.11, are the following:

- low installation costs;
- easy following of load variations;
- short assembling period;
- reduced maintenance costs;
- high-quality fuels required;
- heat available at a high temperature (the temperature of exhaust gases is nearly 500°C).

The heat obtained in the form of steam in the heat-recovery boiler placed downstream the turbogas, can be used for the direct distribution to technological users, for urban heating through a superheated steam/water exchanger, or it can be used to supply a steam turbine that operates an electric generator obtaining in this way the *gas-steam combined cycle*, that shows huge efficiency for both cogeneration and electric power production purposes (see next paragraph).

Considering a cogeneration efficiency of 85% and an electric performance of around 30%, the value of the electric index amounts to about 0.35.

#### 5.2.1.5 Cogeneration with Combined Plants

Combined plants are particularly suitable for the combined generation of electric power and heat. The combined cycle, in carrying out thermodynamic transformations far more efficient than the steam processes traditionally used for cogeneration purposes, is able to generate, given the same quantity of thermal energy, higher quantities of "fine" (electric) energy. Therefore, the efficiency values obtained are far higher than the case of cogeneration with a simple cycle, with considerable savings in terms of primary energy, even more than doubled compared to the steam solution. Another very important advantage offered by the combined cycle is that it keeps high "energy indexes" also in case of a very much variable thermal load: it is possible to use a plant destined to cogeneration for the production of electric power alone, continuing to save remarkable quantities of primary energy, whereas with simple cycles this saving depends on the use of heat.

The main characteristics include:

- high electric efficiency;
- huge working flexibility as concerns the thermal and electric power supplied;

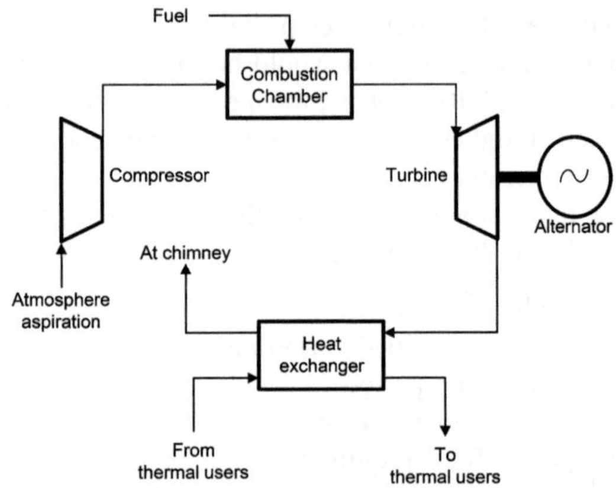


**Fig. 5.11** Scheme of a cogeneration plant with gas cycle

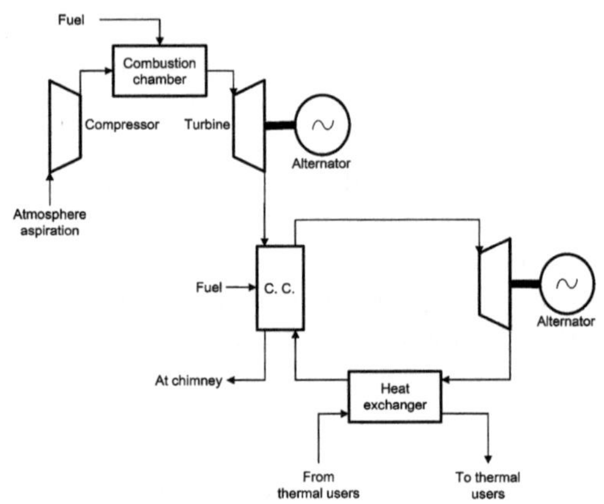


**Fig. 5.12** Scheme of a cogeneration plant with combined cycle

- large-size plants (usually exceeding 10 MW);
- expensive and complex plants.

Figure 5.12 shows a combined cycle cogeneration scheme. If we consider a cogeneration efficiency of 85% and an electric performance of nearly 50%, the value of the electric index amounts to nearly 0.58.

#### 5.2.1.6 Cogeneration with Stirling engines

Stirling machines are potentially very effective cogeneration systems. Their history, their potentialities and the reasons of their limited diffusion up today are referred in [3, 4].

The recent rebirth of such external combustion engines is particularly concentrated in micro-cogeneration applications [5].

## 5.2.2 Typical Indexes of Cogeneration Plants

The performance of cogeneration plants can be defined through some typical indexes that refer to the working of the system at the nominal project conditions: *efficiency of the first* and *second principle*, and *saving of primary energy*.

### 5.2.2.1 Efficiency of the First Principle

It is also called *fuel global utilisation index* and expresses the *ratio between the sum of the thermal and electric powers obtained and the power supplied to the system* by the primary energy source.

Indicating with $P_e$ the electric power, $P_t$ the thermal power and $P_c$ the power supplied to the system by the fuel, we have:

$$\eta_I = \frac{P_e + P_t}{P_c} \tag{5.19}$$

The efficiency of the first principle is nothing else but the already defined *cogeneration efficiency*. This parameter equally measures the thermal and electric power supplied (from which the definition of efficiency of the first principle), and does not consider the level of temperature at which the thermal power is obtained. As shown above, its value is substantially the same for all the cogeneration technologies.

### 5.2.2.2 Efficiency of the Second Principle

It is also called *fuel qualified utilisation index* and expresses the *ratio between the sum of the powers supplied*, where thermal power is translated into conventional mechanical power through the Carnot efficiency, *and the power supplied to the system by the primary energy source*.

Please note that this parameter is more significant than the previous one, in converting the thermal power into conventional mechanical power, differently weights the powers supplied and also considers the temperature at which heat is obtained as a qualifying element (from which the definition of efficiency of the second principle).

If we consider $P_{t,T}$ the thermal power supplied at temperature $T$ and $T_a$ the reference room temperature, the corresponding mechanical (or electric) conventional power $P_{e,T}$ is:

$$P_{e,T} = P_{t,T}\left(1 - \frac{T_a}{T}\right) \tag{5.20}$$

noting that if the temperature at which heat is produced is variable, $T$ is a suitable average value. The equation of the efficiency of the second principle $\eta_{II}$ therefore is:

$$\eta_{\mathrm{II}} = \frac{P_{\mathrm{e}} + P_{\mathrm{e,T}}}{P_{\mathrm{c}}} \tag{5.21}$$

Considering (5.20) and (5.21) can be written as follows:

$$\eta_{\mathrm{II}} = \frac{P_{\mathrm{e}} + P_{\mathrm{t}} - P_{\mathrm{t}}\frac{T_{\mathrm{a}}}{T}}{P_{\mathrm{c}}} = \eta_{\mathrm{I}} - \frac{P_{\mathrm{t}}\,T_{\mathrm{a}}}{P_{\mathrm{c}}\,T} \tag{5.22}$$

Indicating with $P_{\mathrm{tot}}$ the total power (electric + thermal) produced, (5.22) can be written as follows:

$$\eta_{\mathrm{II}} = \eta_{\mathrm{I}} - \frac{P_{\mathrm{tot}} - P_{\mathrm{e}}}{P_{\mathrm{c}}} \cdot \frac{T_{\mathrm{a}}}{T} \tag{5.23}$$

Bearing in mind (5.4) and (5.23) becomes:

$$\eta_{\mathrm{II}} = \eta_{\mathrm{I}} - \eta_{\mathrm{I}}(1 - IE) \cdot \frac{T_{\mathrm{a}}}{T} = \eta_{\mathrm{I}}\left(1 - \frac{T_{\mathrm{a}}}{T}\right) - \eta_{\mathrm{I}} \cdot IE \cdot \frac{T_{\mathrm{a}}}{T} \tag{5.24}$$

(5.24) can be written as follows:

$$\eta_{\mathrm{II}} = \eta_{\mathrm{I}}\left[1 - \frac{T_{\mathrm{a}}}{T}(1 + \mathrm{IE})\right] \tag{5.25}$$

From (5.25) it is possible to infer that *the efficiency of the second principle is always lower than the one of the first principle*. Furthermore, given the same efficiency of the first principle, the efficiency of the second principle increases as the temperature $T$ at which heat is available increases, and as the electric efficiency increases. Therefore, whereas—as already recalled—the efficiency of the first principle is substantially independent from technology, the efficiency of the second principle markedly varies according to the specific characteristics of the technology under consideration, and is high in those cases in which heat is available at high temperatures and/or in case of high electric indexes (namely of high electric efficiencies), as for instance high temperature fuel cells.

#### 5.2.2.3 Primary Energy Saving

The Primary Energy Saving index RI is defined as the *ratio between the consumption of primary energy of the cogeneration process $P_c$ and the consumption of a conventional process $P_c$ that separately produces the same quantities of thermal $P_t$ and electric $P_e$ energy*. Indicating with $\eta_{\mathrm{t}}$ the production efficiency of separate thermal energy, $\eta_{\mathrm{e}}$ the production efficiency of separate electric energy, it follows that:

$$\mathrm{RI} = \frac{P_{\mathrm{c}}}{\frac{P_{\mathrm{m}}}{\eta_{\mathrm{e}}} + \frac{P_{\mathrm{t}}}{\eta_{\mathrm{t}}}} \tag{5.26}$$

This parameter highlights the saving that can be achieved through the cogeneration process, although it does not take into account the level of temperature at which heat is supplied and produced.

## 5.2.3 District Heating

The use of cogeneration in the civil sector mainly consists *of using the heat* produced in *cogeneration plants for the heating and the production of sanitary hot water for domestic use.*

As far as the efficiency of this solution is concerned, some considerations are necessary. Besides the evident advantages entailed by the combined production of electric power and heat which are also highlighted by the parameters defined above, it is useful to examine the efficiency of traditional heat production systems for heating purposes: combustion of fuels in boilers, mostly gasoil and methane. ***Very often efficiency is only considered according to the first law of thermodynamics, that is to say energy conservation***. This means considering as cause of inefficiency only the energy losses in the passage from the combustion to the utilisation phase, that is a problem of combustion efficiency and thermal insulation. Clearly, obtaining good combustion efficiency values and reducing as much as possible heat losses entails primary energy saving but, even accepting heating systems that have an efficiency of the first principle close to 100%, there is in any case energy waste that cannot be avoided, since it is due to the fact that the chemical potential of the fuel (gasoil or methane) is transformed into heat at 40–50°C, as requested by the household comfort.

We better consider also the degradation of the energy level from high temperatures, beyond 1,000°C, supplied by through fossil combustion, to the end use temperatures. This degradation can be identified by considering the *exergy* that, according to the second law of thermodynamics, sets some "quality criteria" of energy. These criteria are the following:

- Heat and work **are not** equivalent forms of energy, since work can be entirely transformed into heat, whereas heat can be transformed into work only in a maximum percentage equal to *Carnot efficiency*. **Therefore, work is a form of energy more "valuable" than heat**;
- Thermal energies quantitatively equal, available at different temperatures, are not equivalent: the one at a higher temperature is more "valuable".

Therefore, while considering the second item, it is possible to understand that the drop of nearly 1,000°C, typical of residential boilers for the production of hot water, originates a thermodynamic "waste" that does not depend on the quality of the system.

### 5.2.3.1 Benefits Obtainable Through District Heating

After making these considerations of merely thermodynamic nature, the advantages of district heating—in particular the primary energy savings—are now taken into consideration.

**Fig. 5.13** An example of thermal needs duration curve

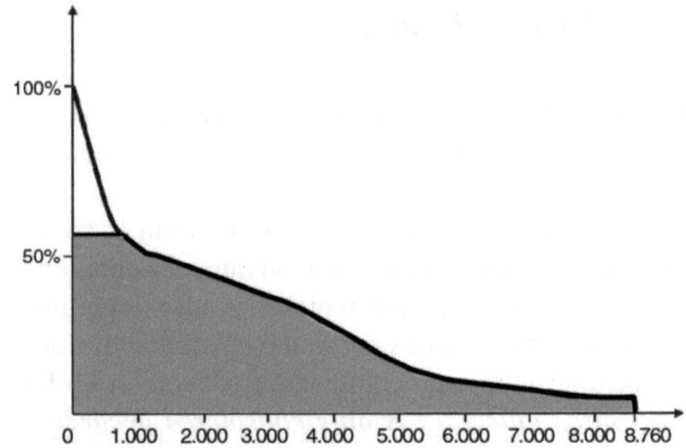


Heat, produced jointly with electric power in cogeneration plants, is distributed to residential users through a distribution network. The energy vector is hot water, usually at a temperature of about 120°C.

The costs, sometimes considerable, of the distribution network and the heat losses in the network itself (5–10%) are offset by the following advantages:

- *fuel saving* by nearly 1/3 compared to the two separate productions, due to the cogeneration and the use of higher efficiency boilers compared to the ones used in buildings;
- *environmental benefits*, since in plants it is possible to obtain technically better combustions with pollutant reduction systems and with higher controls in operations and maintenance;
- *elimination of management and maintenance problems* for users.

Overall, with cogeneration and district heating the costs due to the distance between production and consumption are acceptable, in view of a higher efficiency, allowing energy saving.

For a correct selection of parameters in sizing a cogeneration plant, it is necessary to consider the "*duration curve of thermal demand*". As an example, Fig. 5.13 shows this curve as relating to a typical city of Northern Italy.[9] A peak of thermal power requested for a limited duration, followed by a medium thermal power area relating to heating basic needs, and a final part of low thermal power corresponding to the need for hot water (sanitation use).

The optimisation of the exploitation of cogeneration plants suggests their sizing for a power accounting for 50–60% of the maximum peak. Management tricks (use of the accumulation heat) and the form itself of the duration diagram allow this sizing in terms of power that corresponds to a heat supply to the network whose 85–90% comes from cogeneration. The remaining 10–15% is supplied by integration boilers suitably divided by number and size, in order to coherently

[9] The curve of durations of thermal need for remote heating is clearly linked to the climate characteristics of the area.

meet the integration and reserve functions. These plants, in the implementation by subsequent stages of a general district heating plan, meet the entire heat needs during users' integration transition phase.

After these preliminary remarks, if $F_t$ is the thermal need and $\eta_{CC}$ the average efficiency of building or family boilers of the traditional heating system, the consumption of $E_f$ fuel without cogeneration is:

$$E_f = \frac{F_t}{\eta_{CC}} \tag{5.27}$$

To calculate the consumption in case of cogeneration, it is necessary to consider the technology used and make some general considerations on the entire energy cycle in order to evaluate general and global benefits (primary energy consumption). As an example, let's consider steam (backpressure) and gas cogeneration plants, powered by fossil fuels. In order to assess consumption required to meet thermal needs in case of cogeneration, the following criterion can be applied: considering that a cogeneration plant produces electric power and heat, after calculating the fuel consumed by the cogeneration plant for the production of the heat requested and the corresponding electric power produced, the production of heat requested is attributed to the difference between this consumption and the one that would apply by taking the same quantity of electric power from the network.

As for the consumptions due to the intake of electric power from the network, it is necessary to consider first of all the losses due to the transportation of the energy itself $p_T$ (expressed as a percentage of the energy transported); indicating with $E_p$ the electric power produced and with $E_u$ the one available to users also considering losses, we come to:

$$E_u = E_p - E_p \cdot p_T = E_p \cdot (1 - p_T) = E_p \cdot \eta_{TR} \tag{5.28}$$

where the *transportation efficiency* $\eta_{TR}$ is a complement to 1 of the losses. After calculating $E_p$ it is necessary to consider the percentage produced by non renewable sources ($E_{NR}$) on the basis of the percentage of electric power production $I_{NR}$ of the national electric endowment:

$$E_{NR} = E_p \cdot I_{NR} = E_u \frac{I_{NR}}{\eta_{TR}} \tag{5.29}$$

Indicating with $\eta_m$ the *average efficiency of the production endowment from fossil fuels* (or, more generally, non-renewable sources) at a national level, indicating with $E_{PNR}$ the consumption of non renewable primary energy, it follows[10]:

$$E_{PNR} = \frac{E_{NR}}{\eta_m} = E_u \cdot \frac{I_{NR}}{\eta_{TR} \cdot \eta_m} \tag{5.30}$$

[10] It is also necessary to consider energy consumption due to the extraction, the processing and transportation to the plants of the fuels used: in an approximation analysis, they can be neglected.

Let's now define a fictitious global electric efficiency, geg that besides production and transportation, takes into consideration also the percentage of electric power produced from renewable sources, as follows:

$$\eta_{eg} = \frac{\eta_{TR} \cdot \eta_{m}}{I_{NR}} \tag{5.31}$$

It is evident that such "efficiency" increases as the efficiencies of plants and transportations increase, so as the percentage of electric power produced by renewable sources.

In consideration of (5.31) and (5.30) becomes:

$$E_{PNR} = \frac{E_{u}}{\eta_{eg}} \tag{5.32}$$

Now it is necessary to calculate fuel consumption to cover the thermal needs under consideration through district heating, as well as the corresponding quantity of electric power produced by the cogeneration plant. First of all, it is necessary to consider the efficiency of heat distribution of the district heating network $\eta_{DT}$: the thermal energy $E_{T}$ to be produced by cogeneration plant to cover thermal needs is:

$$E_{T} = \frac{F_{t}}{\eta_{DT}} \tag{5.33}$$

On the basis of the assumptions dealing to the thermal needs duration curve not all the thermal need is covered by district heating. Indicating with $I_{QT}$ the percentage of thermal needs covered by district heating, with $I_{QI} = 1 - I_{QT}$ the one relating to integration boilers, with $\eta_{CI}$ their efficiency and with $\eta_{t}$ the thermal efficiency of the cogeneration plant, it is possible to state that fuel consumption of the cogeneration plant $E_{fc}$[11] is:

$$E_{fc} = \frac{E_{T}QT}{\eta_{t}} + \frac{E_{T} \cdot I_{QI}}{\eta_{CI}} = \frac{\frac{F_{t} \cdot I_{QT}}{\eta_{t}} + \frac{F_{t} \cdot I_{QI}}{\eta_{CI}}}{\eta_{DT}} \tag{5.34}$$

(5.34) can be written by expressing the thermal efficiency according to the first principle and the electric efficiency, as follows:

$$E_{fc} = \frac{F_{t}}{\eta_{DT}}\left(\frac{I_{QT}}{\eta_{t}} + \frac{I_{QI}}{\eta_{CI}}\right) = \frac{F_{t}}{\eta_{DT}}\left(\frac{I_{QT}}{\eta_{I} - \eta_{e}} + \frac{I_{QI}}{\eta_{CI}}\right) \tag{5.35}$$

[11] In this case, the use of fossil fuels is hypothesised both for the cogeneration plant and for the boilers. In general, the consumption of non-renewal primary energy consumption must be considered.

Recalling the definition of electric index $I_e$, and (5.33), the corresponding electric power produced by the cogeneration plant (and therefore available to users) is:

$$E_u = \frac{F_t \cdot I_{QT} \cdot I_e}{\eta_{DT} \cdot (1 - I_e)} \tag{5.36}$$

By replacing (5.36) in (5.32) we obtain:

$$E_{PNR} = \frac{F_t \cdot I_{QT} \cdot IE}{\eta_{DT} \cdot \eta_{eg}(1 - IE)} \tag{5.37}$$

Finally the total consumption for heat production through cogeneration; $E_{c\ tot}$ is:

$$E_{c\ tot} = E_{fc} - E_{PNR} = \frac{F_t}{\eta_{DT}}\left[\frac{I_{QT}}{\eta_I - \eta_e} + \frac{I_{QI}}{\eta_{CI}} - \frac{I_{QT} \cdot Ie}{\eta_{eg}(1 - Ie)}\right] \tag{5.38}$$

The ratio $R$ between the consumption to cover the thermal need without cogeneration and the one with district heating, from (5.38) and (5.27) is:

$$R = \frac{\eta_{DT}}{\eta_{CC}\left[\frac{I_{QT}}{\eta_I - \eta_e} + \frac{I_{QI}}{\eta_{CI}} - \frac{I_{QT} \cdot IE}{\eta_{eg}(1 - IE)}\right]} \tag{5.39}$$

Equation 5.39 shows that the ratio $R$ depends on the characteristics of the cogeneration plant ($\eta_I$, $\eta_e$ and $I_e$), of district heating network ($\eta_{DT}$), on the technology of the plant ($I_{QI}$ and $I_{QT}$), of the traditional heat production systems that is being replaced ($\eta_{CC}$), finally on the characteristics of the national energy system ($\eta_{eg}$).

As an example, let's consider a district heating system with a gas plant ($\eta_e = 30\%$, $\eta_I = 80\%$; therefore $I_e = 0.375$). Then it is possible to assume $I_{QT} = 0.85$, $I_{QI} = 0.15$ and $\eta_{CI} = 90\%$. Furthermore, the heat losses in the district heating network are considered as accounting for 8%, that is to say $\eta_{DT} = 92\%$, while for integration boilers $\eta_m = 90\%$. As concerns the national energy system, by considering the case of Italy, let's assume $\eta_m = 39\%$, the average distribution losses accounting for 5% (therefore $\eta_{TR} = 95\%$) and $I_{NR} = 80\%$ [1]. On the basis of this data we calculate $\eta_{eg} = 46.3\%$. Finally, taken $\eta_{CC} = 80\%$, one comes to $R = 1.72$.

The effectiveness of this solution is linked to the "duration" of heat demand for heating purposes[12]; in fact, during the periods in which there is no heating request, the thermal demand is limited to sanitary hot water needs. Sizing the plant giving priority to heating needs, the plant—as far as heat production is concerned—would

[12] Therefore this solution is particularly suitable for cold climate.

be practically unused.[13] Furthermore, from the economic viewpoint, the use of heat for short periods of time prevents the amortisation of the relevant costs of the heat distribution network. In this connection, it can be the technology of absorption heat pumps look attractive, which use heat for the air conditioning (both winter and summer). In this way, there is a heat demand for long periods also in mild climates in which the request would be limited to a few months for year in winter time.

Another consideration relates to the technologies that can be used: on the basis of the above, it is evident that district heating is energy efficient with the use of mature and tested technologies presently competitive as gas turbine plants. The use of innovative technologies, such as high temperature fuel cells powered through syngas produced locally by the gasification of biomasses (hence from a renewable source) would make this solution even more interesting from the energy and environmental viewpoint. From an economic perspective, however, these technologies are not yet competitive.

In general, as concerns the cogeneration technologies to be adopted for district heating purposes, it is interesting to note the modularity of the entire district heating system (heat distribution network, integration boilers and substations) against the technology used. In other words, it is possible to plan and implement a district heating system by immediately using the most competitive technologies presently available (steam and gas plants), moving on in the future to new and more efficient technologies that avail themselves of renewable sources, by using the same district heating network.

#### 5.2.3.2 Cogeneration in the Industrial Sector

The self-production of electric power by industries–a rather widespread practice nowadays, in stations located near production plants–allows the use of cogeneration in the industrial sector. Although plants devoted to self-production only of electric power exist in large industries, in the vast majority of cases cogeneration plants are used that allow, where the suitable conditions exist, to obtain high overall energy efficiency with subsequent saving of primary energy. Obviously the feasibility and the convenience of cogeneration plants, are reached with a simultaneous demand of electric power and heat. There are many industrial sectors in which this condition takes place, including the following industries:

- extraction;
- iron-and-steel;
- metallurgic;
- mechanical;
- chemical and electrochemical;

[13] Regulations impose minimum limits to the ratio between the heat actually used and the one produced by the plant to "certify" the plant as a cogeneration plant. To comply with these limits, when there is no heat demand for long periods of time, the plant must be stopped.

- oil;
- textile;
- food;
- paper;
- ceramics and glass;
- brick.

For the different industrial sectors, the characteristics of the simultaneous demand for electric power and heat vary on a case-by-case basis: in particular, the quantities of electric and thermal power demanded vary, and consequently the electric power/thermal power ratio, as well as the temperature at which heat is demanded. However, this does not endanger the convenience of cogeneration.

As concerns the variations in the quantity of electric and thermal power, that is to say the electric power/thermal power ratio, it is possible first of all to chose the type of plant that is most suitable for specific users. In general steam plants are suitable for huge power and low electric power/thermal power ratios. In particular, the backpressure plants supply electric power and heat always with the same predetermined ratio, whereas the controlled bled-steam ones allow to largely vary this ratio within given limits. Gas plants are suitable for lower powers and higher electric power/thermal power ratios. Highly important is also the connection of the station to the electric power distribution public network, with the possibility of *purchasing* electric power from the network in the periods of highest consumption and to *sell* to the network itself the electric power, when in excess, at economically convenient conditions. This allows to separate—within given limits—the production of electric power from its consumption and therefore to size and manage the plant with particular reference to the thermal demand having considered that the separate heat production is the most penalising from the energy previously discussed.

As concerns the efficiency viewpoint, at different values of temperature at which heat is demanded in different specific cases, it is possible to meet this need within given limits according to the type of plant being considered. In steam cogeneration plants, for example, both backpressure and controlled bled-steam, it is possible to select in the project design phase the temperature at which the thermal energy is intended to be available, setting the steam conditions at which expansion stops in the former case, or in which spilling is made in the latter one. However, it is necessary to bear in mind that by increasing the temperature at which heat is "recovered", the electric energy production efficiency drastically decreases; as a matter of fact, this limits the arbitrariness in the choice of temperature for heat utilisation purposes.

As we said the maximum temperature of a steam cycle is around 550°C, thus when heat is demanded at very high temperatures, the use of gas systems is necessary in which the temperature of exhaust gases is 400–500°C. Furthermore, as already said, the strong air excess of exhaust gases allows to perform post-combustion with the possibility of obtaining very high temperatures (over 2,000°C). It is clear that, in the case of post-combustion, the heat sent to thermal users is not entirely "recovered", but it is partly produced ad hoc. This implies a decrease in the global efficiency of the cycle as much as higher than the temperature at which heat is demanded (more precisely the difference between the temperature at which heat is requested and the one of exhaust gases). In any case, it is possible to obtain, in this instance as well, a saving of primary energy compared to the case of combustion of air taken from the atmosphere, precisely amounting to the one corresponding to the quantity of heat necessary to bring air—through combustion—from room temperature to the exhaustion temperature of the turbine.

It is interesting to observe how, sometimes, in industrial cogeneration, more generally in industrial energy self-production it is possible to have a further possibility of saving primary energy, namely fuel, burning the residues of the production process; this solution appears particularly interesting in the agricultural sector, despite the low incidence of energy consumption in this sector out of the total, since many remains of agricultural activities can easily be used as fuels.

In order to build these plants, the technology of fluidised-bed combustion chambers is fundamental, since it allow to burn, with very good results both in terms of efficiency and harmful emissions, materials with poor chemical physical characteristics and potentially very much polluting.

## References

1. Ackermann T, Andersson G, Lennart L (2001) Distributed generayion: a definition, Electric Power Systems Research 57 (2001) Elsevier. Also: Sharma D, Bartels R, (1998) Distributed electricity generation in competitive energy markets, The energy Journal, Special issue-Distributed resources: towards a new paradigm of electricity business. The International Association for Energy Economics
2. European Smart Grids Technology Platform (2006) Vision and strategy for Europe's electricity networks of the future—directorate general for research sustainable energy system C.E.C.
3. Walker G (1980) Stirling Engines, Oxford Press, Oxford.
4. Naso V (1991) La macchina di Stirling. In: Masson (Ed) Editoriale ESA, Milano, also in English: (1995) The Stirling Machines, Editoriale ESA, Rome
5. Proceedings of the International Stirling Engine Conferences (2009) Groningen, The Netherlands, http://www.stirlinginternational.org

# Chapter 6
# Energy Useful Effect and End Use

## 6.1 Introduction

The exploitation of energy sources and the entire pathway of energy vectors in time and space have a main purpose: to meet the needs of human beings in order to make their lives more comfortable, more productive and socially active. Apparently human beings do not directly need any energy sources or vectors additional to food. As a matter of fact, they have concrete and easily identifiable needs for better living, and energy is aimed at meeting these needs.

As already highlighted in Chap. 1, the human needs that can be met, thanks to the implementation of a suitable energy flow are grouped under the definition of *useful effects*. As said, human beings need to warm or to cool the rooms in which they live, work or carry out their social activities; they need to operate their vehicles, their household appliances, to move quickly and safely. Furthermore, human beings need to light the rooms in which they live, irrespectively from the availability of natural light, in order to free themselves from the rhythm of the alternation of day and night and seasons. They also need to operate their information technology (IT) and telecommunications equipment, their small and large monitors and displays.

The useful effects from an energy perspective are represented by the reaching and keeping of the desired comfort (temperature and humidity) in our rooms, by the useful result deriving from the operation of machines, household equipment and means of transport, by the correct level of lighting of indoor or open-air places, and the processing, sending or receiving of data thanks to their electronic devices.

From the energy viewpoint, in the end use—the one that directly produces the desired useful effect–men need *mechanic energy*, *thermal energy*, *luminous energy*, and *electric power*—in the last few years, the most required. The entire world energy system must therefore be finalised to the availability and accessibility for each end user, namely for each man living on Earth, of these forms of energy in the time, space, quantity and quality desired.

F. Orecchini and V. Naso, *Energy Systems in the Era of Energy Vectors*,
Green Energy and Technology, DOI: 10.1007/978-0-85729-244-5_6,

**Table 6.1** Energy useful effects and end use

| Useful effect | Energy end use |
|---|---|
| Rotation of the drum of a washing-machine | Mechanic |
| Heating of a room | Thermal |
| Lighting of a hall | Luminous |
| Operation of a computer | Electric/electronic |

The point of contact between energy vectors and useful effects is therefore fundamental for the implementation of the entire energy system. In fact, any system is aimed to the realisation of useful effects, attainable thanks to one of the four energy end uses mentioned above. The correct knowledge of useful effects, and from an energy viewpoint especially of the energy end uses in which they can be translated, allows to identify the appropriate energy vector able to meet such human need, within the level of technological development of the solutions available.

This chapter, therefore, contains an analysis of the energy end uses that is the energy summary of all human needs:

- mechanical:
- thermal;
- luminous;
- electronic/electrical.

These four groups of end uses allow to achieve all the possible useful effects. In fact, *useful effect* is defined as the real and concrete need met, such as the heating or cooling of a room, the lighting of a hall, the rotation of the drum of a washing-machine or of the wheels of a vehicle and the activation of the microchips of a computer, of liquid crystals, or of the light emitting diode (LED) of a video screen.

Therefore the energy end use indicates the form in which energy is used by men at the point of need (Table 6.1).

*Electric/electronic use* means that the form in which energy is used to achieve the useful effect, i.e. electric power, as it happens in computers, printers, faxes, TV sets, etc.

In many other cases, on the contrary, although the equipment placed at the end of the energy chain under consideration is supplied by electric power, it is not the energy form that leads directly to the useful effect. Very often, in fact, electricity is converted in a more or less distant point of the energy chain into another form of energy: mechanic, thermal or luminous. For instance, thinking about home applications, many household appliances convert the electric energy with which they are powered into mechanic energy (vacuum cleaners, washing-machines, etc.). The same happens for the equipment destined to industrial production: in these cases, the form of energy that must be identified with the useful effect is *mechanic energy, not electric energy*. Similarly, many types of equipment (electric fires, ovens for both domestic and industrial use and many others), convert the electric power with which they are powered into thermal energy through the Joule effect. In these cases, the energy end use is the *thermal one*.

As for mechanic use, in the industrial and agricultural sector this form of energy is used in productive processes (process work) and for the handling of goods; in the civil sector, mechanical energy is often used in household appliances. A particularly important sector for the use of mechanical energy is ground transportation of goods and people.

## 6.2 Thermal End Use

### *6.2.1 Overview*

Some exemples of energy thermal end use are the heating of rooms, the supply and implementation of the thermal conditions necessary to carry out several industrial production processes, in addition to cooking and to the production of domestic hot water.

These applications have very different needs in terms of characteristics of the energy flow (the temperature requested in particular) and this determines the choice of the single technologies used in the various cases. The heat requested is presently produced in industrialised countries mainly through the combustion of fossil fuels at the point of need, or by deriving it from the distribution grid of the electricity vector (heat pumps, boilers, electric heaters and electric ovens), and more rarely by using the renewable fuels or solar energy locally available at the point of used.

The thermal uses can be divided as follows:

- *Industrial, agricultural and food-industry use*, including, among the most important ones one finds:
  - Process heat
  - Air-conditioning of greenhouses–Cooking
- *Civil and residential use*, including:
  - Room air-conditioning
  - Production of domestic hot water
  - Cooking

#### 6.2.1.1 Industrial Use

Industrial use means *the use of heat in production processes*, whereas applications for room air-conditioning and for the production of hot sanitary water (also present in the industrial framework) are considered as a civil use. There are many industries in which production processes demand thermal energy; among them:

- extraction
- steel
- metallurgic
- mechanic
- chemical and electrochemical
- oil
- textile
- food
- paper
- ceramic and glass
- brick
- other

The use of heat in production processes is mainly at high temperatures, and the specific characteristics of the heat flow demanded, therefore the usable heat-transfer flow, depend on the specific productive process. In these applications, heat is traditionally produced by the combustion of fuels in boilers.

The analysis of the global energy needs of an industrial site, or of the industrial park in which it is included, shows how to obtain the desired heat flows by strongly improving the energy efficiency of the system thanks to the cogeneration of electricity and heat. This evidence is making industrial cogeneration more and more widespread.

Less widespread, although equally useful, are renewable energy sources and theire conversion technologies. This is evident while moving from a standard efficiency analysis to a more advanced analysis of system efficacy, which also takes into consideration the parameters as the renewability of sources, the decrease in the quantity of emissions produced and the local availability of energy resources.

The two energy vectors mainly requested by the industrial system, that is *heat-transfer fluid* and *electricity*, are both directly producible from the conversion technologies of renewable sources that are almost always available, such as solar radiation, wind energy, hydroelectric energy, biomass and geothermal energy. And perspectively also from cogeneration plants based on the technology of high temperature fuel cells, powered by syngas rich in hydrogen that can be equally produced locally and from renewable sources or distributed with the same procedures as methane or other fuel gases of fossil origin already widespread for energy use.

#### 6.2.1.2 Civil Use

As already mentioned, *thermal energy* in the civil sector is mainly *used for cooking food*, for the *production of hot sanitary water* and for *room air-conditioning*.

Production of Domestic Hot Water

The production of domestic hot water is a *low-temperature application* (which makes the use of solar energy particularly efficient through simple, cost-effective and reliable solar flat-plate collectors); in fact the temperature of use is normally around 40–45°C and usually does not exceed 60°C.

As concerns the temperature of the water system, it obviously depends on the season and the site, and on average is 5–10°C; the temperature increase demand, therefore, varies from 30 to 40°C.

The thermal energy $Q$ necessary to heat a water mass $m$ is:

$$Q = m \cdot c \cdot (T_\mathrm{f} - T_\mathrm{i}) = m \cdot c \cdot \Delta T \tag{6.1}$$

indicating with $c$ the water specific heat, $T_\mathrm{f}$ and $T_\mathrm{i}$ the final and initial temperatures of water. Having considered water specific heat amounts as 4.186 kJ/kg, and considering $\Delta T = 35$°C, the *unitary heat necessary for the production of domestic hot water* is 146.5 kJ/kg.

The daily need for domestic hot water depends on the type of users (such as households, hospitals and offices): based on the average data available in literature, this need ranges between 50 and 200 l/person/day. Tipically the need is of 60 l/person/day and $\Delta T = 35$°C, then *daily energy need relating to the production for one person amounts* to about 8,790 kJ.

The use of heat in the civil sector is typically at low-temperature, with predictable characteristics of the heat flow requested, then the usable heat-transfer fluid is practically always water. In these applications, heat is produced in many countries by electric boilers in single apartments, or by the combustion of fossil fuels (gasoil and methane, more rarely coal) in traditional boilers, directly at the point of need (as district heating in case of compounds or block of flats, but also entirely autonomously for single flats). In other countries, whose urban areas developed simultaneously with the electricity production system or with the industrial system, heat is supplied by cogeneration plants located even at tens of kilometres from the point of use (district heating).

A standard analysis of global energy needs of each single building, or of the residential district in which a building is located, shows how it is possible to obtain the desired heat flows by improving the energy efficiency of the system thanks to the cogeneration of electricity and heat.

Moving from considerations linked to the pre-determination of fossil fuels, electricity from the electricity grid and (where possible) the district heating heat-transfer fluid as the only alternatives available, to more careful considerations of energy analysis, it is then clear in this case as well how many more attractive alternatives are available. A system efficacy analysis, which also takes into consideration the parameters of renewability of the sources, and of lowering of emissions, as well as the local availability of energy resources, certainly leads to resort to renewable sources and to the relevant conversion technologies.

The energy vector requested by civil thermal needs–that is to say a largely available and well-identified heat-transfer fluid as water, mostly at relatively low-temperature (from 40 to 120°C, according to the applications and to the resort to local distribution grids for district heating)–can be produced directly or in cogeneration also by renewable energy conversion technologies as solar radiation, biomass and geothermal energy. Also small residential cogeneration plants can be used, based on low-temperature fuel cell technology, and powered by hydrogen that can be produced also locally from the mentioned renewable energies but also by wind energy and hydro power.

### Room Air-Conditioning

In this type of application, *the useful effect is the thermal comfort of the rooms* in which human activities are carried out. Therefore, we better first define room comfort.

*Room comfort*: an univocal definition of room wellbeing is very difficult, since the concept of wellbeing itself is part of the perception, intrinsically subjective although influenced by objective and measurable magnitudes as temperature and level of humidity. The American Society of Heating, Refrigeration and Air Conditioning Engineers (ASHRAE) defines thermal comfort as *the condition in which men express their satisfaction for the surrounding environment*. It is evident how such definition, although simple and clear, is not objective at all: it stresses, on the contrary, that the comfort is a *personal feeling* and that therefore there is no condition identifiable through objective parameters (temperature, humidity, etc.) that might be rigorously defined as optimal for all from the thermal comfort viewpoint.

In the same environmental conditions, different people can express different opinions on their level of wellbeing. More frequently, however, that are some ranges of climatic conditions in which anybody, or almost anybody, finds his optimal comfort conditions.

The magnitudes that mostly influence these conditions are *temperature* and *relative humidity*. One of the mostly accredited methods for the identification of room comfort conditions is the Fanger method [1].

The human body exchanges heat with the surrounding environment by convection, radiation and evaporation (conduction is little significant); the factors that influence this exchange are the following:

- Environmental factors
  - room temperature[1]
  - radiant average temperature[2]

---

[1] Average temperature measured close to individuals.

[2] It is given by the weighted average of the temperature of walls and of the objects present in a room, where weight is the surface of walls and of objects.

- air temperature
- relative humidity

- Personal factors
  - metabolic activity[3]
  - thermal insulation of clothes

- Physiological factors
  - skin temperature
  - fraction of body area covered with dresses

According to the Fanger model, the feeling of hot, neutral and cold is linked to the *energy balance of the human body* that is, to say the difference between the heat produced by the body metabolism and the heat exchanged by the body with the surrounding environment.

*Three objective conditions* necessary for the achievement of the thermal comfort can be identified:

1. in the human body there shall be neither accumulation nor release of heat;
2. the average skin temperature must be as close as possible to 35.7°C;
3. the thermal exchange through sweating must be limited.

Through the study of significant samples of people it was possible to find a correlation between the predicted mean vote (PMV) and the predicted percentage of dissatisfied (PPD). This correlation (ISO 7730) considers room conditions as acceptable when the PMV is included between –0.5 and +0.5 that is, to say when the percentage of dissatisfaction is lower than 10%.

The ISO regulation 7730 describes the room conditions guaranteeing the thermal comfort:

- The difference between the radiant temperature of the different vertical surfaces must be below 10°C;
- air temperature and speed must be:
  - *winter*

temperature: between 20 and 24°C
speed: below 0.15 m/s

- *summer*

temperature: between 23 and 26°C
speed: below 0.25 m/s

- temperature difference between the heights of 0.1 and 1.1 m from the floor must be below 3°C

[3] It is linked to the physical activity, to motion and to other individual factors.

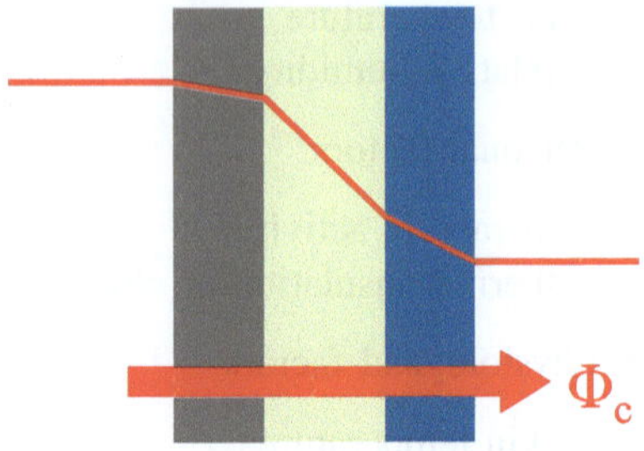


**Fig. 6.1** Heat transfer through a wall

- the floor temperature must range between 19 and 26°C, exception made for radiant floors for which it shall not exceed 29°C.

*Calculation of the energy needs for room comfort*: the thermal energy necessary to keep in the surrounding environment the conditions requested for room comfort is calculated on the basis of the following contributions:

- heat transmission through walls, windows, doors, ceiling and floor;
- energy necessary for the treatment of external air;
- thermal contributions to solar radiation;
- thermal contribution of people and equipment.

### Heat Transmission

Dispersions $\Phi_c$ through walls, ceiling and frames (Fig. 6.1) depend on the difference of temperature ($\Delta T$), the extension of the dispersing surface ($S$) and its thermal characteristics that is, to say the thermal transmittance $k$ (W/m$^2$ K) through the ratio:

$$\Phi_c = S \cdot k \cdot \Delta T \tag{6.2}$$

### Direct Solar Radiation

The contribution of direct solar radiation is calculated according to the maximum thermal flow $I_f$ and to the surface of the glass wall $S_f$. The value obtained must be corrected using of coefficients that depend on room conditions ($\delta_1$), on the characteristics of the glass wall ($\delta_2$) and on the thermal accumulation of facilities ($\delta_3$). The thermal flow $\Phi_i$ for a general section $i$ is then given by the formula:

$$\Phi_i = \delta_1 \cdot \delta_2 \cdot \delta_3 \cdot I_f \cdot S_f \tag{6.3}$$

### Thermal Power Produced in a Room

They concern people who reside in the room to be air-conditioned and all the other sources therein present. An important contribution is given by luminous sources; other sources include machine tools, office machines (photocopiers, printers, PC, etc.). Indicating with $n$ the number of people present in the room under consideration, $\Phi_s$ the thermal flow generated by each individual and $\Phi_i$ the flow of an equipment present, the total flow $\Phi_a$ of the thermal powers produced in the room is given by:

$$\Phi_a = n \cdot \Phi_s + \sum \Phi_i \tag{6.4}$$

### Energy Necessary for the Treatment of External Air

In order to keep the room comfort, it is necessary to continuously renew the air in the rooms, generally with external air, which has to be brought to the temperature and humidity appropriate conditions, with subsequent energy spending.

Energy expenditure can be read in the *psychrometric diagram*: such energy expenditure is partly due to temperature variation (sensible heat) and partly to the variation of relative humidity (latent heat).

As concerns sensible heat, indicating with $\Delta T$ the difference in temperature, with $c_a$ the specific heat of the air (1 kJ/kg) and with $\rho_a$ air density (1.2 kg/m$^3$), with $\dot{m}_a$ the mass flow (kg/s) and with $\dot{V}_a$ the volume flow (in m$^3$/s), the flow of sensible heat is given by:

$$\Phi_s = \dot{m}_a \cdot c_a \cdot \Delta T = \dot{V}_a \cdot \rho_a \cdot c_a \cdot \Delta T \tag{6.5}$$

The range of air renewal is generally expressed as the number of air changes in 1 h ($n$), depending on how many times the entire air volume in the room under consideration is renewed in 1 h. Therefore, indicating with $V$ (m$^3$) the volume of the room, it follows that:

$$\dot{V}_a = \frac{n \cdot V}{3,600} \tag{6.6}$$

Replacing (6.5) in (6.6) and the values of specific heat and air density, the flow of sensible heat is obtained (W):

$$\Phi_s = n \cdot 0.33 \cdot V \cdot \Delta T \tag{6.7}$$

The flow obtained in this way can be referred to the unit of volume and to the unit of temperature drop (W/K m$^3$):

$$\Phi_s^u = n \cdot 0.33 \tag{6.8}$$

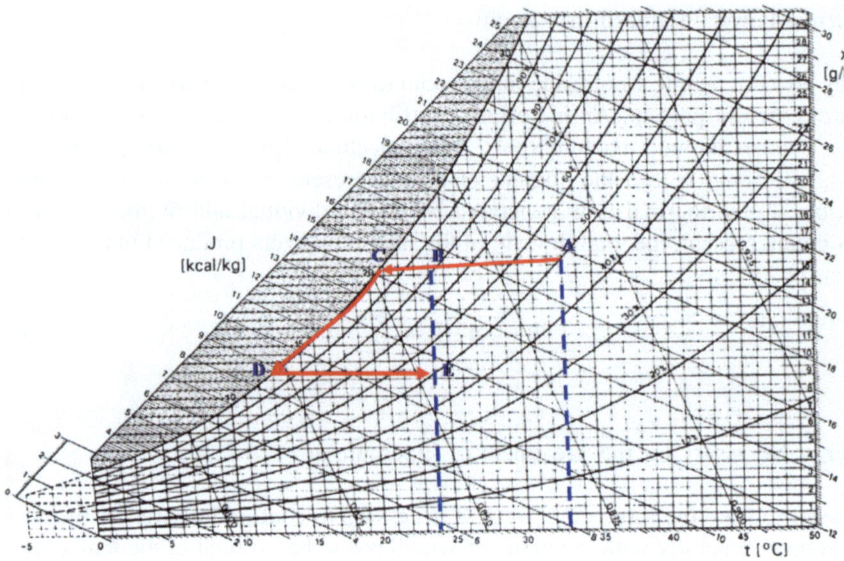


**Fig. 6.2** Energy expenditure for the treatment of 1 kg of external air for summer air-conditioning

For the calculation of the $\Delta T$ it is also necessary to consider the values of humidity relating to the external and internal air (normally around 50%).

In the case of summer cooling (Fig. 6.2), it is possible to consider as an example the external air at 33°C with 50% of *relative humidity*[4] (point A): by cooling constantly (that is to say at a constant absolute humidity) up to the internal desired temperature (24°C in the example) point B is reached, where the relative humidity is about 80%, not corresponding to the conditions required for the room comfort. To dehumidify the air, constant cooling must be continued until reaching the dew point (point C); by continuing to subtract heat, the vapour present in the air condenses and, therefore, the absolute humidity decreases besides temperature. Under the hypothesis that the internal air is at 24°C and with 50% of relative humidity (point E), cooling must be done up to (point D) obtained at the intersection of isoconcentration from point E with the curve corresponding to 100% of relative humidity. Therefore the difference in temperature, having defined $T_i$ the internal temperature and $T_e$ the external one, is:

$$\Delta T = T_A - T_D = T_e - T_D > T_e - T_i \tag{6.9}$$

[4] Relative humidity is defined as the *ratio between the partial pressure of the steam present in the air* (that depends on the quantity of steam present in the air, namely the absolute humidity x expressed in g/kg) *and the pressure of saturated vapour at the air temperature.*

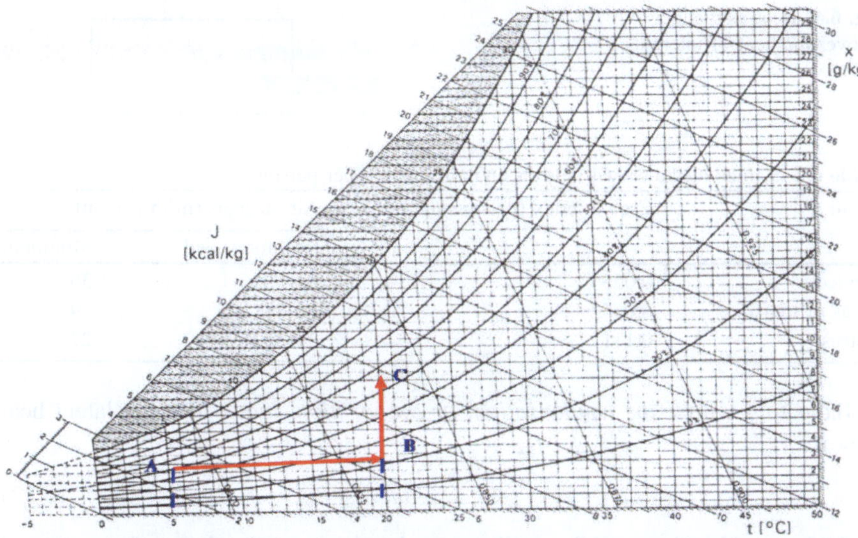


**Fig. 6.3** Energy expenditure for the treatment of 1 kg of external air for winter heating

From the energy viewpoint, the isoconcentration heating from $T_D$ and $T_E$ may not be considered; in fact such heating can take place by using the external air (at higher temperature) without any energy expenditure.[5]

In the case of winter heating (Fig. 6.3) considering, for instance, external air at 5°C with 50% of relative humidity (point A) and internal air at 20°C with 50% of relative humidity (point B), $\Delta T$ is:

$$\Delta T = T_A - T_B = T_e - T_i \tag{6.10}$$

(in the case of *winter heating*). Part of the energy spending relating to sensible heat can be retrieved by heating the input air in a heat exchanger using as *hot fluid* the stale air to be expelled, that is at the internal temperature (in the case described in the example, at 20°C); in the example shown in Fig. 6.4 the internal air is at 20°C and the external one at 0°C. In this case, without heat recovery there would be $\Delta T = 20$°C that, with the exchanger is reduced at only 5°C since air renewal is at 15°C.

Sensible heat must be supplemented by the *latent heat* necessary to steam condensation (in the case of *summer cooling*) or water evaporation (in the case of *summer cooling*). Indicating with *UA* the absolute humidity and with $\dot{m}_a$ the air mass flow, the vapour mass flow, $m_v$, to be generated (in the case of winter heating) is:

$$\dot{m}_v = \dot{m}_a(UA_e - UA_i) = \dot{V}_a \cdot \rho_a(UA_e - UA_i) \tag{6.11}$$

[5] Alternatively, as it usually happens in practice, it cools up to temperature $T_D$ a fraction of the air change that, subsequently mixed with non-cooled air, brings mixed air to temperature $T_E$. From the energy viewpoint, the two processes are equivalent.

**Fig. 6.4** Heat exchanger to recover the heat of internal air

**Table 6.2** Values of the crowd indexes and air changes per person

| Room | Crowd index (persons/m$^2$) | Air changes (m$^3$/h person) | |
|---|---|---|---|
| | | Recommended | Minimum |
| Services | 0.055 | 54–90 | 36 |
| Areas of transit | 0.055 | 12–18 | 9 |
| Offices | 0.111 | 27–45 | 27 |

Indicating with $\lambda$ *the condensation, or evaporation, latent heat* the latent heat flow is:

$$\Phi_l = \lambda \cdot \dot{m}_v \tag{6.12}$$

With considerations similar to the remarks above on the flows of air renewal and considering $\lambda = 2{,}500$ kJ/kg, the latent heat flow is obtained (W):

$$\Phi_l = 833 \cdot n \cdot V \cdot \Delta UA \tag{6.13}$$

Air renewals are provided for by the regulations in force and refer to the air renewals that are necessary for each individual; values depend on the destination of use of the room under consideration. The quantity of air to be changed in a given environment is calculated on the basis of the changes envisaged per person and the value of the crowd index of the room itself; also the values of this index are referred to by technical regulations. As an example, Table 6.2 shows some values of the crowd indexes and the air changes per person.

On the basis of the above, it is evident that the energy need for air-conditioning varies—also markedly—according to the climate of the site, the characteristics of the building and the air changes necessary on the basis of the destination of use of the rooms according to the regulations in force. In order to have an idea on the level of these needs, considering the climate characteristics of Rome, Italy, a building with average size and air changes for "normal" rooms (i.e. offices), the winter heating needs are about 28 W/m$^3$, whereas the needs for summer cooling are slightly higher, amounting to nearly 32 W/m$^3$.

From the energy viewpoint, also air-conditioning end use, as well as the heating of domestic hot water, is low-temperature (hot) or at a temperature that can be reached by simple refrigeration cycles (cold). The heat flow requested leads the heat-transfer fluid traditionally used for heating purposes to often be water, especially in those countries in which room heating became widespread before room cooling.

In these applications, heat is produced by boilers in the single buildings or even in the single apartments, through the combustion of fossil fuels (gasoil and methane), or through a thermal exchange with water vapour distributed by district heating originating from cogeneration plants in the electric sector (electricity and heat) or the use of the waste heat of industrial processes (high-temperature heat for the process, with a cascade distribution of low-temperature heat for residential district heating).

In other Countries, whose electric energy is generally use electricity produced by large-size nuclear power plants (i.e. France), heating mainly depends on the electricity energy vector. The "*all electric*" model as an ideal reference of particular interest from the viewpoint of simplicity and standardisation of end use, in these cases has prevailed over considerations of energy and non-energy nature, bringing electricity to be used for all domestic uses, including food cooking.

The analysis of the global energy needs of each single building, or of the residential district in which it is included, together with a structural (social and cultural) analysis of the energy distribution model used until now (distribution of liquid or gaseous fossil fuels, district heating with distribution of water steam heat-transfer fluid, electricity network) leads to considerations on the most suitable and largely realisable evolution of the system.

In this case, it is possible to carry not the same analysis of the possible solutions relating to the production of domestic hot water. Moving from the pre-determination of fossil fuels, of electricity from the electric grid and, where possible, of the district heating heat-transfer fluid as the only alternatives available, to real considerations of energy analysis, it is evident how there are several and attracting alternatives. A system efficacy analysis, that also takes into consideration parameters dealing with the renewability of sources, the lowering of the production of emissions, and the local availability of energy resources, certainly includes in the scenario the resort to renewable sources and their conversion technologies.

If the energy vector requested heat-transfer fluid is water (at temperatures ranging from 60 to 120°C, according to the applications and to the resort to local distribution networks for district heating), it can be produced either directly or in cogeneration with conversion technologies of available renewable sources as solar radiation, wind energy, hydroelectric power, biomass and geothermal energy. Looking at the future, it is also possible to think about cogeneration plants based on the technology of high or low-temperature fuel cells (according to the size used), powered by gases rich in hydrogen or pure hydrogen, producible from distributable or locally renewable sources.

## *6.2.2 Thermal End Use with Heat Pump*

### 6.2.2.1 Room Air-Conditioning

The use of heat pumps for room air-conditioning is likely to be more and more widespread, as an alternative to the conventional systems made up of a cooler and a boiler. The same equipment in fact, through a simple valve, is able to exchange the functions of the evaporator and the condenser, hence supplying heat in winter and cold in summer (reversible heat pump).

The application of the heat pump to room air-conditioning (heating and cooling) is interesting since it generally entails a cost amortisation time acceptable for end users. In the case of existing buildings, the application of the heat pump for room air-conditioning, both in winter and in summer, could demand an intervention for the restoration of the entire thermal and electric plant, with a subsequent increased cost.

There are different applications of the heat pump in the sectors of services and industry, as in the case of the air-conditioning of sports facilities and venues, low-temperature technological processes and drying in the agricultural and food sector, etc.

#### 6.2.2.2 Heating of Rooms and of Domestic Hot Water

Heat pumps can be used for heat production purposes both for room heating and for domestic (sanitary) hot water, competing with traditional systems as electric or gas boilers and water heaters.

In case of room heating, plants can be:

- *Monovalent*
- *Bivalent*

The monovalent configuration is used when the heat pump is able to entirely meet the thermal needs necessary for heating purposes. If heat pumps use as source the external air, this configuration can be adopted in the climate areas in which the external temperature drops below 0°C. Otherwise, it is necessary to develop a bivalent system, made up of heat pump and of an auxiliary heating system, that is to say a traditional boiler that meets thermal needs when air temperature drops below 0°C.

For the heating of domestic water, storage tanks larger than those used in normal water heaters are needed, since the temperature of the water produced does not exceed 50–60°C.

Types of Plants

For *small powers* (up to nearly 2 kW), also usable for the heating of domestic water, heat pumps can be as follows:

- *Monobloc*, all the components are grouped together;
- *Split*, the plant is made up of:
  - *External unit*, including the compressor and a heat exchanger with the function of evaporator or condenser;

Fig. 6.5 Scheme of a plant for small, medium and large sizes

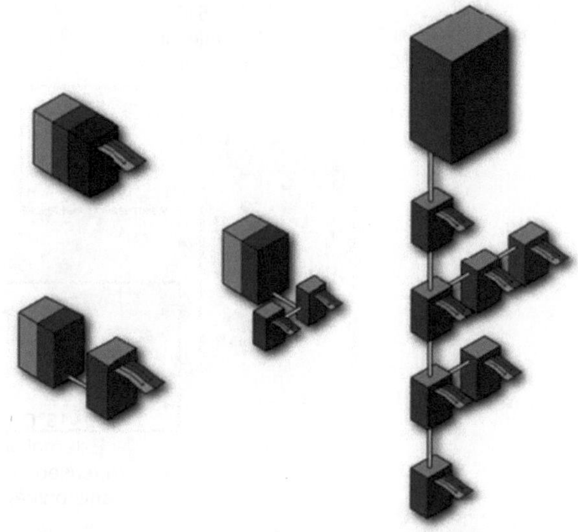


- *Internal unit*, with a cabinet from which hot or cool air is released in the room, depending on the different cases. This system allows to install the noisy components of the plant in an area external to the room to be air conditioned.

In case of *medium power* (up to 10 to 20 kW) suitable for serve several rooms, they can be:

- *Monobloc*, all the components are grouped together;
- *Multisplit*, the plant is made up of:
  - *External units*, including the compressor and a heat exchanger with the function of evaporator or condenser;
  - *Internal units*, made up of several cabinets, regulated individually, from which hot or cool air is released in the room, depending on the different cases. This system allows to install the noisy components of the plant in an area external to the room to be air conditioned.

In case of large power (exceeding 20 kW), destined to several apartments, offices and commercial activities. The plants are made up of:

- *External condensing units*: they produce hot water and cool water;
- *Fan-coils*: internal cabinets, which release in the rooms hot air in winter or cool air in summer, and keep a comfort temperature also in intermediate seasons.

The scheme of the different plant solutions is shown in Fig. 6.5.

In order to improve the efficiency of heat pump plants, very often a device is installed that allows to recover (at least partially) the heat contained in the internal

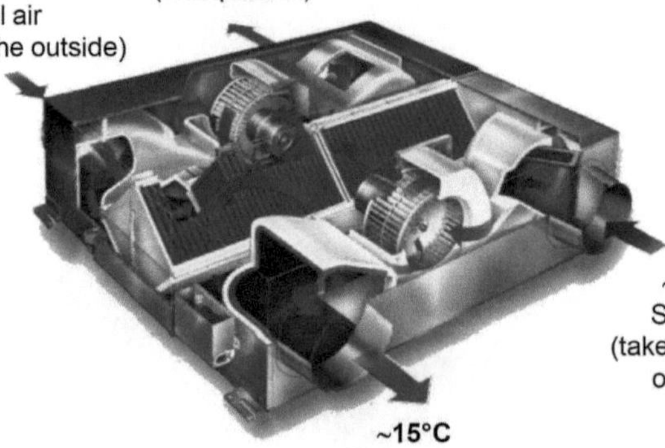


**Fig. 6.6** Internal air/external air heat exchanger

air to pre-heat the air coming from outside. This device is a heat exchanger, an example of which is shown in Fig. 6.6.

## 6.3 Luminous End Use

### *6.3.1 Interior Lighting*

Lighting engineer studies all the aspects relating to lighting, both natural and artificial, in order to guarantee adequate visual conditions to human beings. In particular, in a closed environment, lighting must guarantee the execution of the visual tasks performed and to carry out conditions of visual comfort, that is to say mental conditions of satisfaction expressed vis-à-vis the visual environment. These requirements are met if all the internal elements can be clearly distinguished without any difficulty and tasks performed without any effort whatsoever. To obtain such a result, the lighting to be achieved can be essentially obtained either by exploiting natural lighting or by integrating it with the artificial one.

These two options are not however equivalent: in fact, although artificial lighting, different from the natural one, can be designed with more precise characteristics in terms of quality and quantity, natural lighting has the advantage of a better quality of light (better performance of colours) and the higher levels of lighting that can be reached, which from the quantity viewpoint is practically impossible to achieve with artificial lighting. Furthermore, natural lighting uses solar energy, that is to say the renewable source of lighting *par excellence*, allowing huge energy savings and a neutral impact for the environment.

## 6.3.2 Elements of Photometry

### 6.3.2.1 The Vision in Human Eyes

*Light* is the electromagnetic radiant energy that human eyes are able to perceive; in particular, the white light is given by a mix of all the wavelengths of the visible spectrum and also contains radiations belonging to infrared and ultraviolet rays. The evaluation of the fundamental magnitudes that are used to measure the radiant energy emitted or received by a surface in relation to the feelings produced on individuals through their eyes represents the object of a preliminary part of light engineering that falls within the name of *photometry*.

Photometric Magnitudes

The main photometric magnitudes include:

- luminous flux;
- luminous intensity;
- radiance;
- luminance;
- lighting.

Out of them, the first four characterise the source emission, respectively for **point sources** (*luminous flux and intensity*) and **extended sources** (*luminance and radiance*), whereas the last item refers to the incident light on the receiving surface.

The *luminous flux* represents the luminous power $\varphi$ released by a point source through a solid angle d$\omega$.

*Luminous intensityI* therefore represents the luminous flux released by a point source in a basic solid angle d$\omega$ around a direction *r*:

$$I = \frac{d\varphi}{d\omega} \tag{6.14}$$

It is measured in *candelas* (*cd*).

*Candela* is the fundamental lighting unit in the International System: subsequently, the luminous flux is a derived magnitude, defined through (6.14) as "*the luminous flux emitted into a solid angle of one steradian by an isotropic point source having a luminous intensity of one candela*".

The *radianceM* in a point of a large surface is given by the *ratio between the luminous flux* d$\varphi$ *emitted by a surface element containing the point and the area of the surface itself*:

$$M = \frac{d\varphi}{dA} \tag{6.15}$$

It is measured in lm/m$^2$ (lumen for square meter).

Radiance refers to the visibility curve: it is valid until the source observed has a radiance of 100 lm/m$^2$ (*photopic vision*); in *scotopic vision*, on the contrary, radiance drops below 0.01 lm/m$^2$.

#### 6.3.2.2 Luminance

The *luminance* in a point of a large surface, in a particular direction $\alpha$, is given by the *ratio between the luminous intensity dI emitted in that direction and the emitting surface projected in a perpendicular plan to the direction itself* $dA_n$ (surface apparently seen from direction $\alpha$), that is to say the intensity emitted by the apparent unit of surface d$A_n$.

*Illumination E* in a point of a surface is defined as the *ratio between the luminous flux* d$\varphi$ *incident on a basic surface* dA *which contains the point and the area of the surface itself*:

$$E = \frac{d\varphi}{dA} \tag{6.16}$$

Its unit of measurement is lumen/m$^2$ called *lux* (*lx*).

The measurement of illumination is carried out through *illuminometers*, equipment made up of photovoltaic cells containing a semiconductor sensor in which, if invested by light, an electromotive force is generated, proportional to the power of radiation according to a sensitivity curve that reproduces the normal visibility curve of human eyes.

#### 6.3.2.3 Secondary Sources

A body receiving an electromagnetic radiation sends back a fraction of it amounting to its *reflection coefficient r*.[6] Since the reflection coefficient can take on different values in the different points of a surface, and also depends on the radiation incidence angle, since generally its mean value is referred to.

Furthermore, since *r* is a function of the wavelength, it follows that the incident radiation and the one sent by the surface differ, besides for the power, also for the

---

[6] It must be observed how the reflexion coefficient is different from the one used in the thermal radiation case, since lighting engineering only considers the one relating to the visible band.

composition of the spectrum. In particular, the sending back of the incident light can take place through a *spectrum reflection* (a single *incident ray* produces a single *reflected ray*) as it happens on glossy or smooth surfaces, *diffused* (the presence of infinite infinitesimal surfaces specularly reflecting and oriented to all direction leads the light to be globally reflected randomly), in the case of mat or opaque surfaces that show a certain degree of roughness,[7] *or mixed* (diffuse reflection with a predominant component in the specular direction) on semi-glass surfaces.

The reflecting body can be considered in turn as a luminous source that is generally defined *secondary source*.

Point, Linear and Extended Source

On the basis of their geometry, luminous sources can be divided into:

- point sources;
- linear sources;
- extended sources.

A source can be considered as a *point source* if its size is small compared to the distance from the points to be lighted (the distance must be at least 5 times higher than the largest size of the source). It is conversely considered as *linear* if it shows a size much larger than the others and comparable to the distance from the point to be lighted. It can finally be considered as *extended* if at least two of its sizes are comparable to the distance from the point to be lighted.

## 6.3.3 Natural Light

### 6.3.3.1 Natural Light Sources

The primary light sources are the *sun* and the *celestial vault*. In particular the light can directly come from the sun, although partially absorbed by the atmosphere (*direct light*), or consist in the diffused light of the celestial vault, which is made up of the scattered radiation in the atmosphere by gases, water vapour and particulate, and reflected by the different surfaces of the environment, being it natural (sky, trees, mountains, etc.) and artificial (buildings, parts of the internal environment) (*diffused light*).

The external space performs a role analogous to the one of a lighting equipment (which has the task of filtering and distributing the artificial light) as concerns natural light, scattering, reflecting and transmitting the light.

---

[7] This happens when the size of the surface irregularities is of the same order of magnitude as the wavelength of the incident light.

The entire set of direct and diffused light represents the global daily illumination; the average value of extraterrestrial solar radiation (called *solar constant*) is 1,353 W/m$^2$. The solar radiation, passing through the atmosphere, changes its spectral composition, and the process in general strongly depends on climatic conditions: it is however possible to assume that radiation, as it reaches the earth's surface, shows a spectral distribution that is close to a black body at a temperature of nearly 5,760 K.

The normal curve of visibility of the human eye in a *photopic vision*[8] is suitable for this spectral composition: its peak (at $\lambda = 555$ nm), in fact, corresponds to the maximum of the spectral emittance of the black body at the temperature of 5,225 K. In the field of visible and ultraviolet radiations and in wide bands of infrared radiation, the spectral power of the solar radiation is however lower than the black body, especially near the absorption bands typical of molecules of $O_2$, $N_2$, $H_2O$, $O_3$ and $CO_2$ that is the chemical species. Among those that are present in the atmosphere, that determine a higher absorption of the relevant spectral components.

When sheltered by the direct solar radiation, that is to say when the light arrives only after undergoing the reflection diffused through the atmosphere and on the surrounding bodies, the spectral composition is conversely variable according to the presence of dust and clouds in the air, and is also affected by the presence of buildings, mountains, trees, etc.: in general the diffused solar radiation has a colour tending to light blue.

Obviously, the light on the earth's surface depends on the position of the sun (and therefore on its *height* and *azimuth*) and is variable in its spectral content, in the different months of the year, both in terms of quantity, intensity and quality. The possible extreme conditions are those of *completely bright sky*, without any clouds, and the *uniformly covered sky*. The former case corresponds to a transparent hemisphere, which disperses the light only in small angles (on the basis of this phenomenon, the sky appears *blue* and not *black*), the latter corresponds to a translucent hemisphere that disperses radiation in big angles; all the other conditions can be considered as intermediate between these two.

The *sky* (or *celestial vault*) as light source can be characterised through the distribution of luminance or illumination on a non-obstructed horizontal surface.

## *6.3.4 Artificial Light*

### 6.3.4.1 Artificial Luminous Sources

Artificial light sources are needed to make up for the shortage of natural light. They are generally made up of two parts, *the lighting equipment* and *the lamp*. In particular, in electricity-powered sources, lamps are used for the conversion of

[8] The *photopic vision* is only due to the activity of retina cones. It is the type of vision that occurs when the level of lighting is "normal" (daylight) and allows to identify chromatic differences.

electric power into a luminous flux, whereas the lighting equipment has the task of suitably distributing this flow.

### Lighting Equipment

Lighting equipment represents a fundamental complement for any light source, ensuring a threefold functions: *electric*, *mechanic* and *photometric*. From the manufacturing viewpoint, lighting equipment can be considered as made up of two components:

- the *shell*, with electric and mechanic functions;
- the *optical system*, which has the task of modifying the luminous flux going out of a source and suitably addressing it.

Therefore, they represent technological products (chandeliers, ceiling light, fixtures, spotlights, etc.) that act as a shell for the housing, the mechanic, the thermal and the electric protection of the light source. In particular they include the components that are necessary for the support, the fixing and the protection of the lamp from whatever external agent that might entail a deterioration or a decrease in optical qualities, also including, whenever present, the auxiliary circuits and devices (for instance, transformers) for the connection to the power-supply. From the lighting engineering viewpoint, equipment performs the task of distributing, filtering and/or transforming the light emitted by one or several lamps, exploiting the properties of reflection, refraction and diffusion of the materials composing it.

The *performance*, of an equipment is the ratio:

$$\varepsilon = \frac{\phi_{device}}{\phi_{lam}} \tag{6.17}$$

Generally, the values of $\eta$ are included between 0.4 and 0.8. The fraction of the luminous flux emitted by the source that is absorbed by the equipment is then dispersed in the form of heat.

### Types of Lighting Equipment

The characteristic equipments used in lighting services can be summarized as follows:

*Diffusers*: they have the task of shielding the direct view of the lamp and—thanks to their symmetric geometry, usually vis-à-vis an axis—changing its photometric solid in such a way as the outgoing luminous flux appears uniform in all directions (Fig. 6.7).

They are mainly used for residential exteriors and from their manufacturing viewpoint they are often shaped as *a globe of translucent material* (glass, polymeric material, etc.).

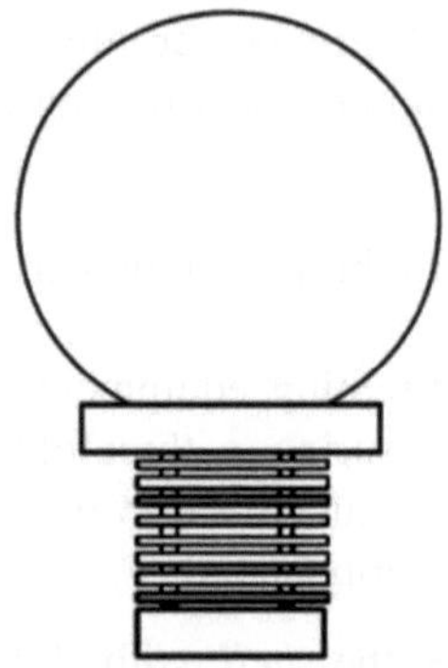

**Fig. 6.7** Diffuser [3]

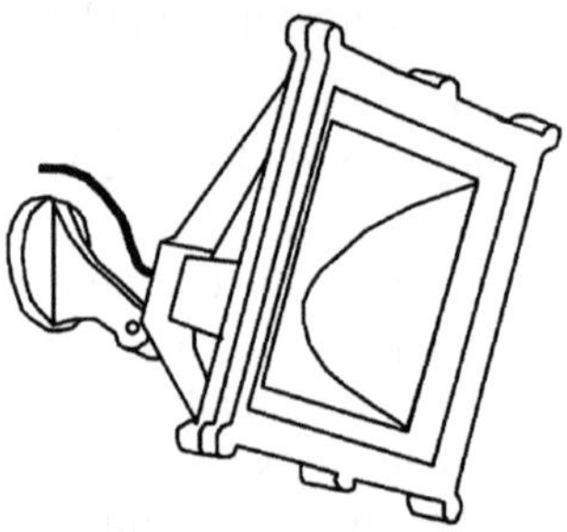

**Fig. 6.8** Reflector [3]

*Reflectors*: they change the photometric solid of the source, exploiting the reflection properties, and inhibit the vision of the lamp only in some directions. From the manufacturing viewpoint, they are made up of *a shell of opaque material* with the internal side treated in such a way that allows to have a high reflection coefficient (Fig. 6.8).

*Refractors*: they are used to modify the emission photometric solid allowing the directional control of the light and avoiding *dazzling* dangers. The change in the spatial distribution of the flux emitted by the source takes place by exploiting in particular the refraction phenomenon. From the manufacturing viewpoint, they are made up of *a glass or plastic shell* that can partially or totally contain the source (Fig. 6.9).

*Projectors*: they have the task of addressing the luminous flux emitted by a source towards a given direction and within a limited solid angle, to particularly light some areas or objects (Fig. 6.10).

Equipment Classification

According to the classification of the Commission Internationale de l'Eclaraige (CIE), lighting equipments can be divided on the basis of the spatial distribution of the luminous flux emitted above and below the horizontal plan passing through their centre. Then one distinguishes the following types (Fig. 6.11):

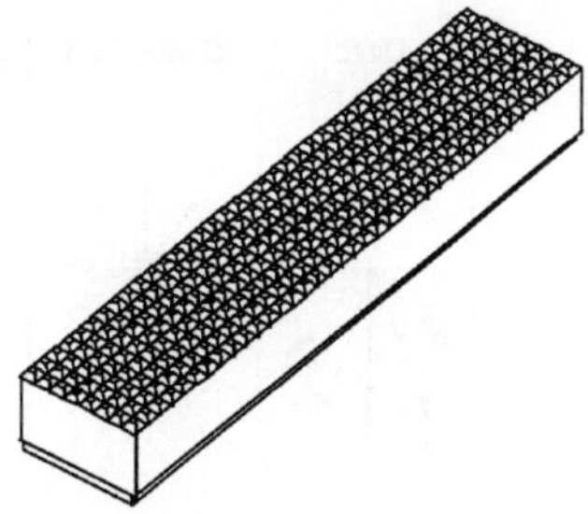

**Fig. 6.9** Refractor [3]

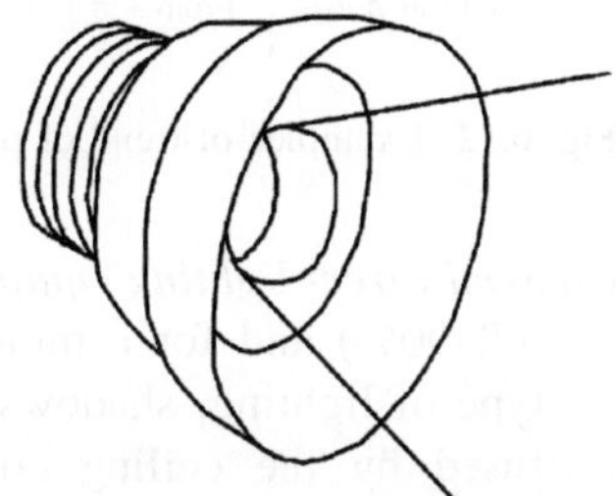

**Fig. 6.10** Projector [3]

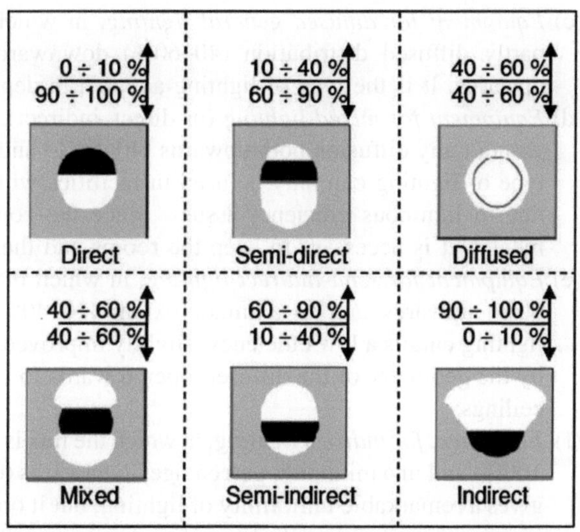


**Fig. 6.11** Main types of lighting

(a) *Equipment for direct light*, in which the flux is mainly emitted downwards (90–100%) and for a minimum part (0–10%) upwards. It is the cheapest illumination, since the lighting equipment is generally simple, cost-effective and high performing. With this type of lighting, the absorption by walls and ceilings is negligible, but shadows are marked and it is advisable to use several luminous sources to reduce them. It is convenient both in exteriors, and in laboratories or offices;

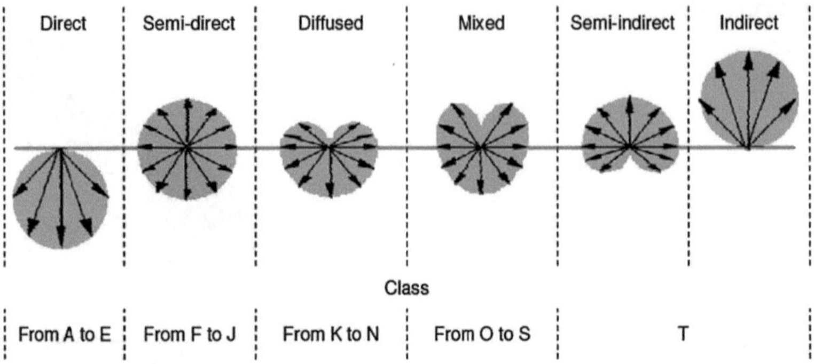


**Fig. 6.12** Examples of trend of photometric curves for the different types of lighting

(b) *Semi-direct lighting equipment*, in which the flux is mostly emitted downwards (60–90%) and for a minor percentage (10–40%) upwards. By adopting this type of lighting, shadows are softened. Furthermore, the part of the flux diffused by the ceiling contributes to increase the global efficiency of the installation and avoids the shadows above the equipment. It is mainly used in offices, houses, canteens, etc;
(c) *Equipment for diffused general lighting*, in which the flux is emitted, with a partly diffused distribution (40–60%) downwards and part of it (40–60%) upwards. It is the type of lighting adopted in department stores;
(d) *Equipment for mixed lighting* (or direct–indirect), in which the flux is emitted without any diffusion both upwards (40–60%) and downwards (40–60%). This type of lighting can only be used in facilities with very much reflecting walls due to luminous efficiency issues. Since the contribution of walls is fundamental, it is necessary to keep the rooms and the ceiling light fixtures clean;
(e) *Equipment for semi-indirect lighting*, in which the flux is mostly emitted (60–90%) upwards and to a minor extent (10–40%) downwards. This type of lighting entails a low efficiency, slightly improved by the direct component and by the geometry of the diffuser, open towards to top; it also requires very light ceilings;
(f) *Equipment for indirect lighting*, in which the flux is mainly emitted upwards (90–100%) and in a minimum percentage downwards (0–10%). This type of lighting gives a remarkable uniformity of lighting, but it obliges to install very powerful lamps, since it generally has a low luminous efficiency. It is used to obtain particular aesthetic results or to highlight architectural details: therefore, it is mostly destined to showrooms, some restaurants, museums, monuments, etc.

Figure 6.12 shows an example of trend of the photometric curves relating to different types of lighting. The same figure also shows a further organisation of the types of equipment in 20 classes, indicated from A to T (ranging from the most concentrated direct to the indirect one), and defined by the CIE to classify the large number of possible spatial distributions of luminous flow.

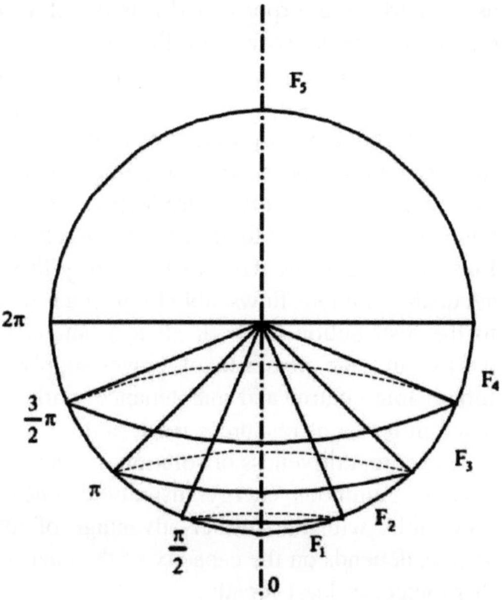


**Fig. 6.13** Solid reference angles for the definition of the class of an equipment

The class of an equipment is defined on the basis of the *flux* $\varphi_i$ *emitted in five different spatial portions*, obtained by dividing the space around the photometric centre into five cones around the axis of the equipment, having a solid angle of $\pi/2$, $\pi$, $3/\pi$, $2\pi$ *ster*, respectively (upper and lower hemispheres) (Fig. 6.13.). The values of the flux emitted in the different spatial portions are provided by manufacturers or can be determined starting from the knowledge of the photometric curve.

By using the values of $\varphi_i$ it is possible to define the *efficiency of the equipment* $\eta_i$, relating to the flow emitted by the source inside and not outside the equipment itself, which is given by:

$$\eta_i = \frac{\varphi_1 + \varphi_2 + \varphi_3 + \varphi_4 + \varphi_5}{\varphi_{tot}} \tag{6.18}$$

where $\varphi_{tot}$ represents the total flow emitted in all directions.

Electricity is the main energy vector for the luminous end use, targeted to meet the useful effect in terms of room luminosity in the absence of natural lighting, up to its completion or independently from its availability.

The main characteristic of each energy vector usable to make the luminous end use possible is its availability at night and in periods of scarce availability of solar energy. Although it might appear trivial at a first sight, this is the key element for the development of energy systems that exploit renewable energies (the main of which is the direct solar source) to supply the energy quantity necessary to meet lighting needs. If electricity is suitable for the achievement of all the applications requested, both in closed rooms and in open ones, and also for large surfaces (as it happens in the case of sports events, social meetings or the lighting of large areas as car parks and airports) and it is therefore the most appropriate vector to reach end users, much space—in the area of energy vectors—remains available to optimise and improve the production procedure and sources, as well as the possibility of more or less distributed storage and close to the point of end use.

The technological progress of electricity-luminous energy conversion systems, in fact, makes it possible for the implementation of energy systems that envisage the integration of diffuse production of electric power as integration or as an alternative compared to the systems based on the storage through the electric grid. Low power and low requests for energy flux in fact increasingly allow to achieve adequate luminous flows, able to meet a growing number of needs. The connection to the distribution network always entails more or less complex infrastructural works, and the presence of power-supply cables translates into a not always foreseeable control and maintenance work, which cannot even be precisely evaluated in terms of resources requested.

The competitiveness of solutions exploiting the sources available at the point of need of luminous energy, usually characterised also by the feature of being renewable, with the further advantage of releasing zero emissions in the atmosphere, depends on the capacity of the energy vector produced of being stored and distributed, at least locally.

For its luminous use, electric power can therefore be certainly produced also by renewable sources, besides from fossil and nuclear sources, the letter are not limited by definition by the availability of the solar source, although under the condition of reaching the point of need for a distribution grid. However, production must occur also through a storable vector, and this opens large perspectives to renewable energy sources based upon the development of electrochemical accumulators:

$$\textit{Electricity production from renewable sources} \Rightarrow \textit{electrochemical storage} \Rightarrow \textit{electricity for lighting purposes}$$

but also for applications that envisage hydrogen as a useful vector:

$$\textit{Hydrogen production from renewable sources} \Rightarrow \textit{hydrogen storage} \Rightarrow \textit{electricity for lighting purposes.}$$

## 6.4 Mechanic End Use

The mechanic end use of energy is aimed at the carrying out of tasks originally performed directly by men with their bodies, with or without the use of any equipment.

The replacement of human work, as well as the speeding up, the increase in precision and repeatability of tasks and works is among the main purposes of the mechanic end use of energy. There are several useful effects in the industrial, agricultural, civil and services fields, ranging from the processing and assembly of industrial components and products, to the performance of household tasks (movement of water and of the drum of the washing-machine), to carry out agricultural tasks (digging, seeding, harvesting, etc.), up to the movement of things and people in the sector of mobility.

All these applications have extremely variable needs in terms of characteristics of the energy flow, and this determines the choice of the single technologies used in the various cases. The mechanic energy demanded is presently obtained in industrialised countries through the distribution network of the electricity vector and by operating an electric power-mechanic power conversion with electric machines and procedures. A specific predominance of the combustion of fossil fuels is present in the sector of transports. In developing countries, animal energy or human energy are still extensively used, with the manual carrying out of tasks that are entirely performed by machines in industrialised areas.

Among mechanic use, it is possible to identify the following:

- *Industrial, agricultural and agro-industrial* uses, among which:
  - Extraction, transportation and handling of raw materials and derivatives
  - Handling and assembly for mechanised production
  - Mechanic processing of parts and components
  - Land cultivation, seeding and collection in agriculture
  - Handling and processing of raw materials and agro-industry products
- *Civil uses and Residential*, among which:
  - Operation of household appliances
  - Operation of lifts, moving pavements, moving staircases
  - Automatic doors
  - Fans and ventilators

At present, the possible solutions for the production of mechanic energy for end use are the following:

- *conversion from chemical energy to mechanical energy* in a thermodynamic cycle that exploits the heat produced through the burning of fuel:
  - gas turbine power plants
  - steam turbine power plants
  - internal combustion (or, more rarely, external combustion) volumetric engines
- *conversion from electric power to mechanic energy*:
  - electric motors

In consideration of their characteristics, gas and steam turbine power plants are used in particular cases in which huge power is required. A typical application of gas turbine power plants for the production of mechanic energy is air transport.

Steam plants are used for the production of mechanic energy in particular industrial sectors; in these cases a part of the mechanical power generated by turbines is normally used, while a part of it is converted into electric power.

As already mentioned internal combustion engines (external combustion ones, of the Stirling type, have a limited use today), are extensively used in the sector of transports and cover almost all transports on road, and part of the maritime and air transport.

The most widespread solution is represented by electric motors; their widespread use is due to various practical factors and to the fact that the global energy efficiency of this solution is in any case similar to the possible alternatives that, however, in many cases are not applicable or show practical problems. Electric motors, in fact, represent now a widely fine-tuned, reliable and cost-effective technology. Furthermore, in the different construction solutions (direct current, alternating current, synchronous or asynchronous) they do not show any limit from the size viewpoint, since they can supply power ranging from a few W to hundreds of kW. An important characteristic is *compactness* that allows to integrate electric motors in the utilisation equipment, being it a household appliance, or an equipment for mechanical working. In many cases it is possible to take electricity directly from the electric grid: alternatively, the electrochemical storage (with a high efficiency) is in any case possible. Another fundamental aspect is represented by the *absence of whatever type of emission* (almost always acoustic), which makes them usable also in closed areas (including domestic use).

As far as the *global energy efficiency* of the solutions mentioned above is concerned, the following must be underlined: in case of gas and steam plants, the values of conversion efficiency (from chemical energy to mechanic energy) are similar to those typical of these plants for the production of electric power (see chapter on technologies). As widely known, electric motors have very high efficiencies (exceeding 90%); for a global energy analysis, however, it is necessary to consider the efficiency of the production of electricity.

As an example, considering a gasoil ICE, it is reasonable to hypothesise a conversion efficiency from chemical energy to mechanic energy amounting to 36%; considering the losses due to fuel extraction, transportation, refining and distribution as amounting to 10%, *a global efficiency of nearly* 32% (from the primary source to the useful mechanic energy) is obtained. Considering, on the contrary, an electric motor, it is possible to hypothesise a conversion efficiency from electric power to mechanic power of 92%; considering the losses due to transportation (5%) and the efficiency of electric power production (40%) as well as the losses for the transportation of the fuel to the plant, *a global efficiency of* 33% is obtained. In terms of primary energy consumption, given the same useful effect, this means that to obtain, for instance, 100 kJ of useful mechanic energy there is a consumption of primary energy of 309 kJ in the case of ICE and of 304 kJ in the case of the electric motor. As in the case of the ICE the primary energy is not renewable (oil), in the case of electric motor a part of the electricity is produced from renewable sources for which, as underlined in several occasions, it is not suitable to talk about "consumption". Assuming that 20% of the electric

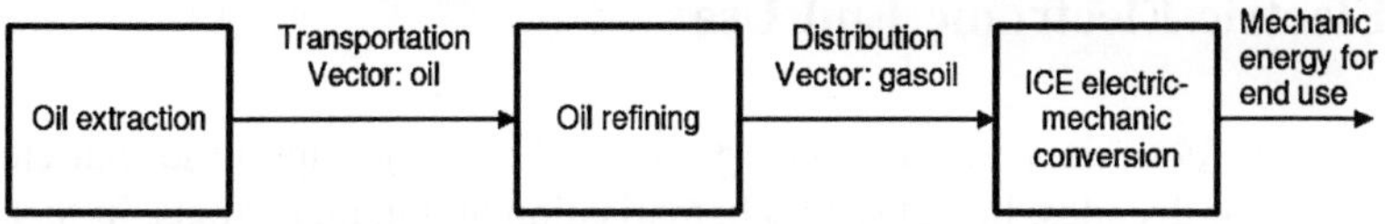


**Fig. 6.14** Production of mechanic energy by ICE: energy cycle from the primary source to end use

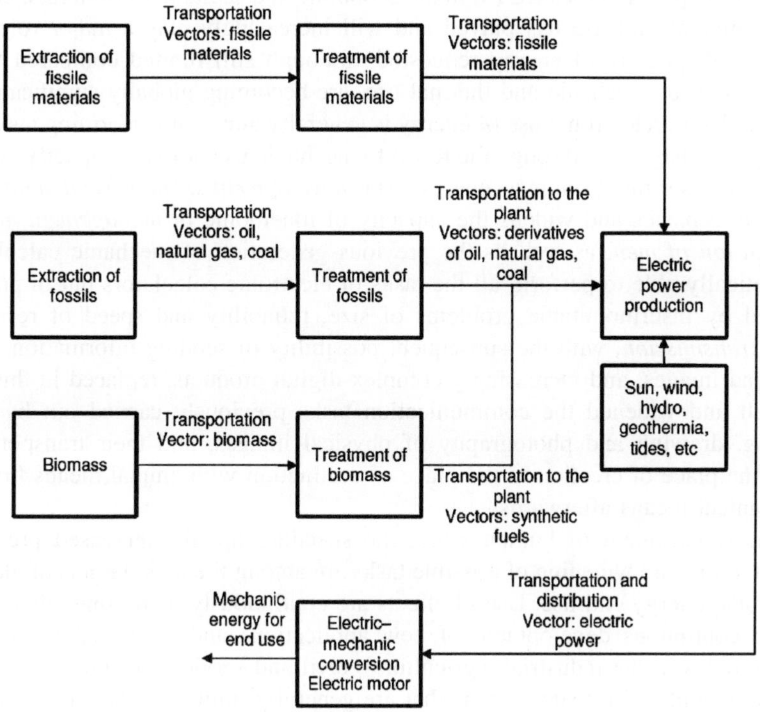


**Fig. 6.15** Production of mechanic energy by electric motors: energy cycle from the primary source to end use

power consumed is produced from renewable sources, the *consumption of non-renewable primary energy* in the case of electric motor is 243 kJ.

Thus, attention has to be paid to the schemes relating to the entire energy cycle (from the source to end use and useful effect). Let's now analyse some cases relating to the technologies devoted to the production of mechanic energy for end use.

Figure 6.14 shows the scheme dealing with the entire energy cycle in the case of production of mechanic energy with ICE fuelled with gasoil, whereas Fig. 6.15 shows the production of mechanic energy through electric motor by taking electricity from the grid.

## 6.5 Electric/Electronic End Use

The *electric/electronic* end use of energy entered in the group of the four end uses only in the last decades. In a classic schematisation and in many texts, in fact, there is still a classification limited to the three groups of *mechanic*, *thermal* and *luminous energy*. As a matter of fact, with the advent of electronics and the explosive growing of information Technologies (ITs) and information and Communication Technologies (ICTs) in the economic and social lives of human beings, this family of end use has played and will increasingly play a major role. The power and quantity of energy requested, although still limited compared to the typical ones of mechanic and thermal use, are becoming globally significant.

The electric/electronic use of energy is generally aimed at performing tasks that were previously made through the resort to mechanic use or to the capacity of men, therefore resorting to human energy. The *development of numerical data management* replaces and widens the capacity of mnemonic or *management manual calculation of men*, as well as the previous generation of mechanic calculators, theoretically able to perform all the tasks of electronic calculators but in practice limited by insurmountable problems of size, reliability and speed of response. *Data transmission*, with the subsequent possibility of sending information, voice data and images, and increasingly complex digital products, replaced in this case as well and widened the communication tasks previously carried out by typewriting, drawing and photography of physical images, and their transportation from the place of creation to the place of destination with animal means first and mechanical means afterwards.

The replacement of human work, the speeding up, the increased precision, repeatability and widening of possible tasks are among the main targets of electric/electronic energy end use. Useful effects are continuously increasing, also thanks to the continuous development of new applications and to the advent of new technologies in the industrial, agricultural, civil and services sectors.

These applications show needs that are generally limited to the energy flow in terms of power and current intensity; this opens large possibilities of choice of energy technologies, of the sources and of vectors that can be used in the various cases, before the necessary conversion into electric power as end use. The electric power requested is obtained today both in industrialised and in developing countries, with the electricity vector mainly being supplied by the distribution grid.

The growing demand for portable applications with end use of electric/electronic energy and the continuous growth of the networks dealing with the distribution, repetition and transmission of signals for their operation have opened the largest market in our history for *energy electrochemical storage systems*. The new and growing demand in this sector, in fact, is the *portability of equipment*, being a computer, a telephone, a sound or video player, a palmtop and much more, having considered the continuous development and the continuous fusion and separation between electronic devices able to perform one or several tasks with this family of applications.

The electric/electronic uses are therefore usable nowadays either statically, with fixed equipment, and dynamically, with portable equipment. Among them, it is possible to identify *working and extra-work use*, including:

- *working and extra-work use*, including:
  - *Calculation and data processing*
  - *Connection to telecommunications networks* (Internet, wireless, telephone and satellite)
  - *Inbound and outbound data transmission* (voice, images and numerical data)
  - *Sound or image play*

The electric/electronic use of energy opens interesting opportunities, in consideration of its characteristics that make it particularly different from the three other end uses, and in particular from the thermal and mechanic ones: the low or very low power at stake, and portability as an element of certain and growing success.

The low power also allows to consider particularly interesting some energy sources that in the past were particularly ignored by classic energy studies, mainly concentrated on the use of industrial and agricultural energy applications, typically involving a high power. Small power and limited quantity of energy pave the way to the possibility of exploiting renewable sources that are practically available all over the world, in the same way as all over the world there are electric/electronic applications that are the symbol of development and of global relations that ask to operate and to be powered. All the conversion technologies of renewable energy sources allow the production of the quantities of electricity requested. Therefore, for the first time all the sources are fully involved in energy competition. In many cases with the advantage (once considered as a disadvantage and still wrongly presented in this way by many authors) of being distributed all over the earth and not focussed in single areas of availability, from which it is necessary to organise a complex, little safe and expensive distribution system.

Conversely, the request for portability paves the way for the development of better and better new solutions of energy storage, which, thanks to their distribution and acceptance in the market of electric/electronic use, show today to the large public and to the global market their existence and their potential usefulness also in other sectors. The characteristic of this sector—new but that can already be considered as dominating from the viewpoint of energy trends—is the precondition for the creation and development of new generations of electrochemical accumulators, increasingly efficient and competitive, in order to allow the access in world markets of new energy vectors that are only competitive in niche applications, such as hydrogen and new chemical, physical or chemical–physical solutions for their storage, and the appearance of mechanical storage systems, easily convertible into electric energy, such as micro and mini flywheels, springs or pressurised fluids.

## References


1. ISO regulation 7730. Moderate thermal environments—determination of the PMV and PPD indices and specification of the condition for thermal comfort
2. Nucara A, Pietrafesa M (2001) Elementi Di Illuminotecnica, Universita' degli studi "Mediterranea" di Reggio Calabria facolta' di Ingegneria Dipartimento Di Informatica Matematica Elettronica e Trasporti
3. Jokem E (2000) Energy End Use Efficiency, World Energy Assesment: Energy and the Challenge of Sustainability, Chapter 6, UNDP and World Energy Council, New York.

# Index




F. Orecchini and V. Naso, *Energy Systems in the Era of Energy Vectors*,
Green Energy and Technology, DOI: 10.1007/978-0-85729-244-5,

MIX
Papier aus verantwortungsvollen Quellen
Paper from responsible sources
FSC® C105338

If you have any concerns about our products,
you can contact us on
**ProductSafety@springernature.com**

In case Publisher is established outside the EU,
the EU authorized representative is:
**Springer Nature Customer Service Center GmbH**
**Europaplatz 3, 69115 Heidelberg, Germany**

Printed by Libri Plureos GmbH
in Hamburg, Germany